METHODS IN MOLECULAR BIOLOGY

Series Editor
John M. Walker
School of Life and Medical Sciences
University of Hertfordshire
Hatfield, Hertfordshire, UK

For further volumes:
http://www.springer.com/series/7651

For over 35 years, biological scientists have come to rely on the research protocols and methodologies in the critically acclaimed *Methods in Molecular Biology* series. The series was the first to introduce the step-by-step protocols approach that has become the standard in all biomedical protocol publishing. Each protocol is provided in readily-reproducible step-by-step fashion, opening with an introductory overview, a list of the materials and reagents needed to complete the experiment, and followed by a detailed procedure that is supported with a helpful notes section offering tips and tricks of the trade as well as troubleshooting advice. These hallmark features were introduced by series editor Dr. John Walker and constitute the key ingredient in each and every volume of the *Methods in Molecular Biology* series. Tested and trusted, comprehensive and reliable, all protocols from the series are indexed in PubMed.

Autophagy in Development and Disease

Methods and Protocols

Edited by

Kursad Turksen

Ottawa, ON, Canada

Humana Press

Editor
Kursad Turksen
Ottawa, ON, Canada

ISSN 1064-3745 ISSN 1940-6029 (electronic)
Methods in Molecular Biology
ISBN 978-1-0716-4267-2 ISBN 978-1-0716-4268-9 (eBook)
https://doi.org/10.1007/978-1-0716-4268-9

This Humana imprint is published by the registered company Springer Science+Business Media, LLC, part of Springer Nature.
The registered company address is: 1 New York Plaza, New York, NY 10004, U.S.A.

Preface

The field of autophagy has experienced huge developments since the publication of *Autophagy in Differentiation and Tissue Maintenance: Methods and Protocols.* The protocols included in this new edition cover both the development and disease state reflecting these developments. Once again, the protocols gathered here are faithful to the mission statement of the Methods in Molecular Biology series: They are well established and described in an easy-to-follow, step-by-step fashion so as to be valuable for not only experts but also novices in the stem cell field. That goal is achieved because of the generosity of the contributors who have carefully described their protocols in this volume, and I am very grateful for their efforts.

My thanks as well go to Dr. John Walker, the Editor-in-Chief of the Methods in Molecular Biology series, for giving me the opportunity to create this volume and for supporting me along the way.

I am also grateful to Patrick Marton, the Executive Editor of Methods in Molecular Biology and the Springer Protocols collection, for his continuous support from idea to completion of this volume.

I would also like to thank David C. Casey, Senior Editor with the Methods in Molecular Biology series, for his outstanding editorial work during the production of this volume.

Finally, I would like to thank the production crew for their work in putting together an outstanding volume.

Ottawa, ON, Canada *Kursad Turksen*

Contents

Contributors

SEVIN ADIGUZEL • *Nanotechnology Research and Application Center, Sabanci University, Tuzla, Istanbul, Türkiye*
HIVA AHVATI • *School of Biology, College of Science, University of Tehran, Tehran, Iran*
AYTEKIN AKYOL • *Department of Medical Pathology, Faculty of Medicine, Hacettepe University, Ankara, Turkiye*
ALAA EMAM ALI • *Department of Pharmacology and Toxicology, Faculty of Pharmacy, Ain Shams University, Cairo, Egypt*
JAVAD ALIZADEH • *Department of Human Anatomy and Cell Science, Max Rady College of Medicine, Rady Faculty of Health Sciences, University of Manitoba, Winnipeg, MB, Canada*
BRIAN J. ANESKIEVICH • *Department of Pharmaceutical Sciences, School of Pharmacy, University of Connecticut, Storrs, CT, USA*
SAMAR S. AZAB • *Department of Pharmacology and Toxicology, Faculty of Pharmacy, Ain Shams University, Cairo, Egypt*
SREEPARNA BANERJEE • *Department of Biochemistry, Orta Dogu Teknik Universitesi, Ankara, Turkiye; Department of Biological Sciences, Orta Dogu Teknik Universitesi, Ankara, Turkiye*
AMIR BARZEGAR BEHROOZ • *Department of Human Anatomy and Cell Science, Max Rady College of Medicine, Rady Faculty of Health Sciences, University of Manitoba, Winnipeg, MB, Canada*
GAMZE BILDIK • *The Graduate School of Health Sciences, Koç University, Istanbul, Turkey; Department of Experimental Therapeutics, The University of Texas MD Anderson Cancer Center, Houston, TX, USA*
EMILIO BOADA-ROMERO • *Department of Immunology, St. Jude Children's Research Hospital, Memphis, TN, USA*
SANDRINE BOURGOIN-VOILLARD • *Université Grenoble Alpes, CNRS, U5525, Grenoble INP, TIMC, EPSP, Grenoble, France; CHU Grenoble Alpes, Institut de Biologie et de Pathologie, PROMETHEE Platform, Grenoble, France*
PAOLA BRUN • *Department of Molecular Medicine, Histology Unit, University of Padova, Padua, Italy*
LIAM E. CARMAN • *Graduate Program in Pharmacology & Toxicology, University of Connecticut, Storrs, CT, USA*
PRASHEN CHELIKANI • *Manitoba Chemosensory Biology (MCSB) Research Group, Winnipeg, MB, Canada; Department of Oral Biology, Dr. Gerald Niznick College of Dentistry, Rady Faculty of Health Sciences, University of Manitoba, Winnipeg, MB, Canada; Department of Biochemistry and Medical Genetics, Department of Physiology and Pathophysiology, Max Rady College of Medicine, Rady Faculty of Health Sciences, University of Manitoba, Winnipeg, MB, Canada*
COURTNEY CLARK • *Department of Human Anatomy and Cell Science, Max Rady College of Medicine, Rady Faculty of Health Sciences, University of Manitoba, Winnipeg, MB, Canada*

MARCO CORDANI • *Department of Biochemistry and Molecular Biology, Faculty of Biology, Complutense University of Madrid, Madrid, Spain; Instituto de Investigación Sanitaria San Carlos (IdISSC), Madrid, Spain*
SIMONE C. DA SILVA ROSA • *Department of Human Anatomy and Cell Science, Max Rady College of Medicine, Rady Faculty of Health Sciences, University of Manitoba, Winnipeg, MB, Canada*
DOAA MOKHTAR ELSHERBINY • *Department of Pharmacology and Toxicology, Faculty of Pharmacy, Ain Shams University, Cairo, Egypt*
YASHAR ESMAEILIAN • *Research Center for Translational Medicine, Koç University, Istanbul, Turkey*
MOHAMMAD FAYYAD-KAZAN • *Department of Natural and Applied Sciences, School of Arts and Sciences, The American University of Iraq – Baghdad, Baghdad, Iraq*
EMEL BAŞAK GENCER AKÇOK • *Faculty of Life and Natural Sciences, Department of Molecular Biology and Genetics, Abdullah Gul University, Kayseri, Turkey*
SAEID GHAVAMI • *Biology of Breathing Theme, Children Hospital Research Institute of Manitoba, University of Manitoba, Winnipeg, MB, Canada; University of Manitoba, Winnipeg, MB, Canada; Biology of Breathing Theme, Children Hospital Research Institute of Manitoba, University of Manitoba, Winnipeg, Canada; Academy of Silesia, Faculty of Medicine, Katowice, Poland; Research Institutes of Oncology and Hematology, Cancer Care Manitoba-University of Manitoba, Winnipeg, MB, Canada; Department of Human Anatomy and Cell Science, University of Manitoba College of Medicine, Winnipeg, MB, Canada; Faculty of Medicine in Zabrze, University of Technology in Katowice, Zabrze, Poland*
JOSEPH W. GORDON • *Department of Human Anatomy and Cell Science, Max Rady College of Medicine, Rady Faculty of Health Sciences, University of Manitoba, Winnipeg, Canada*
DOUGLAS R. GREEN • *Department of Immunology, St. Jude Children's Research Hospital, Memphis, TN, USA*
ALIYE EZGI GÜLEÇ TAŞKIRAN • *Department of Biological Sciences, Orta Dogu Teknik Universitesi, Ankara, Turkiye; Department of Molecular Biology and Genetics, Başkent University, Ankara, Turkiye*
H. HAZAL HÜSNÜGIL • *Department of Biological Sciences, Orta Dogu Teknik Universitesi, Ankara, Turkiye*
ECE ILTUMUR • *The Graduate School of Health Sciences, Koç University, Istanbul, Turkey*
HUIMIN JIANG • *Department of Dermatology, Northwestern University, Chicago, IL, USA; Department of Ophthalmology, The Second Hospital of Anhui Medical University, Hefei, China*
MOHAMMAD HOSSEIN KANANI • *Department of Biology, Faculty of Science, University of Guilan, Rasht, Iran*
NIHAL KAPLAN • *Department of Dermatology, Northwestern University, Chicago, IL, USA*
PRATIMA KHANDAYATARAY • *Department of Biotechnology, Academy of Management and Information Technology, Utkal University, Bhubaneswar, Odisha, India*
AREZOO KHOSRAVI • *Department of Genetics and Bioengineering, Faculty of Engineering and Natural Sciences, Istanbul Okan University, Istanbul, Türkiye*
BINIAM KIDANE • *Department of Surgery, Max Rady College of Medicine, University of Manitoba, Winnipeg, MB, Canada*
MRUNMAYEE KULKARNI • *Developmental Biology Group, MACS-Agharkar Research Institute, Pune, Maharashtra, India; Department of Biotechnology, Savitribai Phule Pune University, Ganeshkhind, Pune, Maharashtra, India*

Robert M. Lavker • *Department of Dermatology, Northwestern University, Chicago, IL, USA*
Andrea Leonardi • *Department of Neuroscience, Ophthalmology Unit, University of Padova, Padua, Italy*
Zhenrui Li • *Department of Immunology, St. Jude Children's Research Hospital, Memphis, TN, USA*
Min Liu • *Department of Dermatology, Northwestern University, Chicago, IL, USA*
Suli Lv • *Department of Biochemistry and Molecular Biology, School of Basic Medicine, Tongji Medical College, Huazhong University of Science and Technology, Wuhan, China*
Joelle Magné • *Department of Immunology, St. Jude Children's Research Hospital, Memphis, TN, USA*
Luigi Mari • *Department of Immunology, St. Jude Children's Research Hospital, Memphis, TN, USA*
Parvaneh Mehrbod • *Department of Influenza and Respiratory Viruses, Pasteur Institute of Iran, Tehran, Iran*
Nazli Şevval Menemenli • *Department of Biochemistry, Orta Dogu Teknik Universitesi, Ankara, Turkiye*
Chandra Sekhar Mohanty • *Plant Genetic Resources and Improvement Division, CSIR-National Botanical Research Institute, Lucknow, Uttar Pradesh, India*
Meesala Krishna Murthy • *Department of Allied Health Sciences, Chitkara School of Health Sciences, Chitkara University, Rajpura, Punjab, India*
Zandile Nxumalo • *Synthetic Nanobiotechnology and Biomachines, Synthetic Biology and Precision Medicine Centre, Council for Scientific and Industrial Research, Pretoria, South Africa*
Ozgur Oktem • *Research Center for Translational Medicine, Koç University, Istanbul, Turkey; The Graduate School of Health Sciences, Koç University, Istanbul, Turkey; Department of Obstetrics and Gynecology, School of Medicine, Koç University, Istanbul, Turkey*
Somtochukwu S. Onwah • *Department of Immunology, Max Rady College of Medicine, University of Manitoba, Winnipeg, MB, Canada*
Göksu Oral • *Department of Biological Sciences, Orta Dogu Teknik Universitesi, Ankara, Turkiye*
Özün Özcan • *Department of Biological Sciences, Orta Dogu Teknik Universitesi, Ankara, Turkiye*
Rojalin Pattanayak • *Department of Zoology, College of Basic Science and Humanities, Odisha University of Agriculture and Technology, Bhubaneswar, Odisha, India*
Han Peng • *Department of Dermatology, Northwestern University, Chicago, IL, USA*
Walid Rachidi • *Université Grenoble Alpes, CEA, Inserm, IRIG-BGE UA13, Grenoble, France*
Mohammad Rasouli • *Department of Biology, Faculty of Science, University of Guilan, Rasht, Iran*
Amir Ravandi • *Department of Physiology and Pathophysiology, Max Rady Faculty of Health Sciences, University of Manitoba, Winnipeg, MB, Canada*
Umberto Rosani • *Department of Biology, University of Padova, Padova, Italy*
Fatemeh Safari • *Department of Biology, Faculty of Science, University of Guilan, Rasht, Iran*
Dibyaranjan Samal • *Department of Biotechnology, Sri Satya Sai University of Technical and Medical Sciences, Sehore, Bhopal, Madhya Pradesh, India*

Michael L. Samulevich • *Graduate Program in Pharmacology & Toxicology, University of Connecticut, Storrs, CT, USA*

Merve Şansaçar • *Graduate School of Engineering and Science, Department of Bioengineering, Abdullah Gul University, Kayseri, Turkey; Faculty of Engineering, Department of Genetic and Bioengineering, Alanya Alaaddin Keykubat University, Alanya, Antalya, Turkey*

Nourhan Sayed • *Department of Pharmacology and Toxicology, Faculty of Pharmacy, Ain Shams University, Cairo, Egypt*

Karan Selarka • *Developmental Biology Group, MACS-Agharkar Research Institute, Pune, Maharashtra, India; Department of Biotechnology, Savitribai Phule Pune University, Ganeshkhind, Pune, Maharashtra, India*

Michel Seve • *Université Grenoble Alpes, CNRS, U5525, Grenoble INP, TIMC, EPSP, Grenoble, France; CHU Grenoble Alpes, Institut de Biologie et de Pathologie, PROMETHEE Platform, Grenoble, France*

Serap Sezen • *Department of Molecular Biology, Genetics and Bioengineering, Faculty of Engineering and Natural Sciences, Sabanci University, Tuzla, Istanbul, Türkiye; Nanotechnology Research and Application Center, Sabanci University, Tuzla, Istanbul, Türkiye*

Shahla Shojaei • *Department of Human Anatomy and Cell Science, Max Rady College of Medicine, Rady Faculty of Health Sciences, University of Manitoba, Winnipeg, MB, Canada*

Bhupendra V. Shravage • *Developmental Biology Group, MACS-Agharkar Research Institute, Pune, Maharashtra, India; Department of Biotechnology, Savitribai Phule Pune University, Ganeshkhind, Pune, Maharashtra, India; Department of Zoology, Savitribai Phule Pune University, Ganeshkhind, Pune, Maharashtra, India*

Nisha Singh • *Manitoba Chemosensory Biology (MCSB) Research Group, Winnipeg, MB, Canada; Department of Oral Biology, Dr. Gerald Niznick College of Dentistry, Rady Faculty of Health Sciences, University of Manitoba, Winnipeg, MB, Canada*

Tanjing Song • *Department of Biochemistry and Molecular Biology, School of Basic Medicine, Tongji Medical College, Huazhong University of Science and Technology, Wuhan, China; Cell Architecture Research Institute, Huazhong University of Science and Technology, Wuhan, Hubei, China*

Lidong Sun • *Department of Biochemistry and Molecular Biology, School of Basic Medicine, Tongji Medical College, Huazhong University of Science and Technology, Wuhan, China; Cell Architecture Research Institute, Huazhong University of Science and Technology, Wuhan, Hubei, China*

Arun Surendran • *Mass Spectrometry Core Facility, Rajiv Gandhi Centre for Biotechnology, Thiruvananthapuram, Kerala, India*

Mutsa M. Takundwa • *Synthetic Nanobiotechnology and Biomachines, Synthetic Biology and Precision Medicine Centre, Council for Scientific and Industrial Research, Pretoria, South Africa*

Deepak B. Thimiri Govinda Raj • *Synthetic Nanobiotechnology and Biomachines, Synthetic Biology and Precision Medicine Centre, Council for Scientific and Industrial Research, Pretoria, South Africa*

Jude E. Uzonna • *Department of Immunology, Max Rady College of Medicine, University of Manitoba, Winnipeg, MB, Canada; Department of Medical Microbiology and Infectious Diseases, Max Rady College of Medicine, University of Manitoba, Winnipeg, MB, Canada;*

Department of Pathology, Max Rady College of Medicine, University of Manitoba, Winnipeg, MB, Canada

RUI VITORINO • *Department of Surgery and Physiology, Faculty of Medicine, University of Porto, Porto, Portugal; Department of Medical Sciences, Institute of Biomedicine iBiMED, University of Aveiro, Aveiro, Portugal*

HAREGEWOIN BEZU WOLDEKIDAN • *Department of Biotechnology, College of Natural and Applied Science, Addis Ababa Science and Technology University, Addis Ababa, Ethiopia; Synthetic Nanobiotechnology and Biomachines, Synthetic Biology and Precision Medicine Centre, Council for Scientific and Industrial Research, Pretoria, South Africa*

ADUGNA ABDI WOLDESEMAYAT • *Department of Biotechnology, College of Natural and Applied Science, Addis Ababa Science and Technology University, Addis Ababa, Ethiopia*

SEVGI YUSUFOGLU • *The Graduate School of Health Sciences, Koç University, Istanbul, Turkey*

ATEFEH ZAREPOUR • *Department of Research Analytics, Saveetha Dental College and Hospitals, Saveetha Institute of Medical and Technical Sciences, Saveetha University, Chennai, Tamil Nadu, India*

ALI ZARRABI • *Department of Biomedical Engineering, Faculty of Engineering and Natural Sciences, Istinye University, Istanbul, Türkiye; Graduate School of Biotechnology and Bioengineering, Yuan Ze University, Taoyuan, Taiwan*

Methods in Molecular Biology (2025) 2879: 1–22
DOI 10.1007/7651_2023_508

Published online: 15 December 2023

Autophagy Analysis: A Step-by-Step Simple Practical Guide for Immunofluorescence and Western Blotting

Meesala Krishna Murthy, Pratima Khandayataray, Dibyaranjan Samal, Rojalin Pattanayak, and Chandra Sekhar Mohanty

Abstract

Autophagy is a vital cellular process responsible for breaking down faulty cellular components and organelles, ultimately routed through lysosomes for degradation. This intricate mechanism involves the translocation of LC3, a cytoplasmic protein, onto the autophagosome membranes. As a result, it becomes feasible to discern cells engaged in autophagy by employing fluorescent markers designed for LC3 or other indicative autophagy markers. Although a variety of techniques such as immunofluorescence and western blotting serve as indispensable tools for assessing autophagy, the definitive confirmation comes from the visualization of autophagosomes using transmission electron microscopy. While numerous protocols for antibody staining can be found in scientific literature and on antibody suppliers' websites, these procedures often demand significant time and financial resources for setup. This chapter endeavors to provide a user-friendly and cost-effective guide for practitioners seeking proficiency in immunofluorescence staining and western blotting techniques.

Key words Immunofluorescence staining, Western blotting, SDS-PAGE, Antibodies, Microscope

1 Introduction

Autophagy, a meticulously regulated and widely preserved mechanism, serves to maintain cellular equilibrium by breaking down accrued proteins and impaired cellular components [1]. This fundamental biological process assumes a pivotal role in sculpting cells throughout their differentiation and developmental stages [2, 3]. Furthermore, it exerts a significant influence on the progression of diverse ailments, including but not limited to cancer [4–6], neurodegenerative disorders [7, 8], inflammatory conditions [9, 10], viral infections [11, 12], and various other medical conditions.

Proteins' localization within fixed biological samples is commonly explored through immunofluorescence (IF) staining, a widely embraced and practical approach [13–15]. This method hinges on the specific interplay between antibodies and antigens and is deployable via two distinct approaches. First, the direct technique involves a primary antibody linked to a fluorophore, allowing it to directly bind to the target antigen, thereby facilitating direct visualization of fluorescence through a microscope. Conversely, the indirect approach employs a secondary antibody marked with a fluorophore, capable of binding to the unaltered primary antibody [14, 16]. This versatile methodology finds utility in the examination of both cells, known as immunocytochemistry (ICC), and tissues, referred to as immunohistochemistry (IHC) [17]. In the context of tissue samples, IHC is applicable to specimens preserved in either paraffin or cryopreservation.

One of the methods employed in scientific research for the separation and identification of proteins is known as the western blot (WB) technique. This method facilitates the segregation of a protein mixture based on their respective molecular weights, using a process called gel electrophoresis. Subsequently, these proteins are transferred onto a membrane, where each protein uniquely manifests itself as a distinct band [18–21]. This membrane is then subjected to treatment with specifically labeled antibodies that selectively bind to the target protein of interest. After rinsing away any excess antibodies, the antibodies that have successfully bound to the target protein are visualized by generating a film. The identification of the target protein is discerned through the presence of a solitary band on the film, with the band's intensity serving as an indicator of the protein quantity within the sample. By employing a standardized reference, the precise protein quantity can be determined.

Many newcomers to the field of research encounter significant hurdles when it comes to identifying and examining their target protein signals, despite the widespread use of these techniques. Frequently, doubts arise regarding the effectiveness and reliability of their antibodies when staining and bands yield subpar results or outright failures. This situation translates into a substantial expenditure of both time and resources as researchers are compelled to seek alternative antibodies. The primary objective of this chapter is to furnish researchers, particularly those new to the realms of IF and WB, with comprehensive and structured procedures. These meticulously outlined protocols aim to facilitate improved outcomes, such as robust and unambiguous signals pertaining to their specific protein of interest. Moreover, they offer the means to diminish undesirable background signals, which often hinder the identification and analysis of the target protein. Adhering to this step-by-step guidance will also empower researchers to produce high-quality images for both fluorescence and blotting applications, serving as valuable visual aids for inclusion in their final research publications.

1.1 Basic Guidelines Before Beginning

(a) Adhering to the standard laboratory safety protocols is crucial. This includes donning appropriate protective attire, such as lab coats, gloves, and masks, and conducting experiments within a well-ventilated environment.

(b) The preparation of cells and tissues must take into account both the timing and the fixation method, particularly in the case of certain antibodies that exhibit high sensitivity to the specific fixation procedures [22, 23].

(c) The primary emphasis should center on the antibodies employed in prior research, when relevant. It is crucial to conduct a thorough assessment of the immunofluorescence staining's efficacy by considering the signal-to-noise ratio.

(d) To locate antibodies directed at a particular protein, individuals can explore commercial antibody manufacturers' websites or consult trusted scientific references like Biocompare1.

(e) Validation experiments play a crucial role in assessing the performance of newly developed commercial or homemade antibodies, ensuring their specificity, selectivity, and reproducibility [23–26]. These experiments involve a range of established techniques such as western blotting, blocking peptides, and negative controls [26, 27]. To affirm the antibody's specificity, it is imperative to employ both positive and negative controls tailored to the specific experimental conditions and available laboratory resources [23–26, 28, 29].

(f) Ensure that you thoroughly examine and record the guidelines pertaining to storage conditions and the anticipated duration of viability for antibodies. Typically, this information can be found within the accompanying documentation provided with the antibodies.

(g) The primary objective of this procedure is to outline the methodology for conducting IHC/IHC on cells/tissues that have been preserved in paraffin. Nevertheless, it is worth noting that this protocol can be adapted for tissues preserved through cryopreservation, requiring the omission of the dewaxing and heat-mediated antigen retrieval stages.

(h) It is generally unnecessary to alter the temperature (25 °C) for the majority of subsequent experiments. Nevertheless, certain antibodies may exhibit improved performance at 4 °C, as suggested by the antibody manufacturer. To investigate this potential effect, one may consider conducting an incubation experiment using two separate samples at both of these temperature settings.

(i) Utilize exclusively ultrapure water, achieved through additional purification of deionized water to attain a resistance of 18 MΩ-cm at 25 °C, in the formulation of all solutions.

Employ reagents of an analytical grade for chemical processes. Keep and employ these reagents under standard room temperature conditions, unless specific instructions dictate otherwise. Adhere strictly to waste disposal regulations when discarding materials. Avoid incorporating sodium azide within the reagent composition.

2 Materials

2.1 Preparation of Solutions for Immunofluorescence Staining Technique

In order to conduct the dewaxing process effectively, it is essential to create roughly 300 mL of the prescribed solutions. This quantity ensures proper filling of every histological staining container, as outlined in the forthcoming guidelines (Fig. 1a):

2.1.1 Solutions for the Dewaxing Step

(a) We obtained pairs of glass containers containing solutions with different concentrations of ethanol, specifically 100%, 95%, and 70% ethanol (chemical formula C_2H_6O; K928.4, Carl Roth GmbH, Germany).

(b) Three clear containers holding xylene C_8H_{10}; 9713.3, Carl Roth GmbH).

(c) An individual glass container containing a mixture of xylene and ethanol in equal proportions (1:1).

The solutions mentioned earlier need to be prepared in advance and then stored in tightly sealed glass containers at room temperature for an extended period. Additionally, these substances can be used repeatedly and only require replacement when they become dirty (see Fig. 1b).

2.1.2 Solution for the Heat Antigen Retrieval (Unmasking) Step

The preparation of the unmasking solution involves mixing 3.6 mL of a citrate-based antigen retrieval solution (pH 6.0) from Vector Laboratories in the United States with 400 mL of deionized water (d-H_2O). It is essential to continuously monitor the pH levels since specific antigens may demand pH levels within the 3–10 range. Nonetheless, for the majority of antigens, a pH of 6.0 is generally adequate [22, 30]. This unmasking solution should be freshly prepared at the start of each experiment and is designated for single use, typically accommodating one batch of paraffin slides, such as 10 or 20 slides consecutively. In scenarios where a larger number of slides are involved, it is advisable to prepare an additional batch of the unmasking solution.

2.1.3 Solutions for the Rinsing and Permeabilization Steps

To carry out the rinsing procedure, a 1000 mL solution of 1X phosphate-buffered saline (PBS) was prepared. This was accomplished by mixing 100 mL of PBS 10X (Product ID: 14190-250, Thermo Fisher Scientific Inc.) with 900 mL of distilled water (d-H_2O). This prepared solution can be conveniently made in

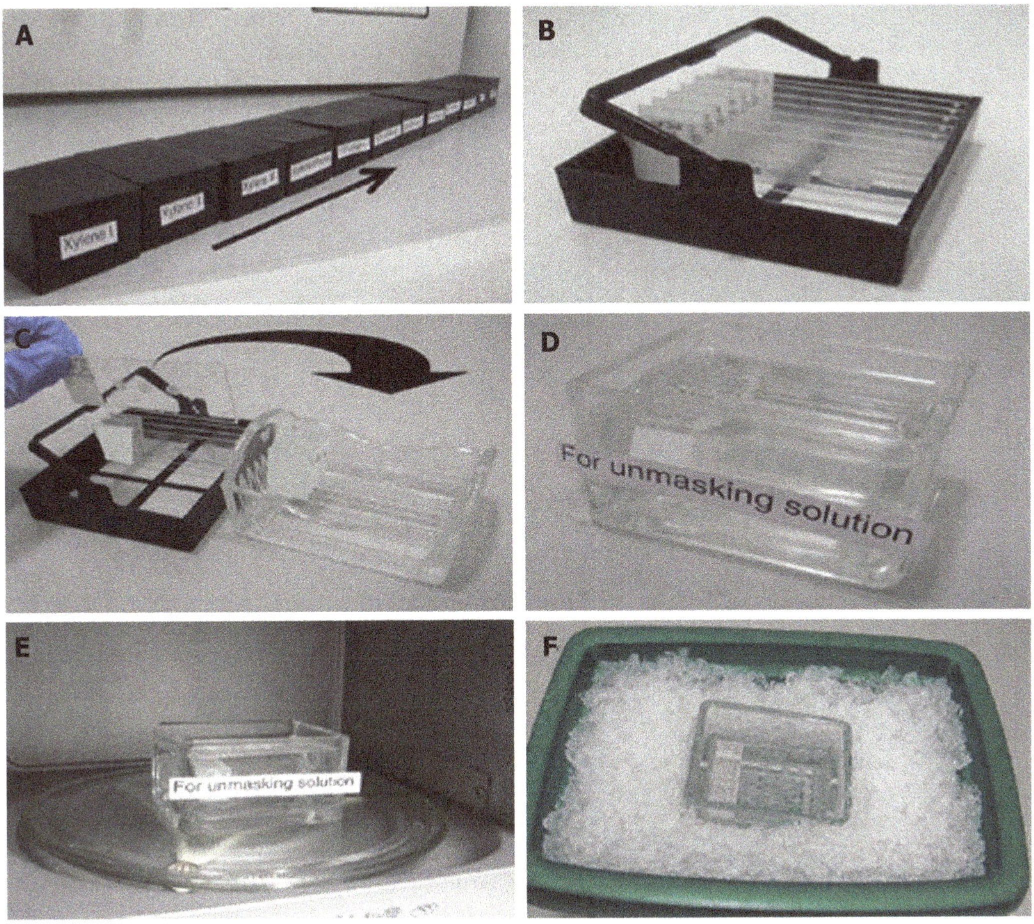

Fig. 1 To prepare histological samples for staining, the following steps are followed: (**a**) Cleaning and proper labeling of histological staining boxes, which are subsequently filled with the required solutions. (**b**) Placing the paraffin section slides onto staining racks. (**c**) Transferring the slides, after dewaxing, onto glass staining racks, and subsequently placing them in glass staining containers. (**d**) Adding the unmasking solution to the containers. (**e**) Subjecting the container with the slides. (**f**) 3 min microwave treatment after boiling, followed by cooling to approximately 50 °C using an ice box [38]

advance and stored at room temperature. It is recommended to use this solution within a 1-month timeframe. To facilitate permeabilization, a permeation-enhancing mixture is formulated by blending 1X PBS, gelatin (0.2% w/v), and Triton X-100 (0.25% v/v). It is essential to store this preparation at a constant temperature of 4 °C, and it is advisable to employ it within a span of 10 days:

(a) To create a 500 mL solution of PBS 1X, the procedure involves the use of a 1000 mL cylinder.

(b) A 500 mL glass beaker was employed to gently warm a mixture composed of 500 mL of distilled water and 2 g of gelatin (product code: 1040781000, manufacturer: Merck KGaA,

Germany) until it reached a temperature of 55 °C, ensuring the complete dissolution of the gelatin.

(c) Afterward, the gelatin solution that has previously been dissolved is added to a 1X PBS solution, previously prepared during the initial phase.

(d) After the solution has undergone the cooling process, we add 2.5 mL of Triton 100X (catalog number 9002931, Sigma-Aldrich, Germany) to the mixture, paying careful attention to its adhesive characteristics.

(e) The final resolution involves a thorough integration process, where plastic-coated magnetic stirring bars are skillfully incorporated.

(f) First, the process involves passing the substance through a filter paper (240 mm; 4.303.240, Neolab, Germany) and subsequently transferring it into a glass container with a pH level set at 7.4 [31].

2.1.4 Solution for the Blocking Step

A 5% bovine serum albumin (BSA) solution is prepared by dissolving 0.25 g of BSA (catalog number A 9647, obtained from Sigma-Aldrich) in 5 mL of a permeabilization solution at a pH of 7.4. This preparation method can be found in the "Solutions for the Rinsing and Permeabilization Steps" section. This solution is specifically designed to reduce nonspecific binding and should be freshly made and stored at 4 °C. It is advisable to use this solution within 48 h of preparation. Additionally, it can be diluted to a 1% concentration for use in primary and secondary antibody dilutions.

2.1.5 Solutions for the Primary and Secondary Antibody Incubation Steps

In a general context, primary antibodies can be divided into two primary categories: monoclonal and polyclonal antibodies. Monoclonal antibodies demonstrate their specificity by binding exclusively to a single epitope, while polyclonal antibodies exhibit sensitivity by binding to multiple epitopes [29]. To minimize the potential for cross-reactivity, which may arise due to a strong affinity between the antibody and a structurally similar antigen, it is advisable to choose a primary antibody generated from a different species than the animal model being studied [22, 32, 33]. An ideal choice for utilizing a rabbit primary antibody would involve its application on mouse cells or tissues. To conduct double or triple IF, it is essential to use primary antibodies sourced from distinct species. This approach enables the separate detection of each primary antibody using secondary antibodies emitting fluorescence at different wavelengths, such as green (488 nm), red (555 nm), and far red (633 nm). As discussed by Marshall and Johnsen in 2017, antibodies labeled with blue or far-red fluorophores are considerably less visible, or sometimes entirely imperceptible, to the human eye, in contrast to antibodies labeled with green-red fluorophores [34]. In specific situations, the use of directly conjugated primary antibodies eliminates the necessity for employing separate secondary antibodies.

The use of a refrigerator is essential for maintaining consistently low temperatures to safeguard the effectiveness and stability of antibodies and their dilutions. The required concentrations of both primary and secondary antibodies are achieved by mixing them in a solution containing 1% BSA, as explained in the "Blocking Step Solution" section. The process of preparing the primary antibody solutions occurs on the initial day of the experiment, particularly during the blocking process detailed in the "Dewaxing Step" section. Conversely, the secondary antibody solutions are prepared on the subsequent day, specifically during the second rinsing step outlined in the "Second Rinsing Step" section.

To initiate the experiment, it is recommended to assess three distinct concentrations of the primary antibody. These concentrations should encompass the manufacturer's recommended dilution or those found in relevant literature, as well as concentrations both higher and lower than the suggested dilution [29]. For instance, if the recommended dilution is 1:300, researchers may experiment with dilutions of 1:100, 1:300, and 1:500. According to the findings of Hoffman et al. (2016), evaluating various concentrations of secondary antibodies is advisable to determine the concentration that yields minimal background noise while maintaining optimal signal strength [35]. In our experimental setting, a concentration of 1:400 has been identified as the optimal choice for most secondary antibodies.

To enhance the detection of nuclei signals, DAPI (1 mL/mL; D9542, Sigma-Aldrich) is added to the secondary antibody solution. The volume of these concentrations is determined based on the number of paraffin microscopic slides requiring staining. Typically, a volume of 200 mL for the antibody dilution per slide is sufficient. However, it is prudent to account for an additional slide during the calculation process to avoid running out of solution for the final slide. For example, to prepare the antibody solution for four slides, it is advisable to create a 1000 mL solution by multiplying five times 200 mL.

In cases involving the use of double or triple IF, the preparation of two or three primary antibodies, respectively, is carried out within the same 1% BSA solution on the first day of the experiment. It is crucial to ensure that secondary antibodies compatible with different fluorescence detection wavelengths are readily available. On the second day of the experiment, the preparation of matching secondary antibodies, along with DAPI, is also conducted in the 1% BSA solution. The antibodies employed for the IF tests in this study are well established and have been previously utilized by other researchers in various cell and tissue studies (see Table 1) [36].

Table 1
List of the primary and secondary antibodies used in autophagy studies [36]

Name	Source	Catalog no.	Dilution	Function
Anti-ULKl	Sigma-Aldrich, Germany	A7481	1:100	Kinase required for formation of the autophagosome
Anti-LC3B	Sigma-Aldrich, Germany	L7543	1:100	Ubiquitin-like modifier involved in formation of autophagosomal vacuoles (autophagosomes)
Anti-LC3	Sigma-Aldrich, Germany	L8918	1:100	Small ubiquitin protein conjugated to PE in autophagic membranes; interacts with proteins containing AIM or LIR motifs
Anti-P62	Abcam, United Kingdom	ab56416	1:100	Ubiquitin-binding autophagy receptor involved in selective autophagy
Alexa fluor 555-labeled donkey anti-mouse IgG	Invitrogen, United States	A31570	1:400	–
Alexa fluor 488-labeled donkey anti-rabbit IgG antibodies	Invitrogen, United States	A21206	1:300	–

2.1.6 Coverslip Preparation

The 22 mm × 22 mm coverslips, with a thickness ranging from 0.13 to 0.17 mm, are placed into a cleaning solution composed of kalium bichromicum, concentrated sulfuric acid, and distilled water, mixed in a ratio of 120 g:200 mL:200 mL. These coverslips soak in the cleaning solution overnight. Once removed from their initial position, they undergo a thorough rinsing with tap water until all traces of the cleaning solution are completely removed. To ensure cleanliness, the coverslips then undergo ten rinses with ultrapure water. Following these cleansing steps, the coverslips are immersed in a 75% ethanol solution for subsequent experiments. In experimental setups, a glass-bottom dish, boasting a 35 mm diameter and a 14 mm well with a glass thickness varying from 0.13 to 0.16 mm, proves invaluable, especially when dealing with challenging cell adhesion or when observing live cells is a priority.

2.2 Preparation of Solutions for Western Blotting Technique

2.2.1 Preparation of SDS (Sodium Dodecyl Sulfate) Polyacrylamide Gel

1. ***Resolving gel buffer***: Prepare an 8.8 pH Tris–HCl solution with a concentration of 1.5 M. Start by adding approximately 100 mL of distilled water to either a 1 L graduated cylinder or a glass beaker (refer to **Note 1** for guidance). Weigh out precisely 181.7 g of Tris–HCl and gently transfer it into the chosen container. Next, add an appropriate amount of water to reach a total volume of 900 mL in the container. Adjust the pH as

needed by carefully introducing hydrochloric acid (HCl) (consult **Note 2** for instructions). Ensure that the final solution does not exceed a volume of 1 L. For optimal preservation, store the solution at a temperature of 4 °C.

2. ***Stacking gel buffer***: Prepare a solution by combining 60.6 g of Tris–HCl with a concentration of 0.5 M and a pH level of 6.8. To prepare this solution with a volume of 1 L, follow the instructions provided in the previous step. It is advisable to store this solution at a temperature of 4 °C for optimal preservation.
3. ***Thirty percent acrylamide/Bis solution (29.2:0.8) acrylamide: bis***: Weigh out 29.2 g of acrylamide monomer and 0.8 g of Bis (cross-linker). Carefully transfer these measured substances into a 100 mL graduated cylinder already containing around 40 mL of water. Introduce a spatula's worth of AG 501-X8 (D) mixed-resin beads into the mixture and agitate it for approximately 30 min. Create a 100 mL solution by diluting with water and then filter it using a 0.45 μm Corning filter (refer to **Note 3**). Store the sample at a temperature of 4 °C in a container wrapped with aluminum foil (refer to **Note 4**).
4. ***Ammonium persulfate (APS)***: A solution containing 10% concentration of the APS in an aqueous medium (refer to **Note 5**).
5. ***N, N, N, N′-Tetramethyl-ethylenediamine***: Please refrigerate at 4 °C (refer to **Note 6**).
6. ***SDS-PAGE running buffer***: In the solution containing 0.025 M Tris–HCl adjusted to a pH of 8.3, combined with 0.192 M glycine, and supplemented with 0.1 % SDS (as described in **Note 7**), a unique blend of components was used.
7. ***SDS lysis buffer (5X)***: The experimental solution employed in this study comprised a mixture composed of 0.3 M Tris–HCl with a pH of 6.8, 10% SDS, 25% β-mercaptoethanol, 0.1% bromophenol blue, and 45% glycerol. It is important to note that one portion of this solution should be maintained at a temperature of 4 °C for instant application, while the rest of the aliquots should be preserved at −20 °C, as outlined in **Note 8** for future use.
8. ***Bromophenol blue (BPB) solution***: To prepare a solution of BPB, blend 0.1 g of BPB powder with 100 mL of distilled water.

2.2.2 Preparation of Western Blotting Apparatus

1. Nitrocellulose membranes.
2. ***Western blot transfer buffer***: Prepare a solution comprising 0.025 M Tris–HCl and 0.192 M glycine, while incorporating 20% methanol, as detailed in **Note 9** for reference.

3. ***Tris-buffered saline (TBS; 10X)***: The solution comprises a mixture containing 1.5 moles of sodium chloride (NaCl) and 0.1 moles of Tris–hydrochloric acid (Tris–HCl) with a pH level adjusted to 7.4.
4. In the solution, TBS with a trace of 0.05% Tween-20 (commonly known as TBST) was used as a mild detergent. This solution is often employed in various experiments to facilitate gentle sample washing and improve the effectiveness of molecular interactions in a controlled environment.
5. ***Blocking solution***: Preserve the solution containing 5% milk in TBS (refer to **Note 10**) by storing it at a temperature of 4 °C.
6. ***Diluent solution***: Prepare a solution by combining 5% milk with TBST as outlined in **Note 10**. Keep this mixture stored at a temperature of 4 °C.
7. Mini PROTEAN® 3 System glass plates.
8. Medium binder clips measuring 1¼ inches in size.
9. Plastic container.
10. Wypall X-60-reinforced paper.
11. ***Nitro blue tetrazolium (NBT)/5-bromo-4-chloro-3-indolyl phosphate (BCIP)***: To prepare the solution, dissolve 1 g of NBT in 20 mL of dimethylformamide (DMF) with a concentration of 70%. Simultaneously, prepare another solution by dissolving 1 g of BCIP in 20 mL of pure, 100% DMF. Before applying this mixture onto the membrane, incorporate 33 μL of BCIP and 66 μL of NBT into 10 mL of alkaline phosphatase buffer. Alternatively, you can opt for the pre-made 1-Step™ NBT/BCIP combination for convenience.

3 Methods

3.1 Immunofluorescence Method

3.1.1 Indirect Immunofluorescence

Unless specified otherwise, it is advisable to conduct all operations under standard room temperature conditions. Now, let us delve into the protocols outlining the steps for performing immunofluorescence staining on adherent cells:

1. Prepare a coverslip by retrieving it from a solution composed of 75% ethanol. Afterward, gently dry the coverslip using a spirit lamp. Following this, carefully position the coverslip inside either a 6-well culture plate or a 35 mm cell culture dish. It is worth noting that this step can be omitted if a dish featuring a glass bottom is employed.
2. Cultivated cell cultures should be allowed to proliferate on coverslips until they reach a cellular density ranging from 60% to 80%.

3. To commence the experiment, it is advisable to remove half of the medium from the petri dish. Following this step, introduce an equal quantity of a 4% paraformaldehyde (PFA) solution into the dish. Allow the resulting blend to incubate at a temperature of 37 °C for a period of 5–10 min.
4. The next step involves replacing the supernatant with 1 mL of a 4% PFA solution within the petri dish. Subsequently, allow for an incubation period of 15 min at a temperature of 37 °C.
5. The recommended procedure involves the elimination of the solution, followed by thorough rinsing of the cells using PBS. This rinsing process should be repeated three times, with each rinse lasting for a duration of 5 min.
6. Replace PBS with 1 mL of a permeabilization solution, following the instructions in **Note 11**. Allow this to occur at room temperature for 15–20 min.
7. The recommended procedure involves eliminating the solution, followed by a thorough triple rinse of the cells using PBS. Each rinse should be carried out for a duration of 5 min.
8. The experimental protocol begins with the introduction of 1 mL of a blocking solution buffer, followed by a 30 min incubation period at ambient room conditions (refer to **Note 12** for details).
9. The recommended approach involves discarding the current solution and proceeding to incubate the primary antibody on the coverslip. This incubation should take place for either 1 h at room temperature or overnight at 4 °C within a humidified chamber (refer to **Note 13** for details). To prepare the primary antibody, it should be appropriately diluted using the specified primary antibody diluent, as outlined in **Notes 14** and **15**. The next step involves the simultaneous co-incubation of the two primary antibodies, which is essential for conducting double immunofluorescence staining. However, it is worth noting that in specific situations, a sequential supplementation may be necessary to achieve optimal results.
10. To ensure proper procedure, it is advisable to warm the humid chamber to room temperature for a period of one h. The initial antibody must be carefully retrieved and stored in a 1.5 mL tube, adhering to the recommended storage conditions. Thoroughly cleanse the cells with PBS, repeating this washing step three times, with each rinse lasting 5 min.
11. The secondary antibodies, labeled with fluorescent markers, are appropriately thinned using a secondary antibody dilution solution, which is elaborated in **Notes 16** and **17**. To ensure optimal binding, the secondary antibody should be gently applied to the coverslip and left to incubate for precisely 1 h

at room temperature, all within a well-maintained humid chamber. In order to perform double immunofluorescence staining, both secondary antibodies are merged and undergo co-incubation. Nevertheless, in specific situations, there may arise a necessity for a sequential supplementation approach. It is advised to carry out the ensuing steps in a softly illuminated setting, unless explicitly specified otherwise.

12. The appropriate course of action involves the disposal of the solution, followed by a thorough rinsing of the cells using PBS for a period lasting 3 sets of 5 min each.
13. In specific situations, DAPI staining finds application in enriching our comprehension of cellular structure. It serves the purpose of differentiation between the morphology and spatial arrangement of cytoplasmic and nuclear regions. Following the provided guidelines, the DAPI solution is appropriately diluted with PBS. The ensuing solution is subsequently administered onto a coverslip and permitted to incubate at ambient room temperature for a period of 5 min.
14. The recommended procedure involves the removal of the solution, followed by three thorough rinses of the cells using PBS. Each rinse should be maintained for a duration of 5 min.
15. To ensure proper slide preparation, a 20 μL aliquot of mounting media is swiftly administered onto the slide's surface. Following this, a cover slip is carefully positioned over the mounting media, and the coverslip is firmly secured in position using clear nail polish.

3.1.2 Image Acquisition

The LAS-AF-Lite_2.2.0_4758 TCS SPF5 II confocal microscope by Leica Microsystems in Wetzlar, Germany, was employed to capture cellular images in every experiment. For the quantification of autophagosome relative average areas, Image-Pro Plus 6.0 was applied following the provided guidelines.

3.2 Western Blotting Methods

Unless explicitly stated otherwise, carry out each procedure under normal room temperature conditions.

3.2.1 Cell Lysis to Extract Protein

Protein extraction is a versatile process applicable to a range of materials, such as tissues and cells. Below, we outline the procedural steps for extracting proteins from adherent cells.

1. To prepare the tissue culture flask or plate for further experimentation, it is essential to cleanse the cells by rinsing them with chilled PBS while gently agitating the container. It is

suggested to disregard any use of PBS in this context. Additionally, maintaining a consistently low temperature during the experiment can be achieved by situating the tissue culture dish on a bed of ice.

2. PBS was introduced into the cell culture, and an implement called a cell scraper was employed to delicately disengage the adherent cells from the culture vessel. The subsequent step involved moving this combination into microcentrifuge tubes through the use of a pipette.
3. The centrifugal procedure ought to be executed at a rate of 1500 rotations per minute (RPM) for a period of 5 min. Subsequently, the liquid component situated above the settled material, commonly referred to as the supernatant, should be extracted and properly disposed of.
4. Combine 20 μL of a fresh protease inhibitor mixture with 180 μL of cell lysis buffer chilled on ice. If the protein concentration remains inadequate after the initial attempt, consider repeating the process with a larger proportion of the protease inhibitor blend.
5. The lysate needs to be kept on ice for 30 min, after which it should undergo a clarification step through centrifugation at 12,000 RPM for 10 min, maintaining a temperature of 4 °C.
6. The clear liquid, also known as the protein mixture, must be carefully moved to a fresh tube and can be preserved by either keeping it on ice or storing it at temperatures as low as −20 °C or even −80 °C.
7. Measure the protein content using a spectrophotometer to obtain accurate quantification.

3.2.2 Sample Preparation

1. To determine the necessary amount of protein extract needed to ensure a consistent 50 μg quantity in each well, one can utilize the formula concentration = mass/volume.
2. To maintain consistency, it is advisable to include 5 μL of sample buffer with the sample. Afterward, balance the volume in every lane by using double-distilled water (dd H_2O). Carefully mix these components. It is suggested to allocate a total volume of 15 μL for each lane.
3. The specimens need to undergo a dry plate heating procedure lasting for 5 min, maintaining a temperature of 100 °C throughout the process.

3.2.3 Gel Preparation

Type of gel	Requirements	Quantity
10% Stacking gel (*see* **Note 11**)	dd H_2O	3 mL
	1 M Tris–HCl	2.1 mL (pH 8.9)
	30%Acr Bis	2.8 mL
	10%SDS	80 μL
	10%APS[a]	56 μL
	TEMED[a]	6 μL
6% Separating gel	dd H_2O	2 mL
	1M Tris–HCl	400 μL (pH 6.7)
	30%Acr Bis	600 μL
	10%SDS[a]	36 μL
	10%APS[a]	24 μL
	TEMED	4 μL

[a]*APS* ammonium persulfate, *TEMED* tetramethylethylenediamine, *SDS* sodium dodecyl sulfate

1. After the formulation of the 10% stacking gel solution, the next step involves setting up the gel solidification rack. It is important to highlight that the inclusion of 10% APS and N,N,N′, N′-Tetramethylethylenediamine (TEMED) causes the solution to solidify. Therefore, you can streamline the preparation process by delaying the addition of these reagents until you reach the final stages for both gels.
2. Exercise caution when adding the stacking gel solution and ensure it reaches the level indicated by the green bar securing the glass plates. For optimal results, it is advisable to wait for a solidification period lasting 15–30 min. An effective method to apply the gel onto the glass plate is to employ a suction pipette.
3. Once the excess water has been eliminated, carefully position the separating gel atop the stacking gel. For effective water removal, it is recommended to tilt the apparatus slightly and employ a paper towel as needed.
4. Place the comb into its designated position with precision, ensuring the complete removal of any trapped air bubbles.
5. It is advisable to allocate an ample amount of time for the gel to firm up. One way to confirm the solidification process is by letting a segment of the gel solution settle undisturbed within a container.

3.2.4 Electrophoresis

1. Transfer the flowing buffer into the electrophoresis apparatus.
2. Place the gel substance into the electrophoretic device and establish a link to a power supply. It is recommended to follow the standard procedure of connecting the positive wire to the corresponding terminal and the negative wire to the matching terminal when establishing the power connection.

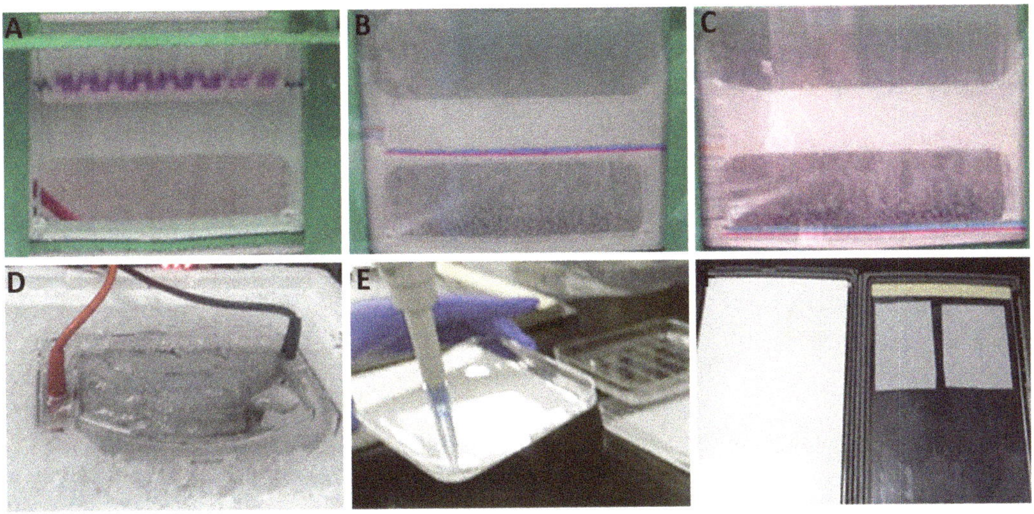

Fig. 2 To achieve optimal results in the process of electrophoresis and western blotting, it is important to follow a set of sequential steps. First, introduce the samples into the stacking gel under lower voltage conditions (**a**). Next, transition to higher voltage settings as the samples progress through the separating gel (**b**). Ensure the gel runs completely to the bottom of the electrophoretic chamber (**c**). For the subsequent transfer step, maintain a cold environment by conducting the transfer on ice (**d**). Afterward, facilitate the membrane's development by incubating it with an ECL mixture, utilizing a 1000 μL pipette to enhance the process (**e**). Finally, when exposing the membrane, employ a suitable cassette within the confines of a dark room (**f**) to obtain accurate and reliable results [39]

3. Make certain that the buffer solution sufficiently envelops the gel, and exercise prudence when extracting the comb.
4. To ensure accurate sample analysis, we recommend the placement of a 6 μL load marker in each well, followed by the addition of 15 μL of the respective samples.
5. The separation gel phase requires the application of a modest 60 V voltage, while the stacking gel phase necessitates a more substantial 140 V voltage, as illustrated in Fig. 2a, b.
6. The gel should undergo an approximate 1-hour electrophoresis process, or until the dye front has entirely reached the lowermost portion of the gel, as illustrated in Fig. 2c. For additional details, kindly refer **Note 18**.

3.2.5 Electrotransfer

1. The process involves trimming six filter sheets and one nitrocellulose/polyvinylidene fluoride (PVDF) membrane to precisely fit the dimensions of the gel, as detailed in **Note 19**.
2. Ensure the sponge and filter papers are dampened using transfer buffer, while moistening the PDVF/nitrocellulose membrane with methanol.
3. The glass plates need to be disassembled to facilitate the extraction of the gel.

4. Permit the gel samples to go through the air-drying procedure for a period ranging from 5 to 10 min, as specified in **Note 20**.
5. To assemble a transfer sandwich, you should arrange the following elements in a precise sequence: start with a sponge, follow it with three filter papers, place a gel layer, position a PVDF/nitrocellulose membrane, and conclude with another set of three filter papers. Ensuring there are no trapped air bubbles between the gel and the PVDF/nitrocellulose membrane is critical. Furthermore, be diligent in removing any excess liquid by gently squeezing it out.
6. Move the sandwich to the specified transfer device, making certain it is placed on a bed of ice to maintain a temperature of 4 °C. Gradually introduce the transfer buffer into the device, ensuring thorough coating of the sandwich. Position the electrodes on the uppermost layer of the sandwich, positioning the PVDF/nitrocellulose membrane between the gel and a positively charged electrode, as illustrated in Fig. 2d.
7. The transfer process typically takes about 90 min. It is important to keep in mind that the time required for this procedure is closely tied to the thickness of the gel. As a result, if you are working with gels that are 0.75 mm thick, it is feasible to reduce the transfer time to just 45 min.

3.2.6 Blocking and Antibody Incubation

1. The membrane underwent a blocking process, which involved immersing it in a solution composed of 5% skim milk in TBST (Tris-Buffered Saline Tween-20) for a period of 1 h.
2. The initial antibody was added to a solution comprising 5% BSA and underwent an overnight incubation at 4 °C on a mechanical agitator.
3. To ensure proper membrane cleansing, it is advisable to rinse it thoroughly using TBST for approximately 5 min. This rinsing step should be repeated three times. For optimal results, it is suggested that all washing and antibody incubation steps take place on a shaker at room temperature. This method guarantees uniform agitation throughout the entire procedure, enhancing its effectiveness.
4. The subsequent antibody must be introduced into a solution comprised of 5% skim milk in TBST, with a subsequent incubation period of 1 h.
5. To ensure the cleanliness of the membrane, it is recommended to rinse it with a solution of TBST for a period of 5 min. It is advisable to repeat this procedure on three times.
6. Prepare the ECL solution by following the manufacturer's suggested ratios of solution A and solution B. Allow the membrane to incubate for a period ranging from 1 to 2 min, as

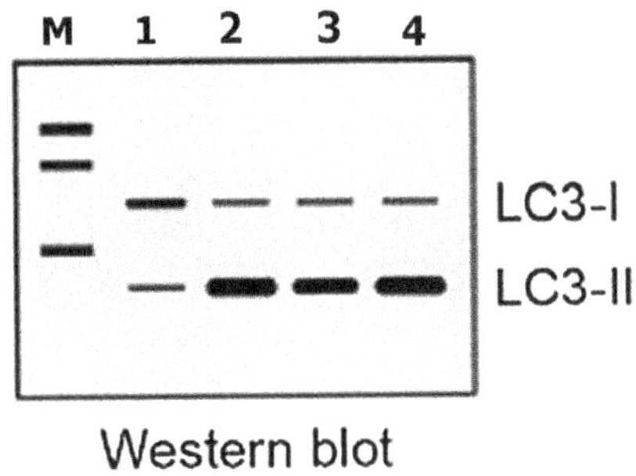

Fig. 3 Detecting LC-3 proteins in HeLa cell lines using western blotting was the primary focus of our study. We utilized a protein marker (M) as a reference point and loaded protein samples into wells labeled 1–4 [40]

illustrated in Fig. 2e. For thorough coverage of both the upper and lower surfaces of the membrane with the ECL solution, it is advisable to utilize a 1000 μL pipette.

7. Take note of the results in the softly illuminated setting, as illustrated in Figs. 2f and 3. An advisable approach could be to reduce the exposure time in cases where the background appears excessively bright.

4 Notes

1. Water's presence at the cylinder's base plays a crucial role in aiding the Tris–HCl dissolution process, thus enhancing the magnetic stir bar's activation efficiency. When opting for a glass beaker, one can accelerate the Tris–HCl dissolution rate by gently warming the water to approximately 37 °C. Nonetheless, it is essential to exercise caution and ensure that the solution attains room temperature before making any pH adjustments.
2. When conducting the acrylamide weighing process, it is advisable to use a protective mask for safety. To reduce potential exposure to acrylamide, consider taking precautionary measures, such as placing a second weigh boat of similar size over the one containing the measured acrylamide during transportation to the fume hood. Inside the fume hood, transfer the measured acrylamide into a cylinder and use a stirrer to mix it. Always exercise caution to avoid direct skin contact with unpolymerized acrylamide, which is known to have neurotoxic properties. For the preparation of acrylamide solution, it is recommended to employ mixed-resin AG 501-X8 (D) with both anion-exchange and cation-exchange properties. This resin effectively removes charged ions, including free radicals, which extends the storage life of the solution. Typically,

investigators store the resulting acrylamide solution along with the resin in a refrigerated environment. However, before storage, it is essential to carry out a filtration process. Dispose of the used composite resin as hazardous waste, as per the manufacturer's guidelines, as it can exhibit explosive properties when combined with oxidizing chemicals. The resin also features a dye that changes color from blue-green to gold once its exchange capacity is depleted.

3. The acrylamide solution should be kept at a temperature of 4 °C for a period of 1 month. When acrylamide undergoes hydrolysis, it leads to the creation of both acrylic acid and ammonia. To ensure consistent reproducibility in daily operations, it is possible to generate sizable quantities of the acrylamide mixture, buffer, and water, dividing them into smaller portions and storing them in frozen aliquots. This approach allows for the continuous use of the prepared solutions, as detailed elsewhere [refer to source 37]. It is advisable to remove the specified quantity, allowing the substance to attain ambient temperature. Subsequently, the remaining components required for the polymerization process should be integrated. In our laboratory, we prepare a fresh acrylamide solution on a monthly basis for the purpose of casting our gel samples.
4. For achieving the best outcomes, it is ideal to prepare new content for every situation.
5. We have found that storing it at a temperature of 4 °C reduces the intensity of its disagreeable odor.
6. One effective method for creating a running buffer involves the following steps: Start by making a tenfold dilution of a native buffer solution that includes 0.25 M Tris–HCl and 1.92 M glycine. To create this dilution, carefully weigh out 30.3 g of Tris–HCl and 144 g of glycine. Mix these two substances together, and then add water to reach a final volume of 1 L. Next, prepare the working solution by diluting 100 mL of a 10× native buffer with 990 mL of water. Finally, incorporate 10 mL of a 10% SDS solution into the mixture, taking care when adding the SDS solution to prevent excessive bubbling.
7. Sodium dodecyl sulfate (SDS) undergoes precipitation at a temperature of 4 °C. Therefore, it is essential to warm up the lysis buffer before using it. To prepare the 10× native buffer, dilute it by adding water until the final volume reaches 800 mL. Additionally, introduce 200 mL of methanol into the solution cautiously. It is advisable not to directly mix methanol into the 10× buffer solution as it may cause the components to precipitate. In such a situation, you can redissolve the precipitate by adding 800 mL of water.

8. To prepare the solution, begin by combining 100 mL of a Tris-buffered saline (TBS) concentrate that is ten times stronger with a graduated cylinder that can hold up to 1 L. Gradually dilute the mixture with water until it reaches an approximate volume of 800 mL. In the same cylinder, introduce 50 g of skim milk powder and stir thoroughly until it completely dissolves. For another solution, add water until you achieve a volume of 1 L. Set aside 500 mL of this solution for use as a blocking agent. In the remaining 500 mL, gently introduce 250 μL of Tween-20 by trimming the tip of the pipette to allow for accurate aspiration. Mix the components until fully blended, and utilize this mixture as a diluting agent.
9. Before the commencement of the run, the centrifugal process applied to the samples plays a crucial role in removing any insoluble debris. This proactive step helps to minimize the likelihood of streaks appearing in the protein lanes during the experiment.
10. To enhance the electrophoretic process's effectiveness, introduce a solitary droplet of BPB with a concentration of 0.1% into the upper chamber's buffer solution. This addition significantly aids in the creation of a stronger and more visible dye front as the electrophoresis proceeds.
11. Triton X-100 is a nonionic surfactant with a chemical composition of $C_{14}H_{22}O\ (C_2H_4O)_n$. It features a hydrophilic polyethylene oxide chain as well as a lipophilic or hydrophobic component made up of an aromatic hydrocarbon. This surfactant demonstrates the ability to solubilize cellular and nuclear membranes by forming pores, which in turn facilitates the passage of antibodies into the cytoplasm or nucleus [33, 34]. Consequently, Triton X-100 is often referred to as a membrane-penetrating agent. A commonly recommended concentration for its use is 0.1% (V/V).
12. In some instances where an antibody demonstrates decreased specificity, extending the blocking period to 1 h could be considered.
13. To reduce the reliance on antibodies, a common approach is to employ a coverslip that entirely envelops a 100 μL solution. A volume of 50 μL is ample to fully overlay all the cells present in a glass-bottom dish.
14. Triton X-100, employed in extremely small quantities in this scenario, facilitates the antibody's penetration into the cellular structure.
15. In this section, we will introduce a range of commonly utilized antibodies. For HeLa cell experiments (as illustrated in Fig. 4), it is advisable to employ a 1:100 dilution of the anti-LC3B antibody (catalog number L7543) from Sigma.

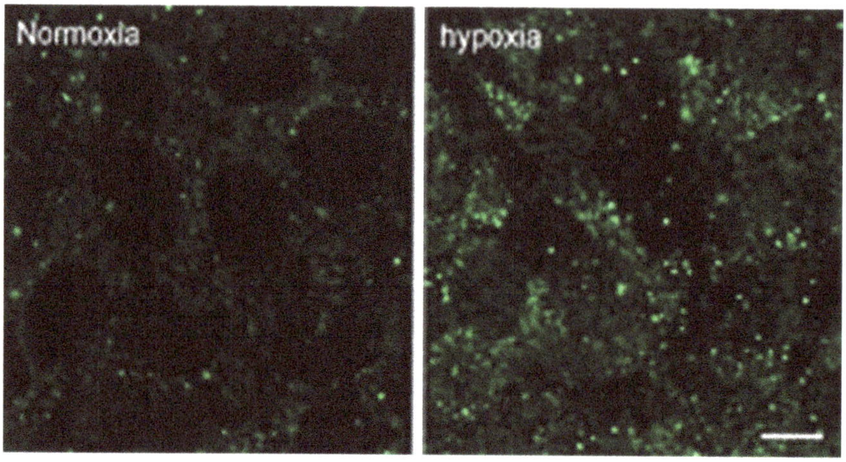

Fig. 4 Detecting LC3 through immunocytochemistry in HeLa cell lines that were fixed with 4% PFA is an essential technique. Additionally, it is crucial to apply this method to both normal and hypoxic HeLa cell lines for comparative analysis. The scale bar, measuring 20 μm, aids in visualizing the results effectively [36]

16. The inclusion of 1% BSA in this context serves the purpose of further reducing the potential for nonspecific interactions with the secondary antibody.
17. The donkey anti-mouse IgG antibodies labeled with Alexa Fluor 555 (product code: A31570, Invitrogen) are prepared by diluting them at a concentration of 1:400. Similarly, the donkey anti-rabbit IgG antibodies labeled with Alexa Fluor 488 (product code: A21206, Invitrogen) are diluted at a ratio of 1:300 for experimental use.
18. Before commencing the run, preparing the samples through centrifugation is essential to remove any insoluble debris, thus reducing the likelihood of streaks appearing in the protein lanes. To enhance the visibility of the electrophoretic process, a small droplet of 0.1% BPB is added to the upper chamber's buffer solution. This addition significantly improves the formation of a strong and distinct dye front during electrophoresis.
19. Hold the top corners of the membranes using both hands. Slowly lower the bottom part of the membrane onto the matching area of the gel, ensuring a gentle release to enable a smooth placement of the entire membrane onto the gel surface. This approach will help prevent the formation of bubbles between the gel and the membrane interface. Before placing the gel-membrane sandwich into the transfer cassette, air bubbles were removed by employing a 10 mL pipette.
20. The effectiveness of the membrane's interaction with the gel experiences a notable boost when the gel is in a completely dry state. Consequently, it becomes essential to expose the gel to a drying procedure that lasts approximately 5–10 min. As a

result, the membrane firmly attaches itself to the gel, facilitating the effortless removal of the gel from the glass plate's bottom through a straightforward lifting of the membrane.

Acknowledgments

The authors are very grateful to Dean and Director of the respective departments.

References

1. Orhon I, Reggiori F (2017) Assays to monitor autophagy progression in cell cultures. Cell 6(3):E20
2. Call JA, Wilson RJ, Laker RC, Zhang M, Kundu M, Yan Z (2017) Ulk1-mediated autophagy plays an essential role in mitochondrial remodeling and functional regeneration of skeletal muscle. Am J Physiol Cell Physiol 312:C724–C732
3. Sbrana FV, Cortini M, Avnet S, Perut F, Columbaro M, De Milito A, Baldini N (2016) The role of autophagy in the maintenance of stemness and differentiation of mesenchymal stem cells. Stem Cell Rev 12:621–633
4. Dower CM, Bhat N, Wang EW, Wang HG (2017) Selective reversible inhibition of autophagy in hypoxic breast cancer cells promotes pulmonary metastasis. Cancer Res 77:646–657
5. Guo JY, White E (2016) Autophagy, metabolism, and cancer. Cold Spring Harb Symp Quant Biol 81:73–78
6. Lei Y, Zhang D, Yu J, Dong H, Zhang J, Yang S (2017) Targeting autophagy in cancer stem cells as an anticancer therapy. Cancer Lett 393: 33–39
7. Menzies FM, Fleming A, Caricasole A, Bento CF, Andrews SP, Ashkenazi A, Fullgrabe J, Jackson A, Jimenez Sanchez M, Karabiyik C, Licitra F, Lopez Ramirez A, Pavel M, Puri C, Renna M, Ricketts T, Schlotawa L, Vicinanza M, Won H, Zhu Y, Skidmore J, Rubinsztein DC (2017) Autophagy and neurodegeneration: pathogenic mechanisms and therapeutic opportunities. Neuron 93:1015–1034
8. Plaza-Zabala A, Sierra-Torre V, Sierra A (2017) Autophagy and microglia: novel partners in neurodegeneration and aging. Int J Mol Sci 18:E598
9. Suh HW, Kim JK, Kim TS, Jo EK (2017) New insights into vitamin D and autophagy in inflammatory bowel diseases. Curr Med Chem 24(9):898–910
10. Zhong Z, Sanchez-Lopez E, Karin M (2016) Autophagy, NLRP3 inflammasome and auto-inflammatory/immune diseases. Clin Exp Rheumatol 34:12–16
11. Polus A, Bociaga-Jasik M, Czech U, Goralska J, Cialowicz U, Chojnacka M, Polus M, Jurowski K, Dembinska-Kiec A (2017) The human immunodeficiency virus (HIV1) protease inhibitor sanquinavir activates autophagy and removes lipids deposited in lipid droplets. J Physiol Pharmacol 68:283–293
12. Feizi N, Mehrbod P, Romani B, Soleimanjahi H, Bamdad T, Feizi A, Jazaeri EO, Targhi HS, Saleh M, Jamali A, Fotouhi F, Nargesabad RN, Abdoli A (2017) Autophagy induction regulates influenza virus replication in a time-dependent manner. J Med Microbiol 66:536–541
13. Zaglia T, Di Bona A, Chioato T, Basso C, Ausoni S, Mongillo M (2016) Optimized protocol for immunostaining of experimental GFP-expressing and human hearts. Histochem. Cell Biol 146:407–419
14. Niedenberger BA, Geyer CB (2018) Advanced immunostaining approaches to study early male germ cell development. Stem Cell Res 27:162–168
15. Smith FW, Gabriel WN (2018) Embryonic immunostaining for the tardigrade hypsibius exemplaris. Cold Spring Harb Protoc 2018: 102343
16. Becheva ZR, Gabrovska KI, Godjevargova TI (2018) Comparison between direct and indirect immunofluorescence method for determination of somatic cell count. Chem Pap 72: 1861–1867
17. Maity B, Sheff D, Fisher RA (2013) Immunostaining: detection of signaling protein location in tissues, cells and subcellular compartments. Methods Cell Biol 113:81–105

18. Towbin H, Staehelin T, Gordon J (1979) Electrophoretic transfer of proteins from polyacrylamide gels to NC sheets: procedure and applications. Proc Natl Acad Sci U S A 76: 4350–4354
19. Kurien BT, Scofield RH (2006) Western blotting. Methods 38:283–293
20. Renart J, Reiser J, Stark GR (1979) Transfer of proteins from gels to diazobenzyloxymethyl paper and detection with anti-sera: a method for studying antibody specifi city and antigen structure. Proc Natl Acad Sci U S A 76:3116–3120
21. Gershoni JM, Palade GE (1983) Protein blotting: principles and applications. Anal Biochem 131:1–15
22. Ramos-Vara JA (2005) Technical aspects of immunohistochemistry. Vet Pathol 42:405–426
23. Shi SR, Liu C, Taylor CR (2007) Standardization of immunohistochemistry for formalin-fixed, paraffin-embedded tissue sections based on the antigen-retrieval technique: from experiments to hypothesis. J Histochem Cytochem 55:105–109
24. Hsi ED (2001) A practical approach for evaluating new antibodies in the clinical immunohistochemistry laboratory. Arch Pathol Lab Med 125:289–294
25. Saper CB (2009) A guide to the perplexed on the specificity of antibodies. J Histochem Cytochem 57:1–5
26. Bordeaux J, Welsh A, Agarwal S, Killiam E, Baquero M, Hanna J, Anagnostou VK, Rimm DL (2010) Antibody validation. Biotechniques 48:197–209
27. Skliris GP, Rowan BG, Al-Dhaheri M, Williams C, Troup S, Begic S, Parisien M, Watson PH, Murphy LC (2009) Immunohistochemical validation of multiple phospho-specific epitopes for estrogen receptor alpha (ERalpha) in tissue microarrays of ERalpha positive human breast carcinomas. Breast Cancer Res Treat 118:443–453
28. Holmseth S, Zhou Y, Follin-Arbelet VV, Lehre KP, Bergles DE, Danbolt NC (2012) Specificity controls for immunocytochemistry: the antigen preadsorption test can lead to inaccurate assessment of antibody specificity. J Histochem Cytochem 60:174–187
29. Magaki S, Hojat SA, Wei B, So A, Yong WH (2019) An Introduction to the Performance of Immunohistochemistry. Methods Mol Biol 1897:289–298
30. Boenisch T (2005) Effect of heat-induced antigen retrieval following inconsistent formalin fixation. Appl Immunohistochem Mol Morphol 13:283–286
31. Gandhi P, Khare R (2018) A unique immunofluorescence protocol to detect protein expression in vascular tissues: tacking a long standing pathological hitch. Turk Patoloji Derg 34:57–65
32. Ivell R, Teerds K, Hoffman GE (2014) Proper application of antibodies for immunohistochemical detection: antibody crimes and how to prevent them. Endocrinology 155:676–687
33. Manning CF, Bundros AM, Trimmer JS (2012) Benefits and pitfalls of secondary antibodies: why choosing the right secondary is of primary importance. PLoS One 7:e38313
34. Marshall J, Johnsen S (2017) Fluorescence as a means of colour signal enhancement. Philos Trans R Soc Lond B Biol Sci 372:20160335
35. Hoffman GE, Murphy KJ, Sita LV (2016) The importance of titrating antibodies for immunocytochemical methods. Curr Protoc Neurosci 76(212):1–37
36. Li W, Li S, Li Y, Lin X, Hu Y, Meng T, Wu B, He R, Feng D (2019) Immunofluorescence staining protocols for major autophagy proteins including LC3, P62, and ULK1 in mammalian cells in response to normoxia and hypoxia. In: Autophagy in differentiation and tissue maintenance: methods and protocols. Humana Press, New York, pp 175–185
37. Harlow E, Lane D (eds) (1988) Electrophoresis. Appendix I. Antibodies. A laboratory manual. Cold Spring Harbor Laboratory, Cold Spring Harbor, p 638
38. Zaqout S, Becker LL, Kaindl AM (2020) Immunofluorescence staining of paraffin sections step by step. Front Neuroanat 14:582218
39. Mahmood T, Yang PC (2012) Western blot: technique, theory, and trouble shooting. N Am J Med Sci 4(9):429
40. Yamamoto-Imoto H, Hara E, Nakamura S, Yoshimori T (2022) Measurement of autophagy via LC3 western blotting following DNA-damage-induced senescence. STAR Protocols 3(3):101539

Methods in Molecular Biology (2025) 2879: 23–32
DOI 10.1007/7651_2024_563

Published online: 10 August 2024

Monitoring Autophagy During *Drosophila* Oogenesis

Mrunmayee Kulkarni, Karan Selarka, and Bhupendra V. Shravage

Abstract

Macroautophagy (autophagy hereafter) is an evolutionarily conserved mechanism that maintains the health of cells by degrading toxic proteins and damaged organelles within the lysosomes. Tissues like ovary are made up of heterogeneous cell types and each cell type has distinct levels of autophagy. Studying autophagy in a cell-type specific manner helps better understand the role of autophagy during oogenesis. Here, we describe assays for monitoring autophagy during oogenesis in *Drosophila* using the two protein markers, Atg8a and Ref(2)P.

Key words Autophagy, *Drosophila*, Nutrient-stress/Starvation, Oogenesis, Ovary

1 Introduction

Autophagy is an evolutionarily conserved and highly regulated intracellular degradative mechanism mediated through lysosomes to maintain cellular homeostasis. Autophagy involves cargo sequestration in a specialized double-membrane organelle called autophagosome (AP), which subsequently fuses with the lysosome, forming autolysosome (AL), wherein cargo gets degraded and recycled back to the cytoplasm. Autophagy is operational at basal levels in all cells, and is upregulated under stress conditions, including nutrient and oxidative stress (Fig. 1a). Thus, autophagy is critical in maintaining cellular homeostasis and cell survival under stressful conditions.

The entire process of autophagy is regulated by a set of genes known as autophagy-related (Atg) genes. Autophagy is under critical regulation of several signaling pathways and can be regulated both cell-autonomously and cell-non-autonomously, as reviewed in [1–5]. Atg proteins are evolutionarily conserved in yeast, flies, and mammals [3, 6]. Studies in *D. melanogaster* over the last few decades have identified and characterized over 18 Atg proteins

The original version of the chapter has been revised. The correction in the sentence "In starvation conditions, the number of AL (yellow puncta) increased, suggesting elevated autophagy flux in germanium" been done and word flux been deleted. A correction to this chapter can be found at https://doi.org/10.1007/7651_2024_572

Authors Mrunmayee Kulkarni and Karan Selarka have equally contributed to this chapter.

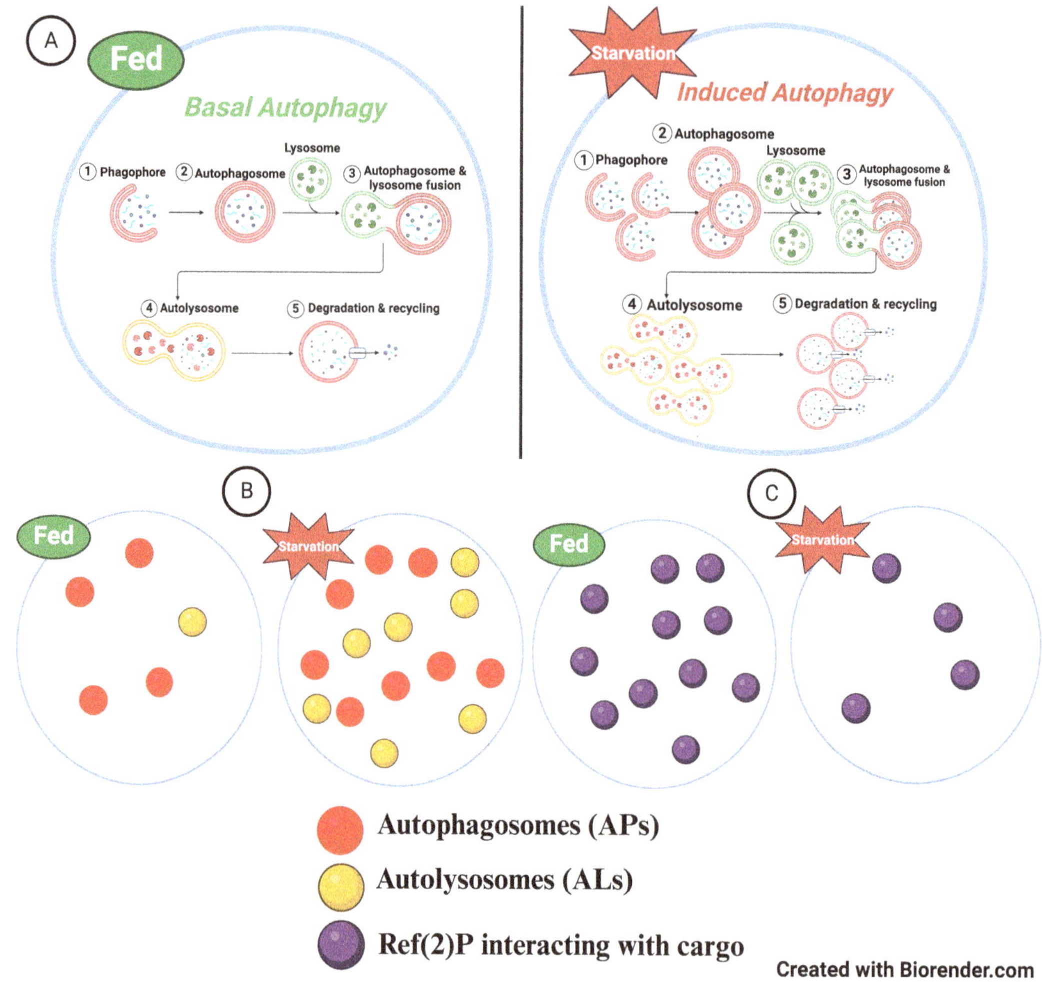

Fig. 1 Illustration showing autophagy status under fed vs starved conditions. (**a**) Under nutrient-replete conditions (fed), basal autophagy is operational (left cell), while upon nutrient-deplete (carbon and nitrogen) condition, autophagy is induced (right cell), which can be illustrated by showing an increased number of autolysosomes (ALs). (**b**) Schematic representation of autophagy flux in fed vs starved conditions by monitoring colocalization of mCherry-Atg8a (autophagosomes: APs in Red) and anti-CathepsinL (lysosome: green) resulting in yellow puncta representing ALs. (**c**) Schematic representation of autophagy cargo degradation in fed vs starved conditions by monitoring autophagy cargo adaptor protein degradation (Ref(2)P: magenta)

and their role in various tissues during development and in disease models [3]. *Drosophila melanogaster* is an ideal model system for autophagy as it offers several advantages—it has a short life cycle of 10 days, cost of maintenance is low, and it is a readily available tool for genetic manipulation [7–9].

D. melanogaster female has a pair of ovaries, each containing 16–18 ovarioles. At the anteriormost end of the ovariole lies the germarium, which houses two to three germline stem cells (GSCs). The GSCs, on their anterior side, are adhered to cap cells (CpCs), which, in turn, connect to the terminal filament cells (TFCs) (CpCs and TFCs: niche cells), as illustrated in Fig. 2. Laterally, the GSCs

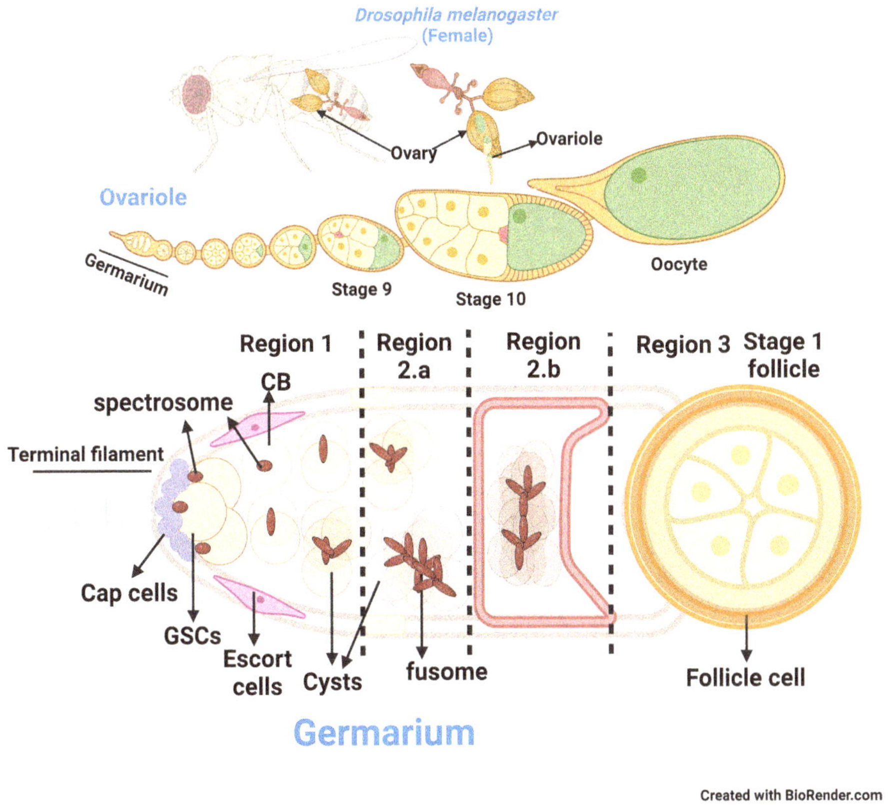

Fig. 2 Overview of *Drosophila* oogenesis. *Drosophila melanogaster* houses a pair of ovaries, of which each ovary is made up of 16–18 ovarioles. Each ovariole mimics an assembly line of egg chambers/follicles with progressively matured eggs at the posterior end. Ovariole at its anterior tip has 2–3 Germline stem cells (GSCs: pale yellow) supported by niche cells, primarily cap cells (CpCs: purple) within the germarium

are surrounded by escort cells, which aid in differentiating GSCs to cystoblasts (CBs). CBs divide four more times with incomplete cytokinesis to produce a 16-cell cyst. One cell from the 16-cell cyst executes the oocyte program, and the rest 15 become nurse cells to support the developing oocyte. The 16-cell cysts are surrounded by a layer of epithelial cells called follicle cells (FCs), which form the developing egg, also called the egg chamber/follicle [10–12]. Egg chambers/follicles are identified in stages 1–14 depending on their progressive age. Stage 14 egg chamber/follicle is the most mature egg chamber. Autophagy has been found to play a role during *Drosophila* oogenesis [13–16].

Over the years, several techniques have been developed to monitor autophagy in *D. melanogaster*, including transmission electron microscopy (TEM), Western blotting, and fluorescence microscopy using fluorescently tagged reporters and/or antibodies [8, 17]. Resources and methods have been developed to study autophagy in the *Drosophila* ovary and the germline [18–20]. Fluorescence reporters and antibodies label/mark specific autophagy proteins, allowing researchers to perform detailed cell biology and biochemical studies during development and in disease conditions. Here, we describe fluorescence microscopy-based assays to monitor autophagy in nutrient stress conditions in the germarium and developing oocytes.

1.1 Monitoring Autophagy by Ref(2)P/p62/Sequestosome-1 (SQSTM-1) Clearance

Ref(2)P (*refractory to Sigma P*), a p62/Sequestosome-1 (SQSTM-1) homolog in *Drosophila*, is an adaptor protein that binds to polyubiquitinated proteins (cargo), interacts with Atg8/LC3 via LC3 interacting motif, and facilitates cargo degradation by autophagy [21]. Upon induction of autophagy, Ref(2)P gets degraded with the cargo within the autolysosomes, resulting in a reduction of Ref(2)P. Thus, monitoring Ref(2)P levels is a reliable assay to measure autophagy activity (Fig. 1c) [22]. Ref(2)P punctae reduce in starved germarium and GSCs (Fig. 3a), cap cells and terminal filament cells (Fig. 3b), follicle cells, and the developing oocyte (Fig. 3c) as compared to fed controls.

1.2 Monitoring the Formation of Autophagosome and Autolysosomes (Autophagy Flux) Using mCherry-Atg8a

Autophagy flux measurement is a quantitative estimation of autophagy degradation activity in the presence and absence of autophagy inhibitors such as chloroquine (CQ) and bafilomycin [22]. In simple terms, it measures the number of autophagosomes and autolysosomes formed in control versus the test conditions. Measurement of autophagy flux is more insightful in the context of oogenesis, considering the dynamic nature of both processes. Atg8a (fly ortholog of mammalian LC3/GABARAP) is a commonly used marker to track autophagy. Atg8a is incorporated into the growing autophagosome membranes and mature APs by covalent conjugation with membrane lipid phosphatidylethanolamine (PE) [18, 23]. This process is also called as lipidation of Atg8/LC3. mCherry-tagged Atg8a (mCherry-Atg8a) has been reliably used to monitor and quantify levels of autophagy in *Drosophila* [17, 18]. We used an antibody against CathepsinL, an acidic hydrolase localized within the lysosomes, to detect lysosomes. These structures are visible as puncta in Fig. 4: autophagosomes are observed as red puncta (mCherry-Atg8a), lysosomes as green puncta (anti-CathepsinL), and autolysosomes as yellow puncta (colocalization of red puncta and green puncta) [18]. In starvation conditions, the number of AL (yellow puncta) increased, suggesting elevated autophagy in germarium (Fig. 4a), GSCs (Fig. 4a-inset), cap cells, terminal filament cells (Fig. 4b), follicle cells (Fig. 4c-inset), and developing egg chamber (Fig. 4c). For CQ treatment refer Nilengekar et al. [18].

2 Materials

1. Drosophila fly line: *yw; +; 3xmCherry-Atg8a (w+)* (Gift from Prof. Gabor Juhasz) [17].

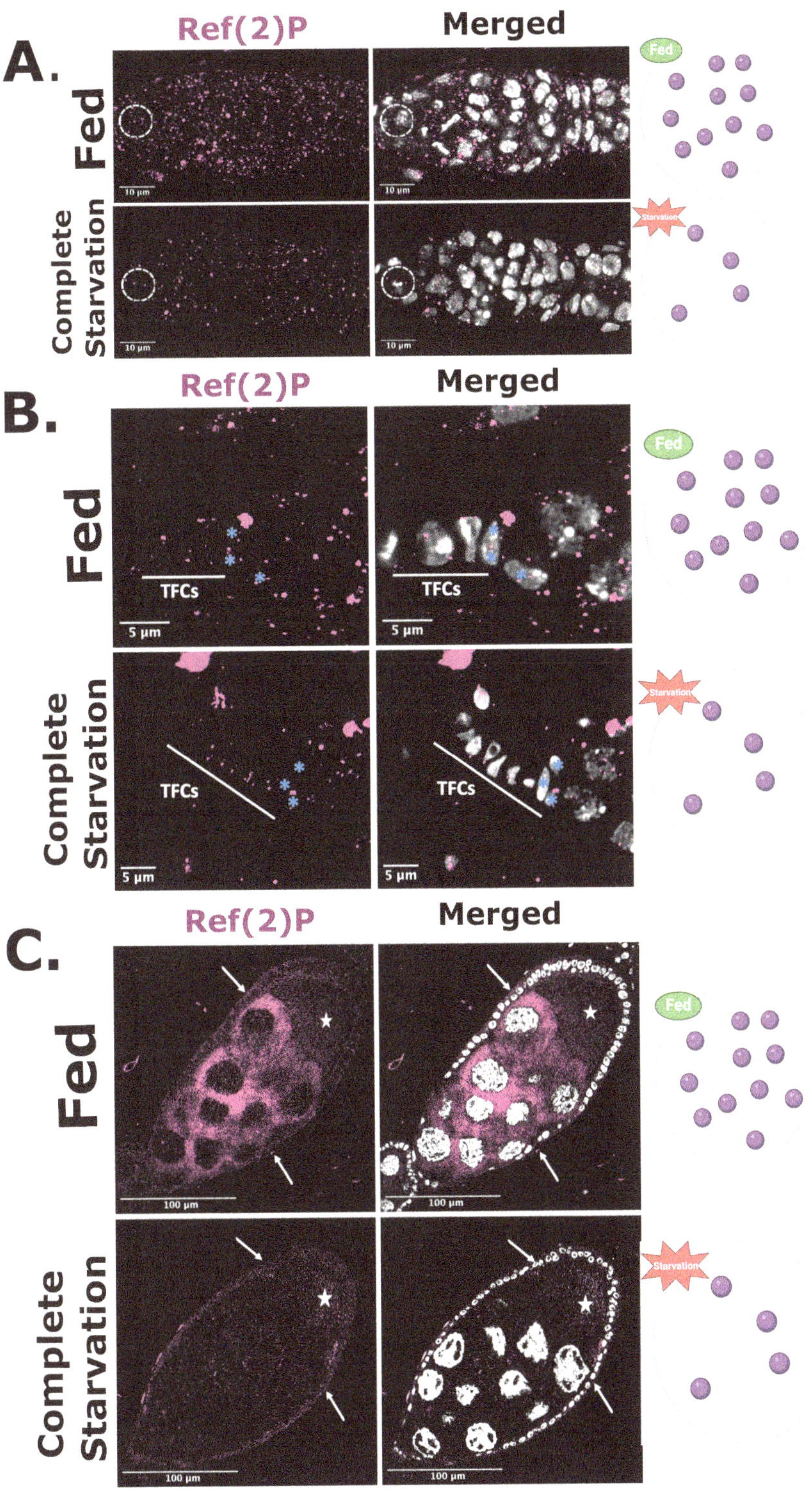

Fig. 3 Starvation induces autophagy cargo degradation as monitored by Ref(2)P/p62/Sequestosome-1 levels. (**a**) Germarium: In nutrient-replete (fed) tissue, Ref(2)P punctae (magenta) are more abundant than in nutrient-deplete (starved) conditions, suggesting that autophagy induction leads to cargo degradation (dotted circles: GSC) (scale bar: 10 µm, DAPI in grayscale). (**b**) Cap cells and terminal filament cells: In nutrient-replete (fed) tissue, Ref(2)P punctae (magenta) are more prominently observed than in nutrient-deplete (starved)

2. Empty vials with tissue paper soaked with sterile 1× PBS (pH 7.4), i.e., carbon and nitrogen starvation.
3. Fixative: 4% Paraformaldehyde in 1× PBS pH 7.4.
4. Grace's medium (GM), 10% BSA, normal goat serum (NGS), Dumont no. 5 forceps, no. 3 forceps, and Minutien pins.
5. PBTx of varying concentrations: Triton X-100 in 1× PBS (pH 7.4), 0.1%, 0.3%, and 1% PBTx.
6. The following primary antibodies were used: anti-CathepsinL (R&D Biosystems: MAB22591) 1:200, anti-Ref(2)P (abcam: AB178440) 1:1000.
7. The following secondary antibodies were used: Alexa fluor 488 goat anti-rabbit 1:250 (RRID AB_143165) and Alexa fluor 647 goat anti-mouse 1:250 (RRID AB_2535805).
8. Mounting medium—80% glycerol.

3 Method

1. Two sets of *Drosophila:* yw; +; 3xmCherry-Atg8a (w+) females were raised on standard diet sprinkled with yeast pellets in 1:1 female to male ratio (maintained throughout the regime) for 2 days. After 2 days, one set of flies was continued to be reared on standard diet sprinkled with yeast pellets and another set was transferred on empty vials with tissue paper soaked with 1× PBS (pH 7.4) for 48 h (vials changed after 24 h) to induce carbon and nitrogen starvation (complete starvation).
2. Dissect ovaries from 10 females in Grace's medium (room temperature) on a cavity slide. Gently separate ovarioles (tease) using Minutien pins (FST, Canada). Immediately transfer ovaries to 1.5 ml microfuge tube using 200 μL pipette tip. Passivate the cut pipette tip with 10% BSA to avoid sticking of ovaries (*see* **Notes 1** and **2**).
3. Carefully remove Grace's medium from the tube, once ovaries are settled at the bottom. Make sure to not aspirate ovarioles. Add 4% PFA for fixing the tissue and keep the tube on a 3D shaker for 20 min at room temperature (*see* **Notes 3** and **4**).

Fig. 3 (continued) conditions, indicating cargo degradation upon induction of autophagy. (line: terminal filament cells; blue asterisk: cap cells) (scale bar: 5 μm, DAPI in grayscale). (**c**) Older egg chamber: In nutrient-replete (Fed) tissue, Ref(2)P punctae (magenta) are more readily detectable than in nutrient-deplete (Starved) conditions, indicating that autophagy activation during nutrient starvation leads to the degradation of autophagic substrates (arrow: follicle cells; star: developing oocyte) (scale bar: 100 μm, DAPI in grayscale)

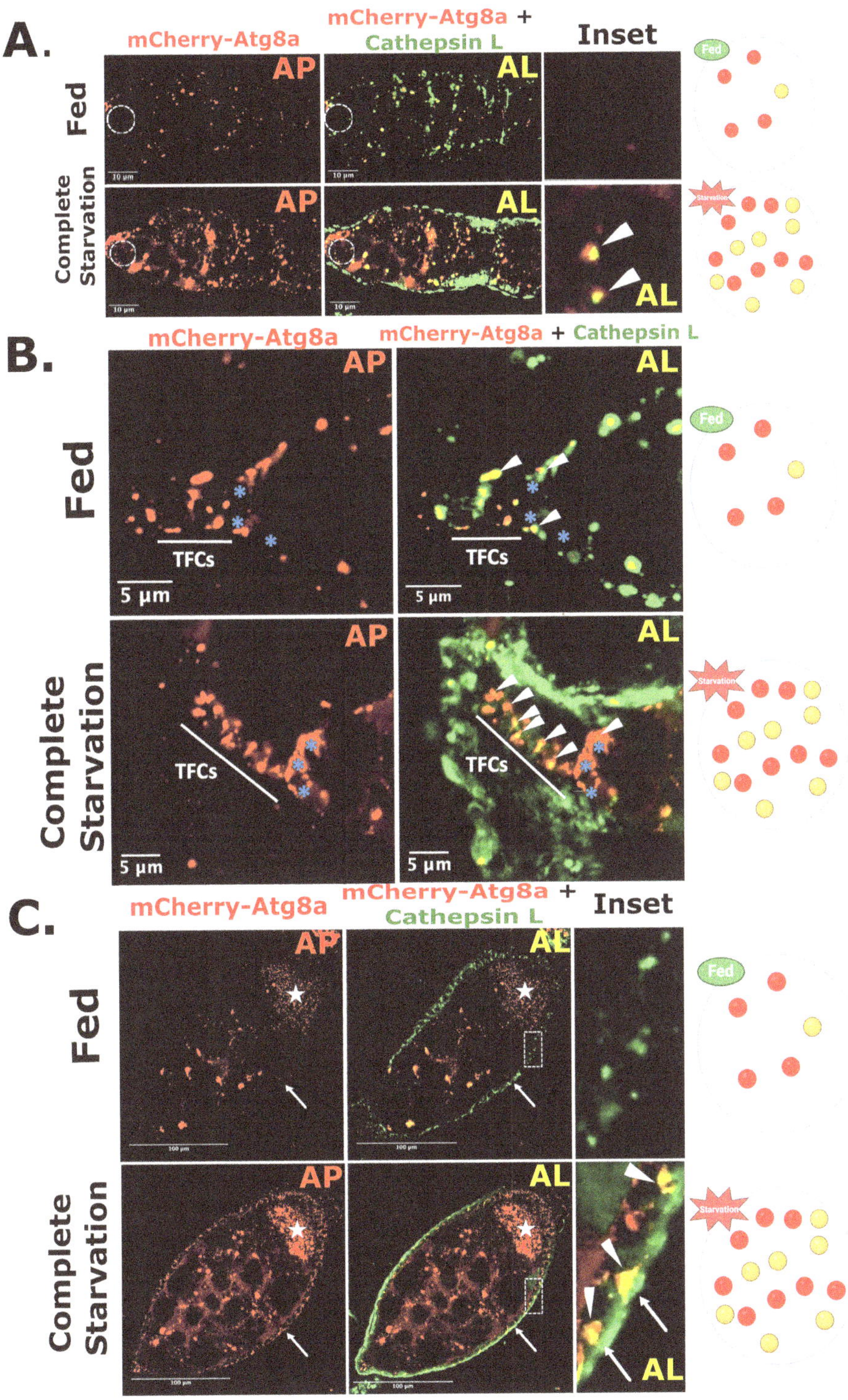

Fig. 4 Starvation induces autophagy as monitored by colocalization of autophagosomes and lysosomes forming autolysosomes. (**a**) Germarium: Nutrient-deplete (starved) tissue exhibits more colocalization (ALs: yellow) of mCherry-Atg8a punctae (APs: red) and anti-CathepsinL (lysosomes: green) compared to nutrient-replete (fed) conditions suggesting increased autophagy flux upon starvation. Inset is shown by a dotted circle and the arrowhead shows colocalization (dotted circles: GSC) (scale bar: 10 µm). (**b**) Cap cells and terminal filament cells: Nutrient-deplete (starved) tissue possesses more autolysosomes (ALs: yellow), colocalized

4. Wash the tissue with 0.1% PBTx, thrice for 5 min on the shaker at room temperature. Block the tissue using 0.5% BSA in 1% PBTx for 1 h on the shaker at room temperature.
5. Further, incubate ovaries in primary antibody solution at appropriate dilution (0.5% BSA in 0.3% PBTx) overnight at 4 °C on a shaker.
6. Wash ovaries thrice with 0.1% PBTx at room temperature and block in 10% normal goat serum in 0.1% PBTx for 2 h on a shaker at room temperature (*see* **Note 5**).
7. Incubate ovaries with secondary antibody solution (10% normal goat serum in 0.1% PBTx) for 2 h on the shaker at room temperature. Protect the tissue from light exposure by covering the tubes with aluminum foil/black paper (*see* **Note 6**).
8. Wash ovaries twice with 0.1% PBTx for 15 min on the shaker at room temperature to remove excess antibodies. Incubate the tissue with DAPI solution (1 μg/ml in 0.1% PBTx) for 15 min at room temperature (*see* **Note 7**).
9. Remove the DAPI solution and wash ovaries thrice with 0.1% PBTx for 5 min at room temperature.
10. Replace PBTx with a mounting medium.
11. Transfer ovaries to a clean glass slide using a wide bore tip and separate egg chambers from ovarioles using Minutien pins. Gently put the coverslip on the slide to avoid bubble formation. Seal the slide using transparent nail varnish.
12. All images were captured using Zeiss LSM900 at 1024 × 1024 resolution, pinhole: 1 A.U., and scan speed 6. Images were captured using 63× oil objective with 1.4, 2.2, and 0.45 zoom factors for germarium, TFCs and CpCs, and older follicles, respectively.
13. FIJI image analysis software (https://imagej.net/software/fiji/downloads) was utilized to edit the images.

Fig. 4 (continued) mCherry-Atg8a (APs: red), and anti-CathepsinL (lysosomes: green) puncta compared to nutrient-replete (fed) conditions suggesting elevated autophagy flux upon starvation. Arrowhead shows colocalization (line: terminal filament cells; blue asterisk: cap cells) (scale bar: 5 μm). (**c**) Older egg chamber: Nutrient-deplete (starved) tissue demonstrates a higher prevalence of autolysosomes (yellow) consisting of mCherry-Atg8a (APs: red) and anti-CathepsinL (lysosomes: green) puncta compared to nutrient-replete (fed) conditions implying higher autophagy flux with starvation. Inset is shown by a dotted box and the arrowhead shows colocalization (arrow: follicle cells; star: developing oocyte) (scale bar: 100 μm)

4 Notes

1. All solutions should be freshly prepared.
2. Ovaries can also be dissected in sterile 1X PBS or 1X Schneider's medium.
3. Ovaries could be fixed overnight at 4° centigrade in 4% paraformaldehyde.
4. The use of freshly prepared 4% paraformaldehyde ensures better fixation of the tissues.
5. Blocking with 1% BSA in 0.1% PBTx before adding a secondary antibody also works.
6. Incubation of secondary antibody can be carried out at 4° centigrade overnight.
7. DAPI can be replaced with Hoescht 33258.

Acknowledgments

We thank the members of the Shravage lab for helpful discussions. We thank Dr. P.K Dhakephalkar, Director, Agharkar Research Institute, Pune, India, and the Developmental Biology fraternity for support and facility access. This work was supported by DST-SERB grant number CRG/2020/002716 and ICMR grant IIRP-2023-1845/F1 to BVS. KS is supported by UGC-JRF fellowship (UGC/NTA Ref No. 191620112609) through CSIR-UGC-NET and is a registered Ph.D. student (UGC-SRF) affiliated to the Department of Biotechnology, Savitribai Phule Pune University (SPPU), Pune, India (Registration No. 196), and MK is supported by CSIR-JRF fellowship (16/06/2019(i)EU-V) through CSIR-UGC-NET and is a registered Ph.D. student (CSIR-SRF) affiliated to the Department of Biotechnology, Savitribai Phule Pune University (SPPU), Pune, India (Registration No. 217). BVS is affiliated to Savitribai Phule Pune University (SPPU), Pune, India, and is recognized by SPPU as a Ph.D. guide in Biotechnology and Zoology.

References

1. Kim YC, Guan KL (2015) MTOR: a pharmacologic target for autophagy regulation. J Clin Invest. https://doi.org/10.1172/JCI73939
2. Yuan HX, Russell RC, Guan KL (2013) Regulation of PIK3C3/VPS34 complexes by MTOR in nutrient stress-induced autophagy. Autophagy 9:1983–1995
3. Mulakkal NC, Nagy P, Takats S et al (2014) Autophagy in drosophila: from historical studies to current knowledge. Biomed Res Int 2014:273473
4. Fenouille N, Nascimbeni AC, Botti-Millet J et al (2017) To be or not to be cell autonomous? Autophagy says both. Essays Biochem. https://doi.org/10.1042/EBC20170025
5. Selarka K, Shravage BV (2024) Illuminating intercellular autophagy: a comprehensive review of cell non-autonomous autophagy.

Biochem Biophys Res Commun. https://doi.org/10.1016/j.bbrc.2024.150024

6. Parzych KR, Klionsky DJ (2014) An overview of autophagy: morphology, mechanism, and regulation. Antioxid Redox Signal. https://doi.org/10.1089/ars.2013.5371
7. Nagy P, Varga Á, Kovács AL et al (2015) How and why to study autophagy in Drosophila: It's more than just a garbage chute. Methods 75: 151–161
8. Lőrincz P, Mauvezin C, Juhász G (2017) Exploring autophagy in drosophila. Cells. https://doi.org/10.3390/cells6030022
9. McPhee CK, Baehrecke EH (2009) Autophagy in Drosophila melanogaster. Biochim Biophys Acta. https://doi.org/10.1016/j.bbamcr.2009.02.009
10. Kirilly D, Xie T (2007) The drosophila ovary: an active stem cell community. Cell Res. https://doi.org/10.1038/sj.cr.7310123
11. Xie T (2008) Germline stem cell niches. In: Stembook. Harvard Stem Cell Institute, Cambridge, MA, pp 1–18
12. Losick VP, Morris LX, Fox DT et al (2011) Drosophila stem cell niches: a decade of discovery suggests a unified view of stem cell regulation. Dev Cell 21:159–171
13. Barth JMI, Szabad J, Hafen E et al (2011) Autophagy in drosophila ovaries is induced by starvation and is required for oogenesis. Cell Death Differ 18:915–924
14. Barth JMI, Hafen E, Köhler K (2012) The lack of autophagy triggers precocious activation of notch signaling during drosophila oogenesis. BMC Dev Biol 12:35
15. Nezis IP, Shravage BV, Sagona AP et al (2010) Autophagic degradation of dBruce controls DNA fragmentation in nurse cells during late Drosophila melanogaster oogenesis. J Cell Biol 190:523–531
16. Nezis LP, Lamark T, Velentzas AD et al (2009) Cell death during Drosophila melanogaster early oogenesis is mediated through autophagy. Autophagy 5:298–302
17. Hegedus K, Takats S, Boda A et al (2016) The Ccz1-Mon1-Rab7 module and Rab5 control distinct steps of autophagy. Mol Biol Cell 27: 3132–3142
18. Nilangekar K, Murmu N, Sahu G et al (2019) Generation and characterization of germline-specific autophagy and mitochondrial reactive oxygen species reporters in Drosophila. Front Cell Dev Biol 7:47
19. Devorkin L, Gorski SM (2014) Genetic manipulation of autophagy in the Drosophila ovary. Cold Spring Harb Protoc 2014:973–979
20. Aishwarya C, Shravage BV (2023) Assays for monitoring autophagy in stem cells. In: Shravage BV, Turksen K (eds) Autophagy in stem cell maintenance and differentiation. Springer International Publishing, Cham, pp 1–34
21. Bartlett BJ, Isakson P, Lewerenz J et al (2011) p62, Ref(2)P and ubiquitinated proteins are conserved markers of neuronal aging, aggregate formation and progressive autophagic defects. Autophagy 7:572–583
22. Klionsky DJ, Abdel-Aziz AK, Abdelfatah S et al (2021) Guidelines for the use and interpretation of assays for monitoring autophagy (4th edition). Autophagy. https://doi.org/10.1080/15548627.2020.1797280
23. Tanida I, Minematsu-Ikeguchi N, Ueno T et al (2005) Lysosomal turnover, but not a cellular level, of endogenous LC3 is a marker for autophagy. Autophagy 1:84–91

Methods in Molecular Biology (2025) 2879: 33–49
DOI 10.1007/7651_2024_561

Published online: 25 July 2024

Assessment of Non-canonical Functions of the Autophagy Proteins in LC3-Associated Phagocytosis and LC3-Associated Endocytosis

Luigi Mari, Emilio Boada-Romero, Zhenrui Li, Joelle Magné, and Douglas R. Green

Abstract

The identification and characterization of noncanonical functions within the autophagy pathway have unveiled intricate cellular processes, including LC3-associated phagocytosis (LAP) and LC3-associated endocytosis (LANDO). These phenomena play pivotal roles in the conjugation of ATG8 with single-membrane phagosomes and endosomes, shedding light on the dynamic interplay between autophagy and cellular homeostasis. Here, we present detailed protocols for both qualitative and quantitative assessment of LAP, including immunofluorescence, flow cytometry, and Western blotting of isolated LAPosomes. Additionally, the protocol for the evaluation of LANDO through immunofluorescent detection of receptor recycling is outlined. The methodologies presented herein serve as a practical guide for researchers seeking to unravel the intricacies of LAP and LANDO. By providing step-by-step instructions, accompanied by insights into potential challenges and optimization strategies, this chapter aims to empower investigators in the exploration of these noncanonical functions of autophagy proteins.

Key words ATG8, Autophagy, Flow cytometry, Immunofluorescence, LANDO, LAP, LC3-associated endocytosis, LC3-associated phagocytosis, Phagosome, Receptor recycling, Western blotting

1 Introduction

Recent progress in autophagy research has unveiled non-canonical functions of the autophagy machinery governing LC3 (ATG8) processing and conjugation [1–3]. Within these functions, a subset of the autophagy machinery is distinctly involved in the lipidation of LC3-II on single-membrane vesicles, a process recently defined as conjugation of ATG8 to single membranes (CASM) [4–6]. Notably, two prominent pathways within this category, LC3-associated phagocytosis (LAP) and LC3-associated endocytosis (LANDO), have emerged. LAP and LANDO utilize elements of the autophagy machinery but diverge in cargo sources, a defining factor setting them apart from classical autophagy. LAP is activated in response to phagocytic cargoes revealing crucial roles in recognizing and clearing various pathogens and dead cells and in regulating inflammatory signaling [1, 7, 8]. In contrast, stimuli such as receptor ligation events trigger endosomal LC3 lipidation in LANDO, modulating vesicle and receptor recycling [2, 9].

For assessing LAP, we describe three separate methodologies (using respectively immunofluorescence microscopy, flow cytometry, and Western blotting) based on the detection of the lipidated LC3-II recruitment to cargo-containing phagosomes (LAPosomes). For LANDO, we describe an antibody-based immunofluorescence protocol to quantify the recycling of internalized antibody-receptor complexes to the plasma membrane during continued rounds of endocytosis. Protocols are organized in the following order:

- Assessment of LAP by confocal microscopy
- Assessment of LAP by flow cytometry
- Assessment of LAP by Western blotting of purified LAPosomes
- Assessment of LANDO by immunofluorescence

2 Materials

Prepare all solutions using tissue culture grade sterile reagents and media. Diligently follow all waste disposal regulations when disposing waste materials.

2.1 Assessment of LAP by Confocal Microscopy

1. Cells: primary macrophages or cell lines (e.g., RAW264.7, ATCC, #TIB-71). It is advantageous, in term of signal resolution, assay timing and costs, if these cells ectopically express fluorescently tagged MAP 1LC3B (LC3b) (*see* **Note 1**).
2. Complete medium: Dulbecco's modified Eagle's medium (DMEM) supplemented with 10% heat-inactivated fetal bovine serum (FBS; v/v), 2 mM L-glutamine, and 100 U/mL penicillin/streptomycin.
3. Zymosan particles fluorescently labeled (e.g., Thermo Fisher, TexasRed #Z2843 or Alexa Fluor 594 #Z23374 or FITC #Z2841;): resuspend at 5 mg/mL in dPBS and store at 4 °C. Because 1 mg of zymosan contains around 2×10^7 particles per mg solid, a 5 mg/mL stock contains 1×10^8 particles/mL (*see* **Note 2**). Alternative LAP-specific phagocytic cargo source can be used, such as dead cells or phosphatidylserine(PS)-coated beads (TIM-4 ligand), immunoglobulin G (IgG)-coupled beads (Fc receptor ligand), or Pam3CysSerLys4 (Pam3csk4)-beads (TLR1/TLR2 ligand) (*see* **step 25**, below, for the Pam3csk4-beads coupling procedure).
4. IBIDI 4 wells chamber slide ibiTreat: #1.5 polymer coverslip, tissue culture treated, sterilized #80426 (*see* **Note 3**).

5. Confocal microscope.
6. Fixing solution: 2% paraformaldehyde (PFA) in 1X dPBS.
7. Nucleic acid staining solution: [1 μg/mL] of Hoechst 33342 or DAPI in 1X dPBS.
8. Permeabilization buffer: TBS (20 mM Tris–HCl, 100 mM NaCl) + 0.1%Triton X-100.
9. Washing buffer: 1X TBS + 0.01% Tween-20.
10. Blocking solution: 1X TBS + 0.05% Tween-20 + 3% bovine serum albumin (BSA).
11. Primary antibody: rabbit polyclonal anti-LC3 (A/B/C) antibody (e.g., MBL, #PM036) if cells do not express fluorescent LC3b.
12. Secondary antibody: e.g., goat anti-rabbit IgG (H + L) highly cross-adsorbed, Alexa Fluor Plus 488 (Thermo Fisher, #A32731) if cells do not express fluorescent LC3b.
13. Post-fixation buffer: 0.25% PFA in 1X TBS.

2.2 Assessment of LAP by Flow Cytometry

1. Cell line or primary cells of interest (*see* Subheading 2.1, **steps 1** and **2** above and **Note 1**).
2. Tissue culture treated multi-well plate.
3. Detaching solution: 2 mM EDTA in 1X dPBS.
4. Deep 96-well plate, 1 mL well (e.g., PlateOne, #1896-1110).
5. V-bottom 96-well plate.
6. Permeabilization buffer: freshly made [200 μg/mL] digitonin in 1X dPBS.
7. FACS buffer: 1% BSA in 1X dPBS.
8. Flow cytometer with 488 and 561 nm lasers.
9. Fc receptor blocking antibody: e.g., Rat anti-mouse CD16/CD32 clone 2.4G2 [10 μg/mL] if cells do not express fluorescent LC3b.

2.3 Assessment of LAP by Western Blot of Purified LAPosomes

1. Cell line or primary cells of interest (*see* Subheading 2.1, **steps 1** and **2** above and **Note 1**).
2. Uncoated 3μm carboxyl-polystyrene beads (Spherotech, #CP-30-10).
3. Pam3csk4 (Invivogen, #tlrl-pms) resupended to 1 mg/mL in ultrapure water.
4. EDC (1-ethyl-3-(3-dimethylaminopropyl)carbodiimide hydrochloride), (Thermo Fisher, #22980).
5. LAP beads: in this example, we used TLR ligands-coupled carboxyl-polystyrene beads (Pam3csk4-coated beads).

Alternatively, phosphatidylserine (PS)-coated beads (Echelon Biosciences Inc., #P-B0PS) or IgG-coated beads can be used.

Bead coupling procedure (scale volume and amounts if needed):

(a) Transfer 1 mL of beads slurry into a glass tube (Pyrex, #9826-16) and pellet the beads (refrigerated benchtop centrifuge, top speed 5 min). Aspirate the supernatant and resuspend the beads in 1X PBS (10 mL) to wash.

(b) Discard supernatant and wash two more times.

(c) Add 1 mL of coupling buffer containing [10 mM] sodium acetate (Sigma, #S8750), pH 5.0, and 5 mg EDC and incubate at RT for 30 min to pre-activate the beads.

(d) Add Pam3csk4 (100 μl per 1 mL of initial slurry).

(e) Incubate at RT for 2–3 h (or overnight at 4 °C) to couple the ligand to the beads.

(f) Wash three times with cold 1X PBS.

(g) Resuspend in 2 mL of cold 1X PBS.

6. Homogenization buffer, 4 °C: [250 mM] sucrose (8%) (Sigma, #S7903) in [3 mM] imidazole (Sigma, #I2399), pH 7.4. A stock solution can be made: 50 mL of [1 M] imidazole, pH 7.4.
7. 10%, 25%, 35%, and 62% (w/v) sucrose solutions, 4 °C:

(a) *Sucrose gradient*: 62%, 35%, 25%, and 10%.

(b) Prepare a 62% sucrose stock solution (e.g., 400 mL, 248 g of sucrose) and dilute with [3 mM] imidazole buffer, pH 7.4, as follows: e.g., for 50 mL.

	35%	25%	10%	8% (250 mM)
Dilution factor	1.77	2.48	6.20	7.75
From 62%	28.23	20.16	8.06	6.45
From 0%	21.77	29.84	41.94	43.55

8. Lysis buffer (1% Igepal-C630). Add protease and phosphatase inhibitors right before use.
9. 15 cm dishes or other appropriately sized vessels.
10. Cell lifter (e.g., Corning, #3008).
11. Conical tubes.
12. Refrigerated bench-top centrifuge.
13. 1 mL syringe with 27-gauge needle.

14. Ultracentrifuge tubes (e.g., 38.5 mL, certified free open-top thin-wall ultra-clear tube, Beckman Coulter, #C13926).
15. Refrigerated ultracentrifuge (e.g., Beckam Coulter, Optima XE) with swinging-bucket and rotors (e.g., SW32Ti #369650).
16. Additional reagents and equipment for Western blotting [10].

2.4 Assessment of LANDO by Immunofluorescence

1. Cell line or primary cells of interest (*see* Subheading 2.1, **steps 1** and **2** above and **Note 1**).
2. IBIDI 8-well chamber slide ibiTreat: #1.5 polymer coverslip, tissue culture treated, sterilized #80806 (*see* **Note 3**).
3. Blocking medium: complete cell culture medium containing 10% of the serum from the same source of the secondary antibody. In our example, we will use complete medium (as in Subheading 2.1, **step 2**) containing 10% donkey serum.
4. Primary antibody labeling medium: complete cell culture medium containing the primary antibody targeting the receptor of interest and 1% of the serum from the same source of the secondary antibody that will be used.
5. Recycling medium: complete cell culture medium containing 10% of the serum from the same source of the secondary antibody.
6. Acid-stripping medium: complete cell culture medium with adjusted pH to 2.0.
7. Secondary antibody labeling medium: complete cell culture medium containing the fluorescent tagged secondary antibody and 1% of the serum from the same source of the secondary antibody.
8. Fixing solution: 4% PFA in 1X dPBS.

3 Methods

3.1 Assessment of LAP by Confocal Microscopy

The principle of this technique is to visually detect cargo-containing phagosomes (LAPosomes) decorated on the membrane with LC3b (Fig. 1).

1. Subculture the RAW264.7 cells in complete medium for 3–4 days in advance of performing the assay. Other cell types may require different, optimal pre-cultures.
2. The day before the assay seed 0.65×10^4 cells in 750 μL medium/well.
3. Remove 150 μL medium from each well to reduce the volume and favor zymosan or bead engulfment.

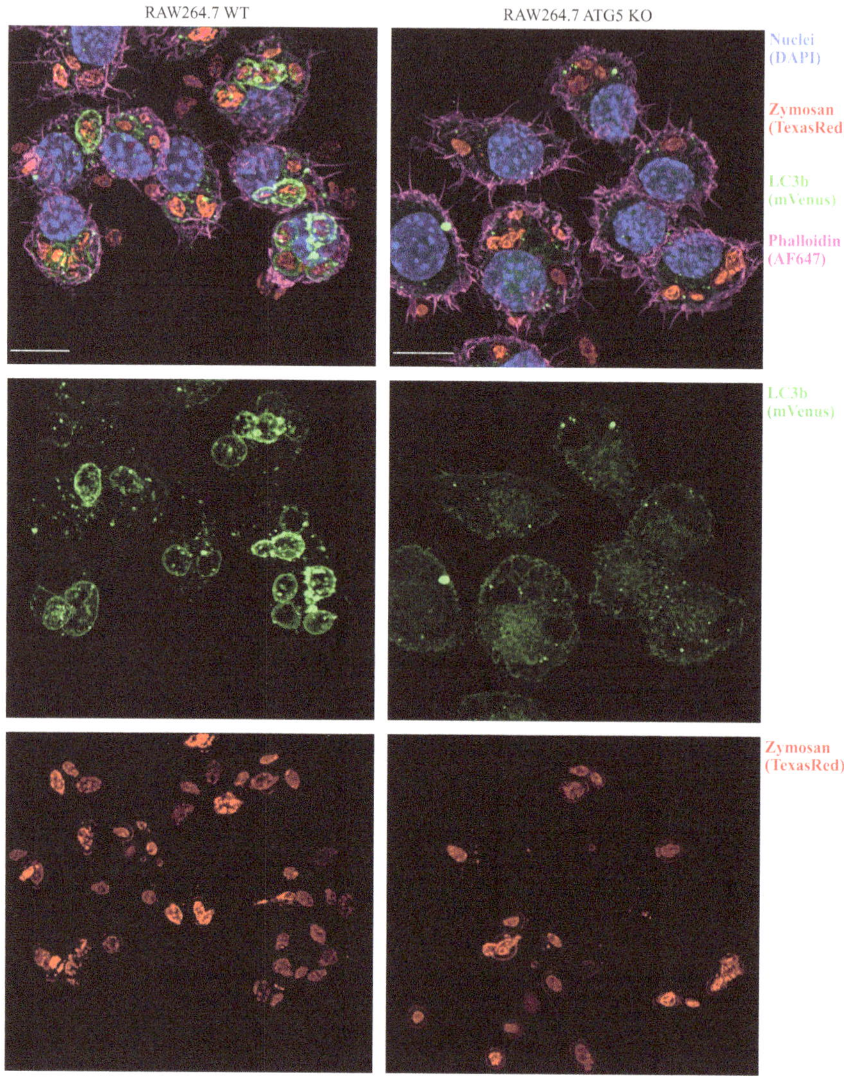

Fig. 1 Assessment of LAP by confocal microscopy in RAW264.7 cells (WT or ATG5 KO) stably expressing mVenus-LC3. Representative confocal images depicting LC3 phagosomal enrichment (green) after 1 h treatment with zymosan-TexasRed (red). Nuclei are stained with DAPI (blue), and actin filaments are labeled with phalloidin staining (pink)

4. Resuspend zymosan (*see* **Note 2**) or beads and add to cells for a final zymosan/bead:cell ratio of 5:1.
5. Incubate cells for 30–60 min in a humidified incubator with 5% CO_2 at 37 °C.
6. Aspirate the medium and gently wash three times with ice-cold dPBS.
7. Add fixing solution to cells for 10 min at room temperature (RT) (*see* **Note 4**).

8. Remove fixation solution and gently wash three times with 1X dPBS. It is possible to stop the experiment here by placing the slide at 4 °C overnight.
9. Incubate with nucleic acid staining solution for 15 min at RT, then wash three times with dPBS.
10. Store the slides at 4 °C and acquire images using a green (mVenus-LC3b), red (zymosan-TexasRed), or blue laser (nuclei).

Alternative protocol for phagocytic cells that do not express fluorescent tagged LC3b:

11. Start after **step 8** as described above.
12. Permeabilize cells with permeabilization buffer for 3 min at RT.
13. Wash three times, 5 min, with washing buffer.
14. Add blocking solution for 30 min at RT.
15. Wash as in **step 13**.
16. Incubate cells with anti-LC3 primary antibody diluted at [1: 300] in blocking solution at 4 °C, overnight (*see* **Note 5**).
17. Wash as in **step 3**. From this step protect the slides from direct exposure to light.
18. Incubate cells in [1:500] Alexa Fluor-488 anti-rabbit secondary antibody diluted in blocking solution for 1 h at RT.
19. Wash as in **step 13**.
20. Add post-fixation solution for 5 min at RT.
21. Wash as in **step 13**.
22. Incubate with nucleic acid staining solution for 15 min at RT, then wash as in **step 13**.
23. Acquire images as described above (Fig. 1).

3.2 Assessment of LAP by Flow Cytometry

The goal of this technique, adapted from [11], is to obtain a more quantitative and less subjective measurement of the cargo-containing phagosomes (LAPosomes) decorated with immobilized LC3-II. This method, however, will also detect canonical autophagy. Therefore, coupling this method with the visualization technique described above provides a robust detection of LAP (Fig. 2).

1. The day before the assay, seed 0.75×10^6 cells in 2 mL medium per well of a 12 multi-well plate (*see* **Notes 1** and **6**).
2. Remove 1 mL medium from each well to favor zymosan uptake by the cells.

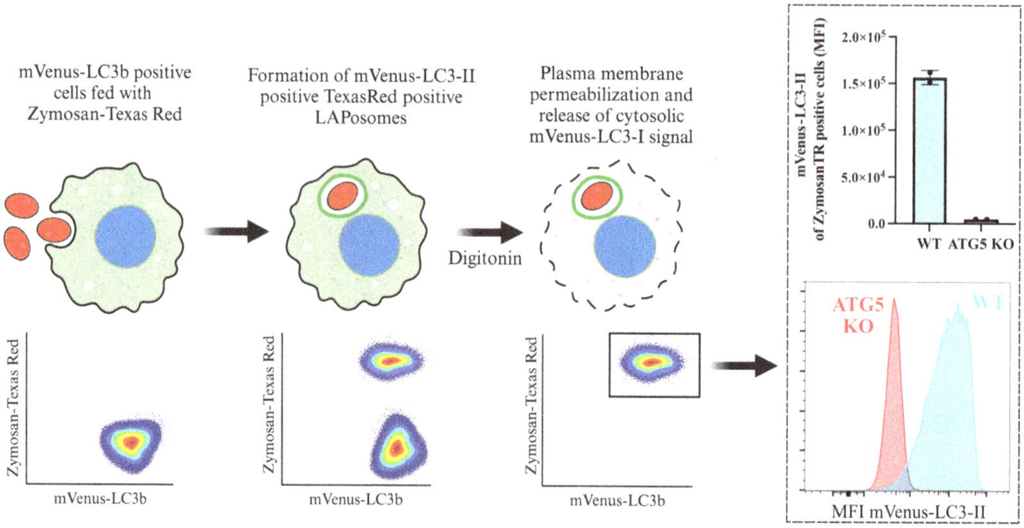

Fig. 2 Assessment of LAP by flow cytometry in RAW264.7 cells (WT or ATG5 KO) stably expressing mVenus-LC3. (Left) Schematic representation of the assay workflow. (Right) Representative quantification of mVenus-LC3 median fluorescence intensity (MFI) levels in LAPosomes from cells treated for 1 h with zymosan-TexasRed following digitonin treatment, as measured by flow cytometry

3. Add zymosan-TexasRed or beads to cells at 5:1 ratio and incubate cells for 30-60 min in a humidified incubator with 5% CO_2 at 37 °C. As a control, add the same amount of medium without zymosan/beads to untreated wells (*see* **Note 2**).
4. Stop phagocytosis by placing the plate on ice and replacing the medium with ice-cold 1X dPBS.
5. Gently wash three times with ice-cold 1X dPBS.
6. Add 0.5 mL/well of ice-cold detaching solution and incubate for 10 min on ice.
7. Collect cells onto an ice-cold deep 96-well 1 mL plate containing 0.5 mL/well of culture medium.
8. Centrifuge at 1800 rpm for 4 min at 4 °C.
9. Resuspend in 150 μL/well dPBS and move cells onto an ice-cold 96-well V-bottom plate.
10. Wash once more with 150 μL ice-cold dPBS and centrifuge as in **step 8**.
11. Allow release of cytoplasmic LC3-I by resuspending the cells in 100 μL/well of digitonin-containing permeabilization solution (*see* **Note 7**) and incubating them for 15 min on ice.
12. Wash twice with ice-cold FACS buffer and centrifuge as in **step 8**.

13. Acquire by flow cytometry using a 561 nm (zymosan-TexasRed) and 488 nm laser (mVenus_LC3b).
14. Assess the mVenus-LC3b median fluorescence intensity (MFI) of the zymosan-TexasRed double-positive events (Fig. 2).

Alternative protocol for phagocytic cells that *do not* express fluorescent tagged LC3b:

15. Start after **step 12** as described above.
16. Block unspecific Fc receptor binding: Incubate cells for 10 min on ice with 75 μL/well ice-cold FACS buffer containing anti-Fc receptor blocking antibody [10 μg/mL].
17. Add 75 μL/well of 2X concentrated anti-LC3 primary antibody [1:300] and incubate cells for 20 min on ice (*see* **Note 5**).
18. Wash twice as in **step 8**.
19. Resuspend in 150 μL/well FACS buffer containing secondary Alexa Fluor-488 anti-rabbit secondary antibody [1:500] and incubate on ice for 20 min.
20. Wash twice as in **step 8** and analyze samples as in **steps 13** and **14** (Fig. 2).

3.3 Assessment of LAP by Western Blot of Purified LAPosomes

The subsequent technique, adapted from [12], aims to analyze latex beads containing LAPosomes. By coating latex-beads with a LAP stimulus, such as Pam3csk4 (TLR2/1 agonist), one can purify the LAPosomes and analyze the protein composition of the vesicles using Western blot (Fig. 3).

1. Plate either cell line or primary cells in approximately 25 mL of cell culture medium at the desired concentration, using 15 cm dishes or other vessels of suitable size. Incubate overnight at 37 °C and 5% CO_2 to facilitate cell adhesion (*see* **Note 8**).
2. Once cells are settled, add uncoated carboxyl-polystyrene beads (control) or LAP beads (PS-coated beads or IgG-coated beads or Pam3csk4-coated beads) at desired bead-to-cell ratio (recommended: 10:1). Incubate at 37 °C and 5% CO_2 for up to 1 h.
3. At the specified time-point(s), remove the medium and wash the cells three times with 20 mL of cold dPBS on ice to eliminate any beads that were not phagocytosed.
4. Remove the final wash by aspiration and add 10 mL cold dPBS to each dish.
5. While keeping each dish on ice, utilize a cell lifter to detach the cells. Collect the cells in dPBS and transfer to a conical tube. Subsequently, centrifuge cells for 5 min at 500 × g, 4 °C.

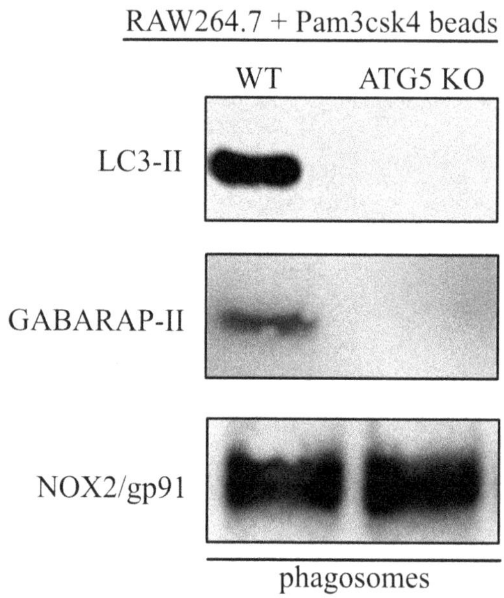

Fig. 3 Assessment of LAP by Western blot of purified LAPosomes in RAW264.7 cells. Representative Western blot showing expression of proteins associated with purified LAPosomes. The levels of LC3-II (Cell Signaling Tech., #2775) and GABARAP-II (Cell Signaling Tech., #13733) are reduced in ATG5-deficient phagosomes. NOX2/gp91-phox (Santa Cruz, #sc-130543) is used here as a loading control

6. Remove the dPBS and wash cell pellet in 10 mL of cold homogenization buffer. Centrifuge the mixture for 5 min at 500× *g* and 4 °C.
7. Discard the supernatant and resuspend the cell pellet in 1 mL of cold homogenization buffer.
8. While keeping each tube on ice, homogenize the cells by repeatedly drawing up and expelling the cell suspension more than 10 times using a 1 mL syringe equipped with a 27-gauge needle (alternatively, glass douncers can be used here).
9. Examine for cell disruption under a light microscope, ensuring that over 90% of the cells are disrupted while maintaining an intact nucleus (*see* **Note 7**).
10. To isolate bead-containing phagosomes through flotation on a sucrose step gradient, continue with the following steps while working on ice. The volumes listed below are appropriate for a 36 mL ultracentrifuge tube.
11. Move the homogenate to a 15 mL conical tube and then add 2 mL of homogenization buffer and 5 mL of 62% sucrose. Thoroughly mix to achieve a final concentration of

approximately 40% sucrose. The final volume of the solution (cells in 40% sucrose) should be approximately 8 mL.

12. Dispense 4 mL of the 62% sucrose solution into an ultracentrifuge tube, specifically a 36 mL ultracentrifuge tube).
13. Carefully and gradually pipette the homogenate (in 40% sucrose) onto the top of the 62% sucrose layer.
14. Carefully and gradually pipette 8 mL of the 35% sucrose solution onto the top of the layer of cells in 40% sucrose containing cell homogenate.
15. Carefully and gradually pipette 8 mL of the 25% sucrose solution onto the top of the 35% sucrose layer.
16. Carefully and gradually pipette 8 mL of the 10% sucrose solution onto the top of the 25% sucrose layer (*see* **Note 9**).
17. Centrifuge the sucrose gradients in a refrigerated ultracentrifuge equipped with a swinging-bucket rotor for 1 h at 100,000× *g* and 4 °C (*see* **Note 10**).
18. Gently collect the band containing bead-containing phagosomes using a 10 mL pipette and transfer it to a new conical tube. Usually, the bead-containing phagosomes band volume is around 5 mL.
19. Wash bead-containing phagosomes with cold dPBS to final 37 mL in a conical tube. Carefully invert the tube several times (do not vortex) to mix well. Transfer the solution to a new ultracentrifuge tube and centrifuge in a refrigerated ultracentrifuge equipped with a swinging-bucket rotor for 30 min at 100,000× *g* and 4 °C.
20. Dispose the supernatant and resuspend the pellet in approximately 50 μL of lysis buffer for Western blot analysis (*see* **Note 11**) (Fig. 3).

3.4 Assessment of LANDO by Immunofluorescence

The goal of this technique, as adapted from [13], is to visualize and obtain a semi-quantitative assessment of the contribution of LANDO in the endosomal recycling of specific membrane receptors. The procedure relies on labeling cell surface receptors using antibodies, triggering endocytosis. Subsequently, any remaining surface-bound tags on receptors are eliminated before initiating the recycling process to return internalized receptors to the cell surface. Receptors carrying the antibody tag that have been recycled to the cell surface are then quantified using a secondary antibody labeled with fluorescence (Figs. 4 and 5a, b).

1. The day before the assay, seed 0.25×10^4 cells in 300 μL medium/well in a humidified incubator with 5% CO_2 at 37 °C.

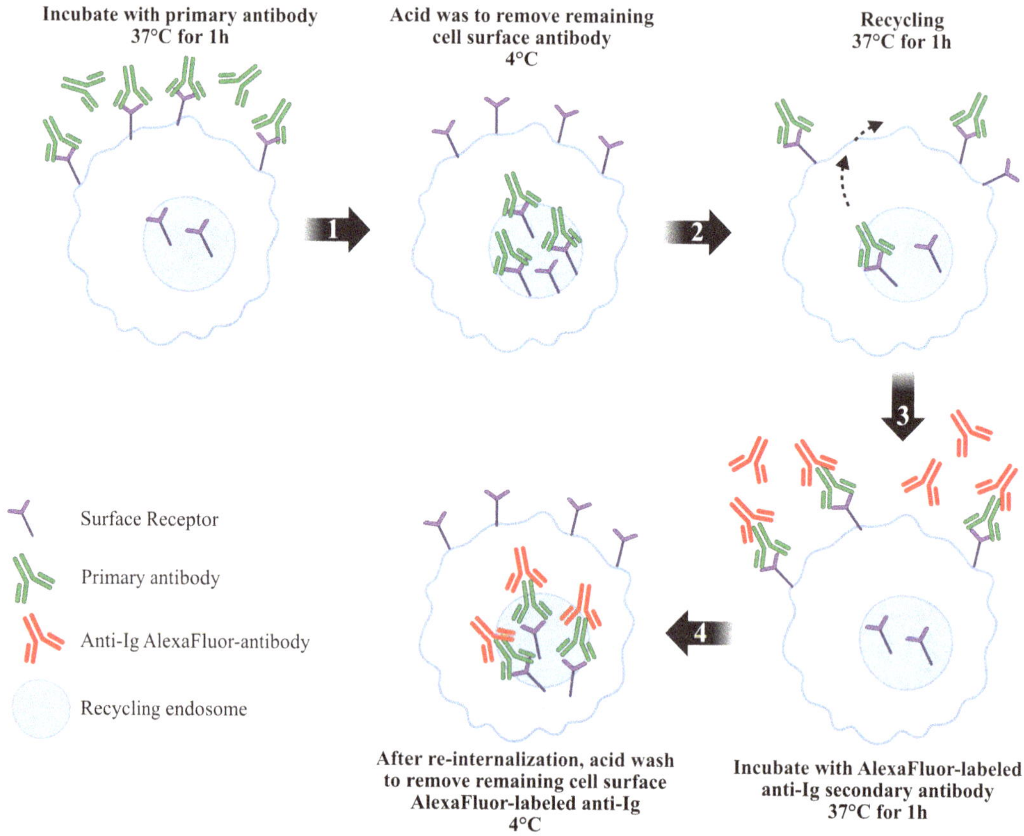

Fig. 4 Schematic representation of the receptor recycling assay workflow

2. Blocking step: After 24 h, replace medium with pre-warmed blocking medium and bring back cells to 37 °C for 15 min (*see* **Note 12**).
3. Labeling receptor of interest: Replace blocking medium with pre-warmed labeling medium (*see* **Note 13**) and bring back cells to 37 °C for 1 h.
4. Strip-off surface-labeled antibodies: put chamber on ice, gently wash once with ice-cold dPBS, and add ice-cold stripping medium for 1 to 3 min (*see* **Note 14**).
5. Receptor recycling: wash cells once with dPBS and incubate cells with pre-warmed recycling medium at 37 °C for 1 h.
6. Label recycling antibody-receptor complexes: replace recycling medium with pre-warmed secondary labeling medium (*see* **Note 13**) and bring back cells to 37 °C for 1 h.
7. Repeat acid stripping as in **step 4**.

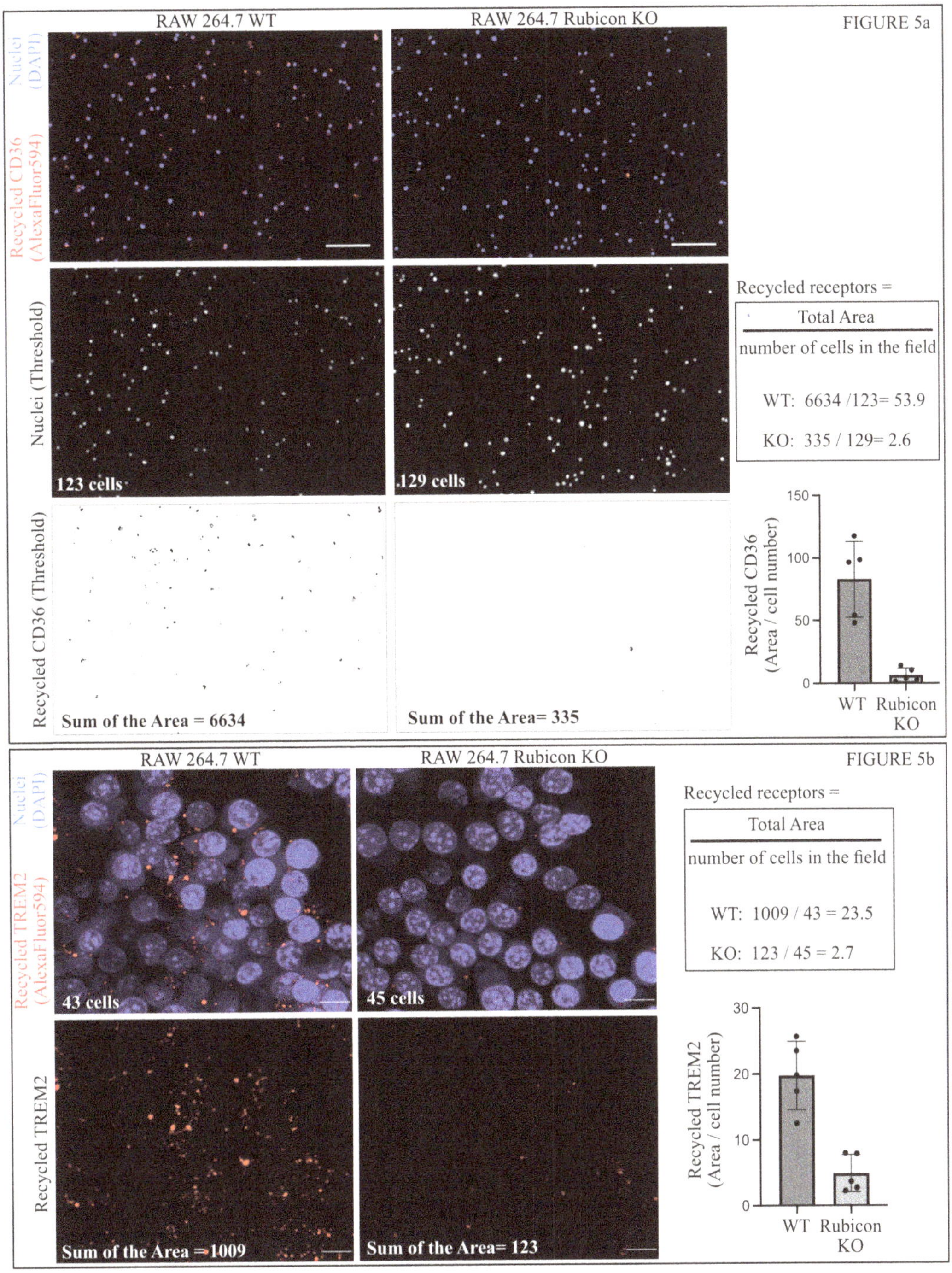

Fig. 5 Assessment of LANDO by immunofluorescence. Representative immunofluorescence images and related quantification strategy of receptor recycling in RAW264.7 cells (WT or Rubicon KO). (**a**) Images showing CD36 receptor recycling (red) as depicted by epifluorescence microscopy. (**b**) Images showing TREM2 receptor recycling (red) as depicted by confocal microscopy. Nuclei are stained with DAPI (blue)

8. Wash cells once with ice-cold dPBS and add fixing solution for 15 min at RT.
9. Wash three times with dPBS.
10. Optional: Incubate with nucleic acid staining solution ([1 μg/mL] of Hoechst 33342 or DAPI) for 15 min at RT.
11. Replace with PBS and image cells using a fluorescence microscope (*see* **Notes 15** and **16**).
12. Quantify receptor recycling as the fluorescent signal coming from vesicles containing recycled receptors divided by the number of cells (nuclei) present in the field (*see* **Notes 17** and **18**) (Figs. 4 and 5a, b).

4 Notes

1. Any phagocytic cell type can be tested here. In representative results related to the described protocols, we used RAW264.7 cells overexpressing mVenus_LC3b. Cells were retrovirally transduced using a pMXs retroviral vector encoding for the RAT_LC3b protein and they were sorted three times to achieve a >98% mVenus-positive cell population. Alternatively, one can use primary murine phagocytic cells, such as bone marrow derived macrophages (BMDMs), obtained from GFP-LC3b expressing mice [14]. As a control, if available, we suggest including in this procedure a cell line in which the LAP pathway is not functional (e.g., lacking the expression of proteins such as ATG5 or Rubicon).
2. Before using the zymosan solution, to circumvent the formation of zymosan aggregates, we highly recommend to vortex the vial containing the zymosan solution and to pass the solution through a 25-gauge needle using a 1 mL syringe several times to break up aggregates.
3. When using different chamber sizes, scale cell numbers accordingly. The assay can be also performed using sterile glass tissue culture coverslip coated with human fibronectin. In this case, a final additional step is needed to mount the cover slip onto imaging slides.
4. Optional before fixation: to reduce fluorescent background signal due to the presence of cytosolic LC3-I, one can permeabilize the cell membranes with a [200 μg/mL] digitonin solution for 10 min at RT, allowing the release of cytosolic LC3-I outside of the cell.
5. Here a primary antibody directly labeled with a fluorophore, e.g., AlexaFluor488, can be used. In this case, the use of a secondary antibody labeling step is avoided.

6. Scale cells and zymosan particle numbers according to the multi-well plate size. Consider plating an extra well of untreated cells for mVenus single color compensation control. The same control of untreated cells can be used to assess mVenus intensity at baseline when comparing different mVenus_LC3b overexpressing cell lines. The baseline intensity will be used to normalize the acquired LAP mVenus-LC3-II median fluorescence intensity (MFI). As an additional control to block canonical autophagy, plate extra wells and treat cells for 6 h with medium containing [300 nM] Torin1 and [100 nM] Bafilomycin A1.
7. We highly recommend preparing fresh digitonin solution for each assay. A [50 mg/mL] stock solution can be prepared in advance and aliquoted at −20 °C. If precipitation occurs, quickly boil the aliquoted solution at 70–90 °C. An efficiency of >90% digitonin permeabilization needs to be achieved which can be monitored by trypan blue microscopy on an aliquot of permeabilized cells.
8. Excessive confluency can lead to reduced engulfment efficiency.
9. Ensure a steady solution flow, avoiding excessive speed. Tilt the tube appropriately to prevent mixing of solutions with different percentages. The different layers of solution should be clearly distinguishable by eye.
10. Following centrifugation, phagosomes float at the interface between the 10% and 25% sucrose layers.
11. The pellet containing the bead-containing phagosomes may be very small, so exercise caution to avoid disturbing it while removing the supernatant.
12. The aim of the blocking medium is to avoid unspecific binding to the large amount of Fc receptors expressed on phagocyte's surface. We highly recommend that one test the efficacy of the blocking medium in a separate pilot experiment. To do so, when working with a new primary antibody, we perform an immunofluorescent staining on the target cells of interest, including in separate wells: 5% of serum from the source of the primary antibody and the Fc receptor blocking antibody rat anti-mouse CD16/CD32 clone 2.4G2 [10 μg/mL]. Keep in mind that if the primary antibody is raised in RAT, the Fc blocking antibody cannot be used; otherwise, it will interfere with the anti-RAT secondary antibody used to detect recycled receptors.
13. It is good practice to perform a preliminary experiment to titrate the antibody concentration to use. As a secondary antibody, a brighter and acidic resistant fluorophore is highly preferred (e.g., Alexa Fluor 594).
14. Usually, 1- to 3-min exposure to acidic stripping medium is sufficient to displace antibodies bound to the cell surface.

Because some cells (e.g., primary cells) are more susceptible than others to the acidic conditions, we highly recommend testing acid stripping efficacy in advance to find the shortest period needed to strip off >90% of receptor from the cell membrane. This can be achieved by performing a preliminary experiment where cells are incubated with a fluorescent tagged version of the primary antibody for 20 min on ice, followed by acidic wash for various times and pH range (2–3.5).

15. Imaging for quantification can be done using an epifluorescence or confocal microscope. For higher-resolution representative images and to clearly visualize the signal of the recycled receptors per cells, we recommend using a confocal microscope.
16. As an alternative acquisition method, the assay can be performed in tissue culture-treated multi-wells, and cells can be harvested and acquired by flow cytometry at the end of the assay.
17. The area of fluorescent signal from vesicle containing recycled receptors can be thresholded using common methods (e.g., ImageJ) or more specialized software (e.g., Imaris imaging analysis software). The area of fluorescent signal is then divided by the total number of cells in the field, resulting in a measurement of total fluorescent area per cell.
18. This protocol is formulated for assessing the endocytic recycling of the receptors of interest. Nonetheless, receptors that have been internalized may not exclusively follow the recycling pathway. For instance, a reduction in recycling rate could stem from heightened intracellular retention or increased degradation. The described assay specifically addresses the measurement of the role of LANDO in receptor recycling and does not examine its impact on receptor degradation or other potential disruptions in endosome trafficking. Therefore, we emphasize the significance of incorporating both recycling and degradation assays to comprehensively observe and comprehend the subcellular routes taken by internalized proteins.

References

1. Sanjuan MA, Dillon CP, Tait SW, Moshiach S, Dorsey F, Connell S, Komatsu M, Tanaka K, Cleveland JL, Withoff S, Green DR (2007) Toll-like receptor signalling in macrophages links the autophagy pathway to phagocytosis. Nature 450(7173):1253–1257. https://doi.org/10.1038/nature06421
2. Heckmann BL, Teubner BJW, Tummers B, Boada-Romero E, Harris L, Yang M, Guy CS, Zakharenko SS, Green DR (2019) LC3-associated endocytosis facilitates beta-amyloid clearance and mitigates neurodegeneration in murine Alzheimer's disease. Cell 178(3):536–551 e514. https://doi.org/10.1016/j.cell.2019.05.056
3. Galluzzi L, Green DR (2019) Autophagy-independent functions of the autophagy machinery. Cell 177(7):1682–1699. https://doi.org/10.1016/j.cell.2019.05.026

4. Durgan J, Lystad AH, Sloan K, Carlsson SR, Wilson MI, Marcassa E, Ulferts R, Webster J, Lopez-Clavijo AF, Wakelam MJ, Beale R, Simonsen A, Oxley D, Florey O (2021) Non-canonical autophagy drives alternative ATG8 conjugation to phosphatidylserine. Mol Cell 81(9):2031–2040. e2038. https://doi.org/10.1016/j.molcel.2021.03.020
5. Green DR (2021) Alternative careers at the autophagy factory. Trends Cell Biol 31(8): 613–615. https://doi.org/10.1016/j.tcb.2021.04.006
6. Wang Y, Ramos M, Jefferson M, Zhang W, Beraza N, Carding S, Powell PP, Stewart JP, Mayer U, Wileman T (2022) Control of infection by LC3-associated phagocytosis, CASM, and detection of raised vacuolar pH by the V-ATPase-ATG16L1 axis. Sci Adv 8(43): eabn3298. https://doi.org/10.1126/sciadv.abn3298
7. Grijmans BJM, van der Kooij SB, Varela M, Meijer AH (2021) LAPped in proof: LC3-associated phagocytosis and the arms race against bacterial pathogens. Front Cell Infect Microbiol 11:809121. https://doi.org/10.3389/fcimb.2021.809121
8. Cunha LD, Yang M, Carter R, Guy C, Harris L, Crawford JC, Quarato G, Boada-Romero E, Kalkavan H, Johnson MDL, Natarajan S, Turnis ME, Finkelstein D, Opferman JT, Gawad C, Green DR (2018) LC3-associated phagocytosis in myeloid cells promotes tumor immune tolerance. Cell 175(2):429–441 e416. https://doi.org/10.1016/j.cell.2018.08.061
9. Magne J, Green DR (2022) LC3-associated endocytosis and the functions of Rubicon and ATG16L1. Sci Adv 8(43):eabo5600. https://doi.org/10.1126/sciadv.abo5600
10. Ni D, Xu P, Gallagher S (2017) Immunoblotting and immunodetection. Curr Protoc Cell Biol 74:6 2 1–6 2 37. https://doi.org/10.1002/cpcb.18
11. Eng KE, Panas MD, Karlsson Hedestam GB, McInerney GM (2010) A novel quantitative flow cytometry-based assay for autophagy. Autophagy 6(5):634–641. https://doi.org/10.4161/auto.6.5.12112
12. Desjardins M, Huber LA, Parton RG, Griffiths G (1994) Biogenesis of phagolysosomes proceeds through a sequential series of interactions with the endocytic apparatus. J Cell Biol 124(5):677–688. https://doi.org/10.1083/jcb.124.5.677
13. Lucin KM, O'Brien CE, Bieri G, Czirr E, Mosher KI, Abbey RJ, Mastroeni DF, Rogers J, Spencer B, Masliah E, Wyss-Coray T (2013) Microglial beclin 1 regulates retromer trafficking and phagocytosis and is impaired in Alzheimer's disease. Neuron 79(5):873–886. https://doi.org/10.1016/j.neuron.2013.06.046
14. Mizushima N (2009) Methods for monitoring autophagy using GFP-LC3 transgenic mice. Methods Enzymol 452:13–23. https://doi.org/10.1016/S0076-6879(08)03602-1

Methods in Molecular Biology (2025) 2879: 51–61
DOI 10.1007/7651_2023_501

Published online: 28 October 2023

Monitoring Autophagy with GFP-LC3 Reporter

Lidong Sun, Suli Lv, and Tanjing Song

Abstract

Autophagy is a key process that maintains cellular homeostasis. Autophagy contributes to various physiological and pathophysiological processes. Development of methodologies for autophagy detection has greatly facilitated the research on autophagy. Among these methodologies, GFP-LC3 reporter has been popularly used in the literature. In this chapter, we will detail step-by-step the GFP-LC3 reporter protocol we have adapted in our lab. This protocol begins with the generation of lentivirus expressing GFP-LC3. Then, the cells are transduced with titrated virus. After selecting the positive cells, single colonies are isolated, characterized, validated, and used in further study.

Key words Autophagosome, Autophagy, GFP, LC3, Lysosome

1 Introduction

Autophagy packs intracellular material in double-membrane structure called autophagosome and then delivers it to lysosome for degradation [1]. Autophagy removes damaged proteins and cellular organelles and maintains cellular homeostasis [1]. Meanwhile, autophagy recycles nutrients to support cell survival and proliferation, especially under starvation [1], and it involves orchestrated function of many proteins. LC3 is an evolutionarily conserved protein. Mammalian LC3 is the homolog of yeast ATG8 [2]. LC3 can be conjugated to phosphatidylethanolamine in a process similar to ubiquitination. This process is also known as lipidation of LC3. Lipidated LC3 is called LC3-II, as opposed to non-lipidated LC3-I. Lipidated LC3 is critical for autophagosome initiation and expansion [2]. Various types of autophagy are now recognized, and not all of them are involved in the formation of autophagosome [3]. GFP-LC3 reporter does not apply to all forms of autophagy; throughout this chapter, autophagy refers to macroautophagy.

Abnormality in autophagy contributes to various human diseases, including neurodegenerative diseases, immune malfunction, and cancer [4–6]. As a result, recent years have seen an explosion in autophagy study. In 2022 alone, about 10,000 papers were found in PubMed, with key term "autophagy" present in title or abstract. This explosion in autophagy study is made possible by development of methodologies for autophagy detection [7]. A classical method is transmission electron microscopy, which allows direct observation of cellular ultrastructure, including autophagosome [7]. But the equipment and techniques are not commonly available. In addition, under high magnification, each photo contains much less cells. As a result, it is also time consuming to take enough photos in order to document statistically meaningful result. Like many other processes, molecular markers have been used instead to monitor the level of autophagy. The level of LC3-II as detected by Western blot has been used for such purpose [7]. Alternatively, puncta of LC3 in live cells can also serve as marker for autophagosome [7]. The philosophy underneath is non-lipidated LC3 diffuses in the cell, while lipidated LC3, LC3-II, selectively enriches autophagosomes, forming visible puncta under fluorescent microscopy. While puncta formed by endogenous LC3 theoretically can be visualized with immunofluorescence, the specificity and sensitivity of LC3 antibody may not be sufficient. In addition, endogenous LC3 is subject to transcriptional regulation by various stimuli, which might compound the readout [8, 9]. As a result, exogenous GFP-LC3 puncta has proven to be a convenient surrogate for autophagosome. In the literature, GFP-LC3 is either transiently or stably expressed in the cell. We choose stable expression here as transient transfection is often associated with variable transfection efficiency and variable expression level among cell population. For stable expression, either cell population or single colonies have been used. Here, we choose single colony as we find that single colony gives more uniform GFP-LC3 expression and puncta pattern, which makes comparison between different experimental groups more amenable [10]. Conventional non-virus vectors suffer from extremely low efficiency of integrating into genomic DNA after transfection. Accordingly, generating stable colonies with these vectors also faces extremely low efficiency. So here we choose to use lenti-virus based GFP-LC3 expression vector. Collectively, with our protocol, multiple single colonies can be generated within 3 weeks with minimal hands-on time. After generating GFP-LC3 reporter cells, we typically perform two types of analysis on these cells. One is to analyze the number of GFP-LC3 puncta with fluorescent microscopy and the other is to measure the turnover of GFP-LC3 by Western blot.

2 Materials

2.1 Required Equipment

1. Biosafety cabinet (e.g., Thermo MSC-advantage).
2. CO_2 incubator (e.g., Thermo BB150).

3. Fluorescent microscope with camera (e.g., Zeiss-A1 Axiovert A1 fluorescence microscopy equipped with 63× oil objective lens with QImaging Retiga R6 Monochrome camera).
4. Mini Gel electrophoresis (e.g., Mini-PROTEAN® Tetra from Bio-rad).
5. Wet-transfer apparatus (e.g., Mini Trans-Blot Electrophoretic Transfer Cell #1703930).
6. Imaging Machine (e.g., ChemiDoc Imaging System from Bio-Rad).

2.2 Required Reagents and Solutions

2.2.1 Cell Culture and Virus Packaging

1. Complete cell culture medium (e.g., DMEM supplemented with 10% FBS).
2. 0.45 μm filter: Millipore CAT# SLHP033RB.
3. 0.22 μm filter: Millipore CAT# SLGP033RB.
4. PEI (Polyethylenimine): Polysciences #23966, 0.323 mg/mL in deionized water, filter through 0.22 μm to sterilize.
5. Polybrene: Sigma CAT# H9268, 8 mg/mL in deinonized water, filter through 0.22 μm to sterilize.
6. Plasmid: psPAX2, pMD2.G, pLenti-GFP-LC3.

2.2.2 Immunostaining

1. Anti-fade fluorescence mounting medium (Abcam #ab104135).
2. DAPI (4′,6-Diamidino-2-phenylindole dihydrochloride): Sigma #D8417, dissolve in water at 5 mg/mL.
3. BCA protein quantitation kit: Beyotime #P0012.
4. PBS (phosphate buffered saline): take KCl 1 g, KH_2PO_4 1.2 g, NaCl 40 g, Na_2HPO_4 7.2 g, dissolve in water, adjust pH to 7.4 with HCl, bring volume to 5 L with water.
5. 4% paraformaldehyde: 2 g paraformaldehyde dissolved in 50 mL PBS.
6. PBST: PBS supplemented with Tween-20 at 0.2% final concentration.
7. 0.5% Triton-X100: 0.5 mL Triton-X100 dissolved in 49.5 mL PBS.

2.2.3 Western Blot

1. 1× loading buffer: 62.5 mM Tris–HCl pH = 6.8, 2% SDS, 10% glycerol, 5% β-2-mercaptoethanol, 0.002% bromophenol blue [11].
2. 1× SDS lysis buffer: 62.5 mM Tris–HCl pH = 6.8, 2% SDS, 10% glycerol, 5% β-2-mercaptoethanol [11].
3. GFP antibody (e.g., ProteinGene #2057 RRID: AB_2924213).

2.3 Software

1. FIJI (ImageJ 1.53q) (National Institute of Health).

3 Methods

3.1 Virus Packaging

1. Day1: Seed HEK-293T cells into 6 cm dish at 50% density.
2. Day2: Transfection: Mix 2.6 μg psPAX2, 1.4 μg pMD2G, and 4.0 μg pLenti-GFP-LC3 plasmid in a 1.5 mL micro-centrifuge tube. Mix 532 μL 150 mM NaCl and 56 μL PEI in another 1.5 mL micro-centrifuge tube. Then add PEI to plasmid, mix well, and let sit for 30 min at room temperature. Afterward, add plasmid-PEI complex gently to cell culture, and shake the cell culture dish gently to mix (*see* **Note 1**).
3. Day3: Change culture medium.
4. Day4: Collect the cell culture supernatant which contains virus. Autoclave and discard the cell culture dish as biohazard. Filter the virus suspension through 0.45 μm filter. The virus suspension can be used right away or frozen in −80 °C freezer in aliquot for later use (*see* **Note 2**).

3.2 Transduction and Selection

1. Day1: Plate cells in 3.5 cm dishes at 30% confluency.
2. Day2: Add increasing amount of virus suspension (e.g., 0.1 μL, 1 μL, 10 μL, 100 μL) to each dish respectively. Add polybrene to culture medium to final concentration 8 μg/mL. Shake culture medium gently to mix (*see* **Note 3**).
3. Day3: Trypsinize the cells and plate one-third of them into 6 cm dishes. Meanwhile, seed a portion of cells in each dish to cover slides for photographing (Fig. 1) (*see* **Note 4**).
4. Day4: Add blasticidin at 10 μg/mL final concentration (or other antibiotics applicable) to the culture medium to initiate selection.
5. Day5–Day12: Continue selection, maintain the cells below 80% confluency. Passage the cells if necessary to prevent over-confluency (*see* **Note 5**).

3.3 Single Colony Isolation (Key Step)

1. Examine GFP-LC3 signal of selected cell population under fluorescent microscope. Among dishes with observable GFP-LC3 signal, choose the one from transduction with lowest virus titer (Fig. 2).
2. Trypsinize cells in the selected dish and dilute the cell suspension to 5 cells/mL. Seed 0.1 mL medium into 96-well plate. Contain 10 μg/mL blasticidin final concentration in culture medium (*see* **Note 6**).
3. Let the cell expand. Examine the cell and change the medium once a week if necessary.

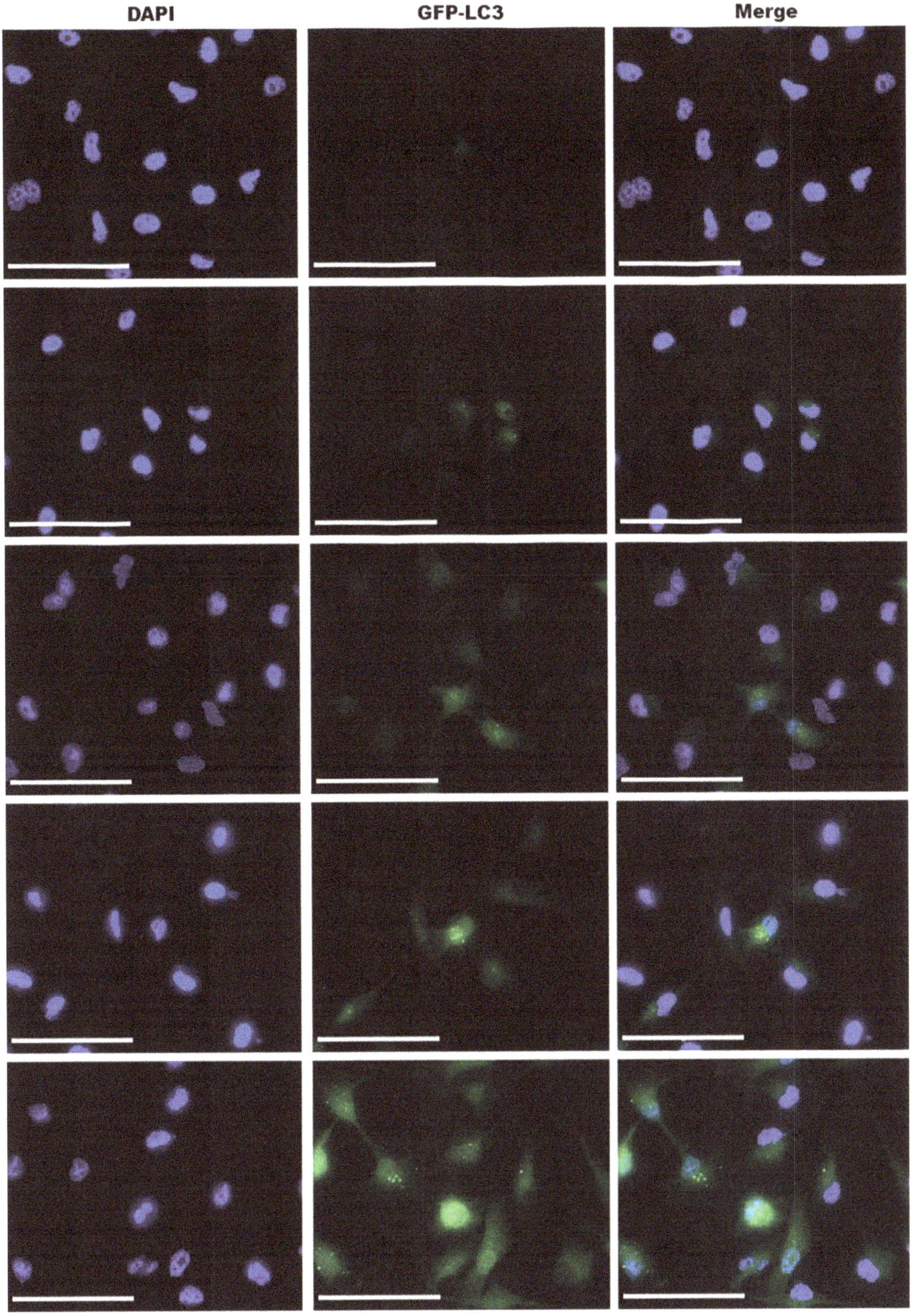

Fig. 1 High virus titer increases GFP-LC3 expression level. Photos show GFP-LC3 signal with increasing virus titer. From top to bottom: 1 μL, 10 μL, 20 μL, 50 μL, and 100 μL virus suspensions were used to transduce cells respectively (scale bar = 80 μm)

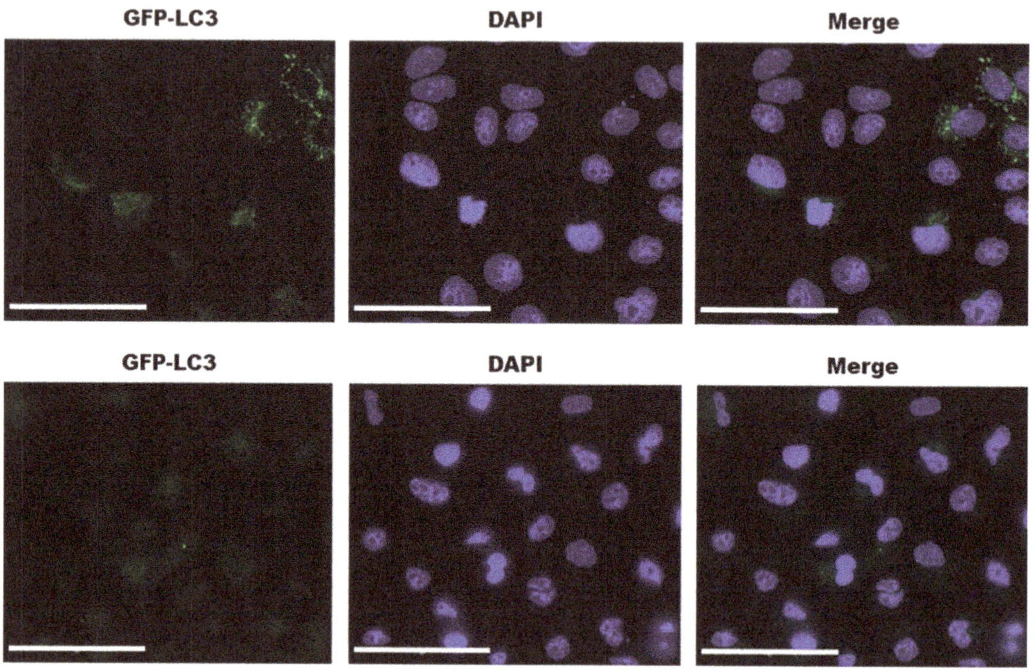

Fig. 2 Antibiotic-resistant cell populations have heterogeneous GFP-LC3 signal. Top: photos show a population of HeLa cells infected with GFP-LC3-expressing virus after blasticidin selection for 2 weeks (scale bar = 80 μm). Bottom: photos show a population of Caki-1 cells infected with GFP-LC3-expressing virus after blasticidin selection for 2 weeks (scale bar = 80 μm)

4. By 1 week generally, colonies form in some of the wells. Pick 5 to 10 wells that contain one and only one round colony. Trypsinize the cell and plate each colony into one 3.5 cm dish.
5. Propagate the cells further into 6 cm dish.

3.4 Single Colony Characterization

Examine GFP-LC3 puncta pattern: Seed cells from each colony onto cover slides and examine GFP-LC3 under fluorescent microscopy. Keep colonies with uniform GFP-LC3 signal and similar GFP-LC3 puncta pattern (Fig. 3).

3.5 Readout

Now the cells can be used for various treatments, including gene knockdown, gene overexpression, and drugs. Afterward, examine GFP-LC3 puncta with fluorescent microscopy or measure GFP-LC3 turnover with Western blot.

3.5.1 Examine GFP-LC3 Puncta Number

Part 1: Prepare the Sample

1. Seed cells onto cover slides and let grow.
2. Fix the cells with 4% paraformaldehyde for 15 min in room temperature.
3. Wash with PBST (PBS + 0.2% Tween-20) for three times 5 min each time.

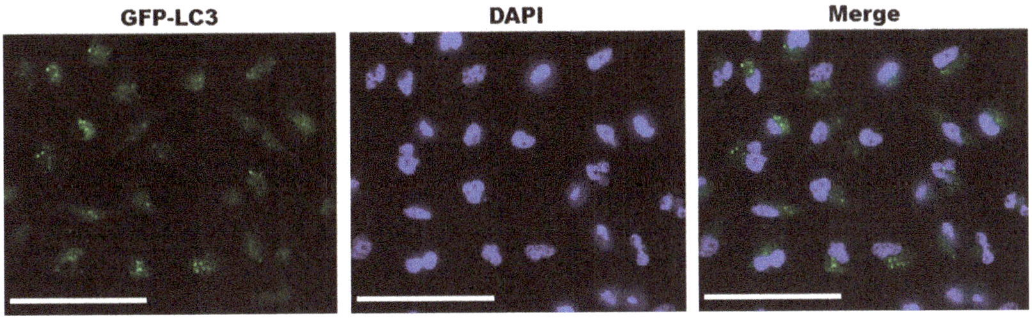

Fig. 3 Cells from single colony have more uniform GFP-LC3 signal. Photos show Caki-1 cells propagated from a single colony (Scale bar = 80 μm)

4. Incubate slides with PBST containing 1.5 μM DAPI for 5 min to counter-stain the cell nuclei.
5. Wash with PBST (PBS + 0.2% Tween-20) for three times 5 min each time.
6. Mount slides onto glass slides with mounting medium.
7. Wait for 6 h for the mounting medium to solidify. Take pictures for randomly selected fields under fluorescent microscope equipped with a 63× object lens. Keep exposure time and all other parameters consistent across the same batch of experiment (Fig. 4).

Part 2: Analyze the Photos with FIJI to Automatically Count Puncta Number in Each Cell

8. Open the image: "File" --> "Open" (Fig. 5a–c).
9. Change image to 8 bit: "Image" --> "Type" --> "8 bit".
10. Subtract background: "Process" --> "Subtract background with rolling ball radius" (set at "27 pixel") (*see* **Note 7**) (Fig. 5d).
11. Turn the photos into binary: "Image" --> "adjust" --> "threshold" --> set threshold at 18/255, choose "Dark background," click "Apply" (*see* **Note 8**) (Fig. 5e).
12. Split merged particles: "process" --> "binary" --> "Wastershed" (Fig. 5f).
13. Call GFP-LC3 particles with "Analyze Particle" function: "Analyze" --> "Analyze particle" --> "size" set at 0.20–6.50 μm^2, "circularity" set at 0–1 (Choose all options except "include holes" in case other parameters of the puncta such as their size are also of interest.) (*see* **Note 9**) (Fig. 5g).

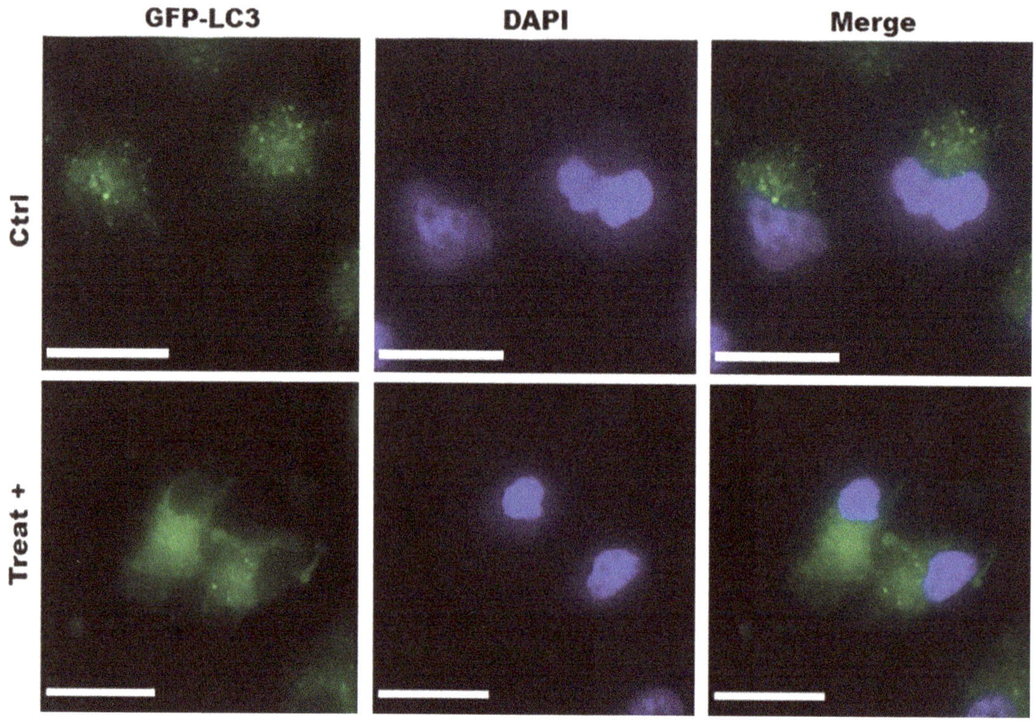

Fig. 4 Number of GFP-LC3 puncta decreases after TFE3 knockdown. Photos show GFP-LC3 expressing-Caki-1 cells. The top panel shows control cells. The bottom panel shows TFE3-KD cells (scale bar = 20 μm)

3.5.2 Measure Turnover of GFP-LC3 with Western Blot

Collect the cells for Western blot analysis. (Use any regular Western blot process. Following is just an example of what we do.)

1. Aspirate culture medium completely. Wash the cells twice quickly with PBS. Aspirate PBS completely.
2. Add 1× SDS lysis buffer to the cell and scrape cell.
3. Sonicate the cell lysate with a probe sonicator at 30 W, 5 s on/ 5 s off, 6 cycles.
4. Heat the sample at 99 °C for 5 min.
5. Centrifuge at 15,000 g for 10 min.
6. Take the supernatant as whole cell lysate.
7. Measure the protein concentration with BCA method.
8. Take 50 μg whole cell lysate, and bring sample volume equal with 1 × SDS loading buffer.
9. Load samples onto 10% SDS-PAGE gel (*see* **Note 10**).
10. Perform Western blot with GFP antibody following a regular Western blot protocol.
11. Acquire data for Western blot (e.g., with ChemiDoc Imaging System from Bio-Rad) (Fig. 6).

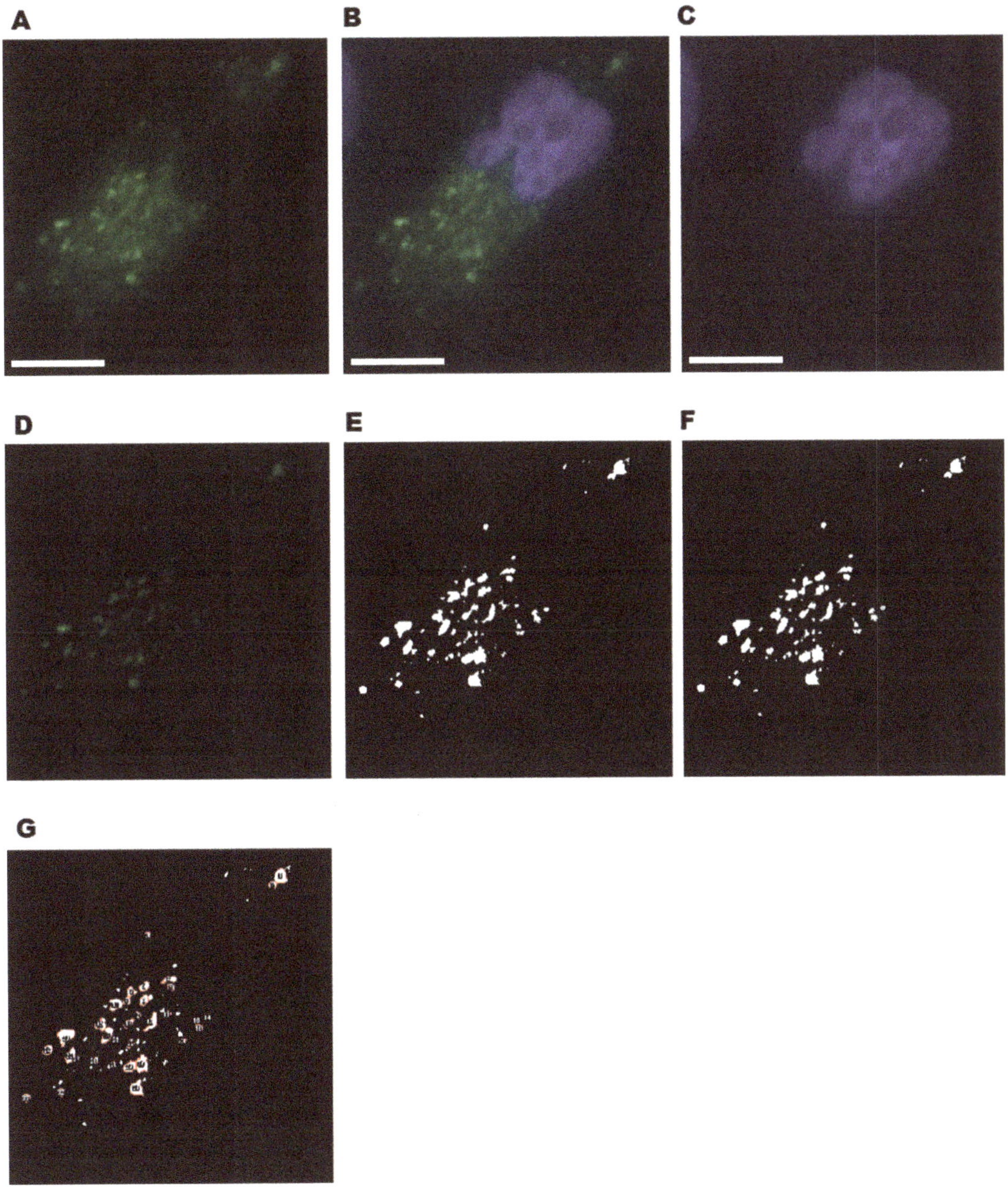

Fig. 5 Counting GFP-LC3 puncta number with FIJI. (**a**) Original photo for GFP-LC3 in a cell. (**b**) Original photo for DAPI in the same cell as shown in (**a**). (**c**) Merge of GFP-LC3 and DAPI. (**d**) GFP-LC3, after "Subtract background." (**e**) GFP-LC3, after "Turn the photos into binary." (**f**) GFP-LC3, after "Split merged particles." (**g**) GFP-LC3, after "Analyze Particle"

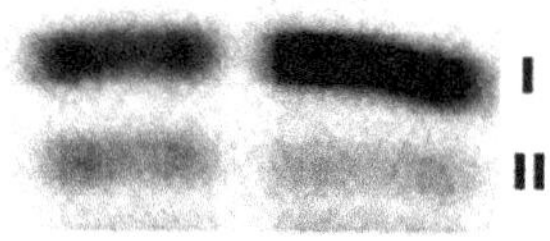

Fig. 6 Change in GFP-LC3 level could be detected with Western blot. Whole cell lysates of control (lane 1) or TFE3-KD cells (lane 2) were analyzed with Western blot [10]

4 Notes

1. Transfection with other protocols should also work. The PEI method used here is just one example.
2. In our practice, virus does not need to be concentrated since very little virus suspension is needed for this purpose. If the transduction efficiency is too low, concentration may help.
3. Since transduction efficiency varies significantly among cells, we typically performed transduction with a series of virus titers.
4. Adequate expression of GFP-LC3 is critical for optimal result. Our experience is that expression level too high may lead to high background and sometimes artificial signal in the nucleus.
5. Selection time varies with the antibiotics. If puromycin is used instead, shorter selection time is feasible.
6. The cell number seeded at each step can be modulated so cells will not be over-confluent at any step.
7. This is for photos with scale 13.75 pixel/μm, so scale up or down accordingly. In addition, radius may be fine-tuned based on the result, but be consistent across the batch.
8. Threshold may be fine-tuned based on the context, but be consistent across photos in one experiment).
9. Exact parameters may be fine-tuned, but be consistent across photos in one experiment.
10. Gel with too high concentration may fail to differentiate LC3-I and -II of GFP-LC3.

Acknowledgments

This work was supported by grants from the National Natural Science Foundation of China (32071296 and 31871284 to L.S, 31971149 and 31800641 to T.S.) and Natural Science Foundation of Hubei Province of China (2022CFB251 to T.S).

References

1. Dikic I, Elazar Z (2018) Mechanism and medical implications of mammalian autophagy. Nat Rev Mol Cell Biol 19:349–364
2. Schaaf MB, Keulers TG, Vooijs MA, Rouschop KM (2016) LC3/GABARAP family proteins: autophagy-(un)related functions. FASEB J 30: 3961–3978
3. Bednarczyk M, Zmarzly N, Grabarek B, Mazurek U, Muc-Wierzgon M (2018) Genes involved in the regulation of different types of autophagy and their participation in cancer pathogenesis. Oncotarget 9:34413–34428
4. Debnath J, Gammoh N, Ryan KM (2023) Autophagy and autophagy-related pathways in cancer. Nat Rev Mol Cell Biol:1–16
5. Themistokleous C, Bagnoli E, Parulekar R, Muqit MMK (2023) Role of autophagy pathway in Parkinson's disease and related genetic neurological disorders. J Mol Biol:168144
6. Santovito D, Steffens S, Barachini S, Madonna R (2023) Autophagy, innate immunity, and cardiac disease. Front Cell Dev Biol 11: 1149409
7. Klionsky DJ et al (2021) Guidelines for the use and interpretation of assays for monitoring autophagy (4th edition) (1). Autophagy 17: 1–382
8. Fullgrabe J, Klionsky DJ, Joseph B (2014) The return of the nucleus: transcriptional and epigenetic control of autophagy. Nat Rev Mol Cell Biol 15:65–74
9. Baek SH, Kim KI (2017) Epigenetic control of autophagy: nuclear events gain more attention. Mol Cell 65:781–785
10. Song T et al (2023) TRIM28 represses renal cell carcinoma cell proliferation by inhibiting TFE3/KDM6A-regulated autophagy. J Biol Chem 299:104621
11. Laemmli UK (1970) Cleavage of structural proteins during the assembly of the head of bacteriophage T4. Nature 227:680–685

Methods in Molecular Biology (2025) 2879: 63–82
DOI 10.1007/7651_2024_525

Published online: 06 March 2024

Investigating Protein-Protein Interactions of Autophagy-Involved TNIP1

Michael L. Samulevich, Liam E. Carman, and Brian J. Aneskievich

Abstract

Myriad proteins are involved in the process of autophagy, which they participate in via their protein-protein interactions (PPI). Herein we outline a methodology for examining such interactions utilizing the case of intrinsically disordered protein (IDP) TNIP1 and its interaction with linear M1-linked polyubiquitin. This includes methods for recombinant production, purification, immuno-identification, and analysis of an IDP associated with autophagy, its ordered binding partner, and means of quantitatively analyzing their interaction.

Key words Bacterial expression, Heat lysis, Intrinsically disordered protein, Microscale thermophoresis, Nickel affinity chromatography, Recombinant protein

1 Introduction

Protein-protein interactions (PPI), in addition to protein-membrane interactions, are key to the initiation and organization of the autophagosome highlighting the need for identifying and assessing such assemblies [1]. At the core of autophagosome assembly are numerous autophagy-related (ATG) proteins that are conserved across species as diverse as yeast and humans [2, 3]. Expanding and specializing ATG-mediated assembly are selective autophagy receptors (SARs) especially given their discriminating delivery of target proteins destined for breakdown and in some cases, themselves being degraded [4–7]. Nevertheless, the varying biophysical nature of the protein partners involved in autophagosome organization complicates in vitro analyses. Individual proteins often present unique challenges for their expression, purification, yield, and possible labeling for interaction assay throughput. Studies of PPI to model such partner protein assembly can be done via diverse in vivo (e.g., cell-based two-hybrid), in vitro (e.g., protein microarrays), and computational (e.g., PPI databases) approaches [8–10]. Each of these has advantages and limitations

Authors Michael L. Samulevich and Liam E. Carman have equally contributed to this chapter.

dependent on individual protein characteristics, throughput scale, and available instrumentation [11–14]. Thus, methodology for procuring protein(s) and studying PPI needs to be under constant development.

Here we present select expression and purification protocols needed for two proteins, TNIP1 and polyubiquitin, destined for in vitro modeling of their interaction as a potential SAR [15, 16] and partner protein [17, 18], respectively. Nevertheless, they required differing production and purification approaches, some of which could be turned to advantage at specific points along protein preparation. Specifics we present on their individualized handling can alert researchers to expect the likely need of tailoring protocol steps to their protein of interest. Also, for PPI such as that between TNIP1 and polyubiquitin, we highlight the emerging technique of microscale thermophoresis (MST) in part because of its low consumption of purified protein relative to other in vitro assays while returning literature-consistent measurements of interaction. MST is increasingly being employed for quantitation of macromolecular protein-protein and protein-ligand interactions for which others have provided excellent detailing of the thermodynamics involved and breadth of applications [19–22].

Among its many functions, TNFα-induced protein (TNFAIP) 3-interacting protein 1 (TNIP1) is a non-enzymatic repressor of cytoplasmic, NF-κB-mediated inflammatory signaling and nuclear receptor-mediated transcriptional activity [17, 18, 23–25]. TNIP1 is an intrinsically disordered protein (IDP) [18, 23], meaning it transits within an ensemble of conformations rather than realizing one fixed tertiary structure thus providing both challenges and opportunities for its recombinant expression and purification (see [26] and Subheading 3 method steps). This conformational flexibility enables its interaction with multiple partners as reflected by its independent discoveries via PPI with cytoplasmic TNFAIP3, HIV proteins, and nuclear receptors (see [17, 18, 27, 28] for reviews). The malleable nature of TNIP1 cuts across the typical structured lock-and-key PPI paradigm. As with some other IDPs [29, 30], its final conformation may indeed be instructed by interaction with specific partners enabling it to be a lock or key in differing interactions. As an IDP, it may be more appropriate to consider TNIP1 a hub protein central to overlapping or independent PPI networks mediated by its direct binding to other proteins as well as its recognition of polyubiquitin conjugated to other proteins [17, 31, 32] and reflected by its extensive interactome [16]. Spokes on the TNIP1 hub include its direct interaction with the LC3 protein, itself central to autophagy, positioning TNIP1 as a target and separately a SAR [16, 33, 34]. It is clear from investigations cited above and now these studies that TNIP1 PPI are central to its multiple roles and thus in need of new methodology for their study. The utility of the protocols provided here is demonstrated by

TNIP1:polyubiquitin interaction values similar to previously established values [31, 35] notwithstanding modified protein preparations and the very different biophysical approach of MST.

2 Materials

2.1 Bacterial Expression of Recombinant Proteins

1. Lysogeny broth (LB): 10 g tryptone, 5 g yeast extract, 5 g sodium chloride, 1 mL 1 M sodium hydroxide, fill to 1 L with type 1, 18.2 MΩ water (Nanopure) (*see* **Note 1**), autoclaved.
2. Chloramphenicol (37 mg/mL in ethanol, 1000× stock, 0.2 μm filter sterilized, stored at −20 °C).
3. Kanamycin (50 mg/mL in water, 1000× stock, 0.2 μm filter sterilized, stored at −20 °C).
4. Isopropyl β-D-thiogalactopyranoside (IPTG) stock (100 mM in water, 0.2 μm filter sterilized, stored at −20 °C).
5. Bacterial colonies expressing protein of interest, BL21 or equivalent (e.g., Rosetta 2 DE3, Sigma-Aldrich; Cat# 71397).
6. Incubator shaker (optimally two) capable of 16–37 °C range and 250RPM.
7. 250 mL flasks.
8. 1 L flasks, preferably baffled.
9. 500 mL centrifuge bottles.
10. Floor centrifuge capable of 20,000× *g* and Beckman JA-10 and JA-20 fixed angle rotors or equivalents.

2.2 SDS-PAGE

1. 5× Laemmli sample buffer: 0.3125 M Tris–HCl, pH 6.8 (dilute from 1 M stock), 10% sodium dodecyl sulfate (SDS), 50% neat glycerol, 0.005% bromophenol blue, and 10% β-mercaptoethanol (added just before use) (*see* **Note 2**).
2. 10× running buffer: 0.25 M Tris–HCl, pH 8.3, 1.92 M glycine, 1% SDS.
3. Destain/fixing solution: 7% acetic acid, 10% methanol in water.
4. Staining solution: 0.05% Coomassie Brilliant Blue, 10% acetic acid, and 50% methanol in water.

2.3 Western Blotting Analysis

1. Nitrocellulose membranes, blot filter paper, foam pads.
2. Western blot transfer buffer: 0.025 M Tris–HCl, 0.192 M glycine, 20% methanol, 0.5% SDS, chilled before use (*see* **Note 3**).
3. 10× Tris-buffered saline (TBS): 1.5 M NaCl, 0.1 M Tris–HCl, pH 7.4, 0.2 μm filter sterilized.

4. 1× TBS containing 0.05% Tween 20 (TBS-T), 0.2 μm filter sterilized (*see* **Note 4**).
5. Blocking solution: 5% BSA in TBS-T, 0.2 μm filter sterilized.
6. Primary antibody (e.g., anti-His6; BioLegend, Cat# 652502).
7. Secondary antibody (e.g., IRDye 800CW goat anti-mouse IgG secondary antibody, Li-Cor; Cat# 926-32210).
8. Protein Ladder (e.g., Chameleon Duo Pre-Stained, Li-Cor; Cat# 928-60000).
9. Li-Cor Odyssey CLx with Image Studio Software for imaging and analysis, or equivalents (*see* **Note 5**).

2.4 Immobilized Metal Affinity Chromatography (IMAC) Purification

1. Thermostable protein lysis buffer: 20 mM sodium phosphate, 500 mM NaCl, 20 mM imidazole, pH 7.3 (dependent on protein isoelectric point). Store at 4 °C until use. Before use supplement with 5 mM β-mercaptoethanol, 0.25 mg/mL lysozyme, 1 tablet/30 mL Roche cOmplete EDTA-free protease inhibitor tablet, 1 mM PMSF.
2. Ordered protein lysis buffer: 4 mL of B-PER (ThermoFisher; Cat# 90084) per gram of bacterial pellet, 0.25 mg/mL lysozyme, 0.2 μL Pierce Universal Nuclease, 1 Roche cOmplete EDTA-free protease inhibitor tablet (*see* **Note 6**).
3. Pierce Universal Nuclease (*see* **Note 7**).
4. Binding Buffer: 20 mM HEPES, 400 mM NaCl, 20 mM imidazole, 10% glycerol, pH 7.3, adjusted to 5 mM β-mercaptoethanol immediately before use.
5. Wash Buffer: 20 mM HEPES, 400 mM NaCl, 40 mM imidazole, 10% glycerol, pH 7.3, adjusted to 5 mM β-mercaptoethanol and 1 mM PMSF immediately before use.
6. Elution Buffers (EB1-EB4): 20 mM HEPES, 300 mM NaCl, imidazole (EB1-EB4; 50, 100, 150, and 300 mM, respectively), 10% glycerol, pH 7.3, adjusted to 5 mM β-mercaptoethanol and 1 mM PMSF immediately before use (*see* **Note 8**).
7. Final Buffer: 20 mM HEPES, 150 mM NaCl, 5% glycerol, pH 7.3, adjusted to 1 mM DTT immediately before use.
8. Glass column for gravity flow (e.g., Econo-Column®, BioRad; Cat #7371512).
9. Nickel-NTA resin (e.g., HisPur™ Ni-NTA Resin, Thermo Scientific™; Cat #88221).
10. 10 mL borosilicate glass tubes treated with SigmaCote (Sigma-Aldrich; Cat #SL2).
11. 100 °C water bath.

12. High-speed PPCO centrifuge tubes (e.g., Nalgene™ Oak Ridge, Thermo Scientific™; Cat #3119-0010).
13. Desalting column (e.g., Sephadex G-25 in PD-10, Cytiva; Cat #17085101).

2.4.1 Thrombin Cleavage on Nickel Resin

1. Lyophilized thrombin from bovine plasma (e.g., Sigma-Aldrich; Cat# T4648-1KU).
2. Calcium chloride ($CaCl_2$) (*see* **Note 9**).

2.5 Microscale Thermophoresis (MST)

1. His-tagged target protein, 200 nM.
2. Ligand protein stock, concentration contingent on predicted K_D (*see* **Note 10**).
3. His-Tag Labeling Kit RED-tris-NTA 2nd Generation dye (NanoTemper; Cat# MO-L018). Kit includes NTA dye, PBS-T with 0.05% Tween 20, and His tag control peptide.
4. Standard Monolith Capillaries (*see* **Note 11**) (NanoTemper; Cat# MO-K022).
5. Monolith X Instrument and related software MO.Affinity Analysis (MO.AA) and MO.Control Software (NanoTemper).
6. 96-well plate (*see* **Note 12**).
7. Aluminum foil (*see* **Note 13**).

3 Methods

3.1 Bacterial Expression of Recombinant Proteins

Bacterial expression systems provide a well-established and reliable method of producing recombinant proteins for downstream use in a wide variety of assays. The *Escherichia coli* expression system remains the chief method of bacterial protein expression [36]. Initially, a limitation of this system was the incongruence between the codon usage preference in the desired human recombinant proteins and the commonly available tRNAs native to *E. coli* [37]. This limitation has been overcome via specialized BL-21 derivative bacterial strains containing a plasmid prior to recombinant plasmid transformation; the plasmid encodes a variety of tRNAs that are uncommon to *E. coli*, thus allowing a protein sequence that has not been optimized for bacterial expression to still be effectively produced by *E. coli*. An example of this modification can be seen in Rosetta™ 2 DE3 *E. coli* which contains the pRARE plasmid encoding low abundance tRNAs [38].

IDPs lack fixed or ordered secondary and/or tertiary structure [39]. Pragmatically, intrinsic disorder also makes proteins like TNIP1 more sensitive to proteolysis under instances of endogenous function and recombinant expression [26, 40]. These TNIP1 traits, unexpected at initial reports of its PPI, lead us to encourage researchers to assess their protein(s) of interest for possible intrinsic disorder and predicted protease cleavage sites before attempting bacterial expression as we and others have previously described [41–43] to be better informed in optimizing recombinant expression protocols. In-depth reviews and methods [e.g., [44, 45] are available to broadly cover challenges in recombinant production of IDPs. Potential solutions include engineering inclusion body and/or secretion tags into plasmids containing the desired protein's coding sequence [46, 47]. Below, we outline simpler options such as reduced temperature and increased time of induction to enhance levels of intact recombinant protein.

1. Prepare 250 mL starter flasks with 40 mL of LB with 40 μL of 1000× chloramphenicol and 40 μL of 1000× kanamycin. Inoculate flasks with individually selected colonies and place in an incubator shaker set to 37 °C and 250 RPM overnight (16 h).
2. Prepare 1 L culture flasks of 800 mL LB containing 800 μL 1000× chloramphenicol and 800 μL 1000× kanamycin for scale up. Take 1 mL from culture media and set aside to use as a spectrophotometry blank. Record OD_{600} of overnight cultures as an internal growth consistency check, then add it to culture flask. Record OD_{600} of inoculated culture media as a reference density to later time points, then split sample into two 400 mL aliquots in two 1 L culture flasks. Load culture flasks into incubator shaker set to 37 °C and 250RPM.
3. Measure OD_{600} of each culture flask after 1 h and every subsequent 30 min until OD_{600} reaches a value between 0.400 and 0.600. Collect a 1 mL sample of each culture immediately prior to induction to pellet and lyse with 200 μL 1× Laemmli buffer.
4. Initiate IPTG induction. For disordered proteins (e.g., TNIP1), bring to 1 μM IPTG (4 μL 100 mM IPTG in 400 mL bacterial culture) and place in incubator shaker at 16 °C and 250RPM for 24 h. For ordered proteins (e.g., triubiquitin), bring to 100 μM IPTG (400 μL 100 mM IPTG in 400 mL bacterial culture) and place in incubator shaker at 30 °C and 250RPM for 5 h (*see* **Note 14**).
5. Remove culture flasks from incubator shaker and take 1 mL culture samples to pellet and lyse with 200 μL 1× Laemmli buffer. Take OD_{600} of the final culture, then spin down cultures at 5000× *g* for 15 min at 4 °C in 500 mL centrifuge bottles.
6. Pour off supernatant leaving bacterial pellet. Transfer bottle to a −20 °C freezer overnight. For long-term storage, transfer pellet to a −80 °C freezer the next day.

3.2 SDS-PAGE

SDS-PAGE is typically performed at multiple steps during production of recombinant proteins, most often for analyzing the results of induction or different steps during column purification. An essential method in proteomics, the basic protocol outlined here can be adapted as needed. The protocol below is optimized for 4–20% Gradient Polyacrylamide Gels (BioRad, Cat# 4561095) with 20 μL sample wells.

1. Prepare 20 μL samples in Eppendorf tubes using 5× Laemmli buffer added to a final concentration of 1× (*see* **Note 15** for details on normalization and sample preparation).
2. Prepare ladder samples at 1× Laemmli (*see* **Note 16**).
3. If using precast gels, remove gels from packaging and put in cassette, using a buffer dam if running an uneven number of gels. Secure using the clamps and insert into box, removing combs from gels once secure (*see* **Note 17**).
4. Dilute 100 mL of 10× running buffer to 1× in 900 mL of water, both pre-chilled, to a final volume of 1 L.
5. Pour running buffer into space between gels and let the excess flow over until the line of the water is just under gel wells. If bubbles obstruct view, a Kimwipe can be used to gently remove them.
6. Using gel loading tips, fill wells of the gel. Fill box to the top with remaining buffer, and secure lid on top connecting the electrodes to the correct terminal.
7. Run for 1 h at 130 V. This step can be optimized for better resolution or for different gel recipes.
8. When gel is done, remove gel from plastic or glass plates. Nick the corner of the gel by lane 1 to maintain orientation and place it in a suitable container for staining. If performing a blot, proceed to **step 1** of Subheading 3.3.
9. Fix the gel for 10 min in destain solution on a platform rocker.
10. Remove destain solution and add staining solution. Rock for 30 min.
11. Remove staining solution and add destain solution. Add one or two Kimwipes, rolled up and knotted to absorb excess stain. Rock overnight for best results.
12. Compare and contrast Coomassie-detected bands as to those in the uninduced versus induced bacteria lysates against molecular weight marker standards. The band(s) of induced protein of interest should be rigorously quality controlled for their overall abundance relative to host cell proteins to assess induction success based on IPTG concentrations and incubation times as described above. Their apparent molecular weight on the gel as determined by the migration marker standards

contrasted with predicted size from amino acid content should be calculated to assess possible breakdown products. Coomassie stained SDS-PAGE and/or western blotting (Subheading 3.3) should be employed as analytics throughout protein production and purification to detect breakdown, assess column binding efficiency, and determine recovery in eluted fractions (Subheading 3.4).

3.3 Western Blotting Analysis

When optimizing induction protocols, an objective method to compare results of manipulating variables such as time, temperature, and inducer concentration is needed. Accordingly, it is most effective to perform immunoblots on samples of lysed bacteria at different time points during induction. It is useful to normalize sample load volumes for each time point to the OD_{600} of the culture at said time points (*see* **Note 15**).

1. Carefully remove completed SDS-PAGE gel from cassette and place into a suitable container for buffer incubation steps. Add a sufficient volume of chilled transfer buffer and equilibrate gel on rocker for 15 min.
2. During equilibration, put foam pads as well as nitrocellulose membrane into a plastic tray with chilled equilibration buffer to pre-soak.
3. Assemble transfer sandwich paying special attention to the migration direction of protein (*see* **Note 18**).
4. Run for 1 h at 100 V and 4 °C (*see* **Note 19**).
5. When transfer is complete, disassemble cassette and remove membrane with forceps from gel. Move this to a suitable container and add blocking solution; if using a fluorescently labeled ladder, protect from light from this point forward. Rock for 30 min at room temperature.
6. In a 50 mL conical tube, combine 3 μL of anti-His6 primary antibody and 3 mL of blocking solution (1:1000; for the listed antibody, this is a final concentration of 500 ng/mL). Carefully remove blocked membrane from box and put into conical tube, protein side facing inward. Cap and place on the rotator overnight at 4 °C and make sure the pool of antibody solution contacts the whole area of the membrane in a rotation.
7. Remove membrane and place it into a clean container. Rinse quickly with TBS-T and pour off liquid. Then perform five 3-min TBS-T washes on rocker, removing as much liquid as possible after each wash.
8. Combine 1 μL secondary antibody and 10 mL TBS-T in the same container (1:10,000 dilution; for the listed antibody, this is a final concentration of 100 ng/mL). Place on the orbital

shaker for 45 min; if using a fluorescently labeled secondary antibody, protect from light from this point forward.

9. Rinse blot as in **step 6**.
10. Image blot as soon as possible. It is best to keep blot wet in TBS-T (*see* **Note 5**).

3.4 Immobilized Metal Affinity Chromatography (IMAC) Purification

IMAC column resins provide for selective adsorption of recombinant proteins carrying a polyhistidine epitope tag, typically comprising six histidine residues, designed in-frame with the protein's amino or carboxyl terminus. IMAC resins available with different metal cores have a long history of use [48] although divalent nickel or cobalt cores are the most commonly employed. As solubilized proteins flow along the resin, the polyhistidine tag ionically interacts with the nickel ion presented by a nitrilotriacetic acid (NTA) moiety on the surface of the supporting matrix. Preferential interaction is achieved by the polyhistidine epitope; proteins with lower affinity for the resin are mostly delivered to the initial flow-through volume and/or removed by subsequent wash steps. Tagged proteins are released from the column by sequential elutions containing stepwise increased concentrations of imidazole which at a high enough strength will outcompete the recombinant protein's polyhistidine tag. IMAC purification is typically followed by a size exclusion chromatography "polishing" step. In our experience, protein recovered from IMAC alone can reach greater than 90% purity, as determined by SDS-PAGE.

Most recombinant protein expression vectors have engineered in-between the protein and the polyhistidine tag a linker region containing an enzymatically cleavable site, e.g., for thrombin, a serine protease. The investigator will need to ensure it does not by chance occur in the protein of interest. By loading previously purified protein onto nickel resin along with active thrombin, one is able to remove the polyhistidine tag and selectively recover the cleaved protein with a compatible buffer such as low concentration imidazole elution buffer, with imidazole being removed in downstream steps. The resin can be processed for reuse by clearing remaining polyhistidine tags (cleaved or still attached to the undigested protein) with a high concentration imidazole elution buffer and water washes before storage.

1. Begin thawing frozen pellets. Supplement chilled lysis buffer (contingent on disordered or ordered protein purification, see above in Methods) according to the recipe and add to pellet before it is fully thawed. Resuspend using a pipette (Fig. 1a).
2. Transfer volume to 50 mL Falcon tubes and place onto the end-over-end rotator for 30 min at room temperature (Fig. 1b). If preparing a disordered protein, proceed to **step 3**. If preparing an ordered protein construct, proceed to **step 4**.

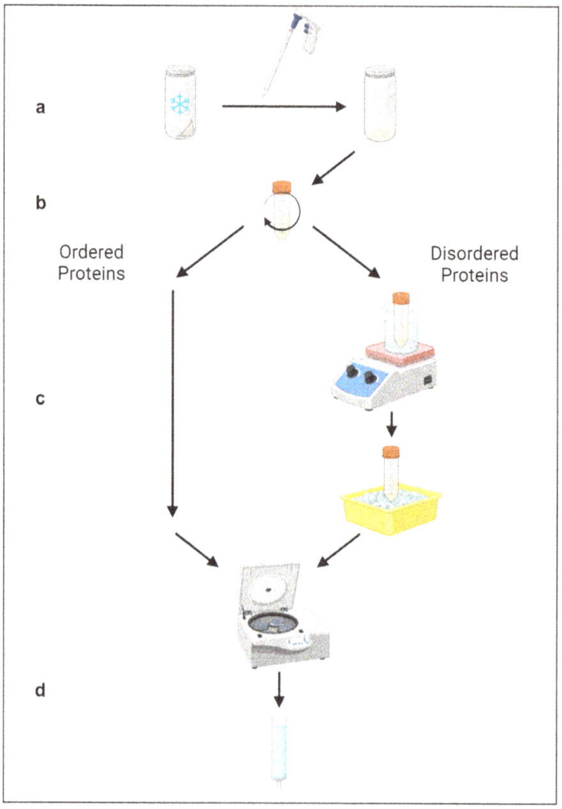

Fig. 1 Schematic for preparation of frozen bacterial pellet. (**a**) Resuspension in respective lysis buffer; (**b**) 30 min tumble at room temperature on rotator; (**c**) direct centrifugation for ordered proteins vs. 20 min boiling followed by 10 min on wet ice for preparation of disordered proteins; (**d**) clarification of lysate at 20,000× *g* for 20 min (*see* **Note 20**) followed by application to the nickel-NTA resin. Citation: Created with BioRender.com

3. For disordered protein constructs, heat the lysate in a boiling water bath for 20 min before transferring to wet ice for 10 min (Fig. 1c).
4. Centrifuge the bacterial lysate at 15,000× *g* for 15 min at 4 °C (Fig. 1d; *see* **Note 20**).
5. Decant the protein-containing supernatant off the waste pellet (*see* **Note 7**).
6. Equilibrate the nickel-NTA resin with 10 column volumes (CV) of chilled Binding Buffer with the flowthrough going to waste. From this point forward, all buffers should be chilled (4 °C or on ice).
7. Begin collecting flowthrough and load the supernatant.
8. Wash the column with 10CV of Wash Buffer.

9. Begin collecting 1CV fractions and add 4CV of EB1 to the column.
10. Add 4CV of EB2 to the column and collect in 1CV fractions.
11. Add 4CV of EB3 to the column and collect in 1CV fractions.
12. Add 4CV of EB4 to the column and collect in 1CV fractions (*see* **Note 8**).
13. Quit collecting fractions and wash the column with 10CV water. The column can be stored at 4 °C in water being sure to maintain several milliliters of water above the resin to keep it submerged.
14. After determining which fractions contain the purified protein (via SDS-PAGE or spectrophotometric protein assays), pool them and transfer into Final Buffer for long-term storage using a PD-10 desalting column following the supplier's protocol.

3.4.1 Thrombin Cleavage on Nickel Resin

1. Prepare column with 10CV of Binding Buffer.
2. Load purified His-tagged protein with thrombin cleavage site.
3. Load 1CV of Binding Buffer containing 10 units/mg of protein with a thrombin-cleavable tag and supplemented with $CaCl_2$ to a concentration of 10 mM (*see* **Note 9**).
4. Carefully allow thrombin-supplemented Binding Buffer to permeate the nickel-NTA resin without allowing the buffer meniscus to drain below the top of the resin. Cover the top of the resin with 1CV of normal EB1 and cap the column.
5. Incubate at room temperature overnight (16 h).
6. Apply 8CV of EB1, collecting each as a separate fraction. Repeat with 8CV of EB4. The EB1 samples should now contain bioactive protein cleaved from its His tag (*see* **Note 21**).
7. Prepare the protein for long-term storage as in **step 14** of Subheading 3.4.

3.5 Microscale Thermophoresis (MST)

MST detects protein movement in response to laser-induced heating [19–22], akin to protein movement in response to applied electrical current. Of particular investigational value is when the rate of that movement along a thermal gradient changes due to interaction between the target protein of interest, dye-labeled for fluorescent detection (as is required in NanoTemper Monolith X), and a candidate binding partner protein allowing extrapolation to binding affinities. In comparison, isothermal titration calorimetry (ITC) is a label-free method. However, ITC use for the PPI we describe here or as seen in the literature at concentrations ranging 10–1000 μM would typically consume total amounts of protein many-folds more for the ITC approach because of the volume of the analysis cell in the instrument being much larger than the microcapillary tubes used for MST [51–53]. Lower consumption of absolute amounts of purified protein partners facilitates high-throughput investigation of PPI conditions (salt concentration, pH, etc.) and/or testing of site-directed mutants of either protein to help document amino acid residues central to the partners associating. From the protocols described here, binding results with linear triubiquitin and the human TNIP1 amino acids 417-509 (hTNIP1$^{417\text{-}509}$) ubiquitin-binding subdomain of TNIP1 we previously characterized [26] show that MST-derived K_D values are nevertheless consistent with those acquired by more commonly reported methods such as ITC [31]. Additionally, they support that IDP obtained via heat lysis, an approach distinct from previous ubiquitin-binding protein investigations [49, 50], remain biophysically active and relevant to downstream studies. For instance, we illustrate the applicability and sensitivity of this method for mapping PPI via introduction of an established [17, 54] TNIP1:polyubiquitin interaction-negating mutation, D472N (*see* Fig. 2).

1. Prepare 100 μL (*see* **Note 22**) of 200nM His-tagged target protein hTNIP1$^{417\text{-}509}$ via dilution in PBS-T (*see* **Note 23**). Combine 2 μL of NTA dye solution and 98 μL PBS-T producing 100 μL of 100 nM NTA dye solution (*see* **Note 24**). Prepare 1:1 solution of 200 nM protein and 100 nM dye and incubate at RT for 30 min, protecting from light.
2. Prepare a 16 step twofold serial dilution of the ligand protein, M1-linked triubiquitin, in a 96-well plate by placing 20 μL of 40 μM triubiquitin (*see* **Notes 10** and **25**) in the first well, 10 μL of PBS-T in the following 15 wells, and diluting as shown in Fig. 3.
3. Add 10 μL of labeled protein to each well and incubate at room temperature for 1 h, protecting from light (*see* **Note 26**). Final triubiquitin concentrations range from 20 μM to 610 pM with a constant hTNIP1$^{417–509}$ concentration of 50 nM.
4. Using MO.AA software, create file for determination of binary interaction binding affinity with red excitation channel. Set excitation power to 100%, MST power to 40%, and temperature to 23 °C (*see* **Note 27**).
5. Place capillaries into each sample well (*see* **Note 28**). Gently tip the plate to ~60° angle allowing samples to fill each capillary. Load capillaries Monolith X instrument and begin the assay with the MO.AA software (*see* **Note 29**).
6. Following analysis, create a brief report in the MO.AA software. Analyze the MO.AA file with MO.Control Software for additional data analysis.

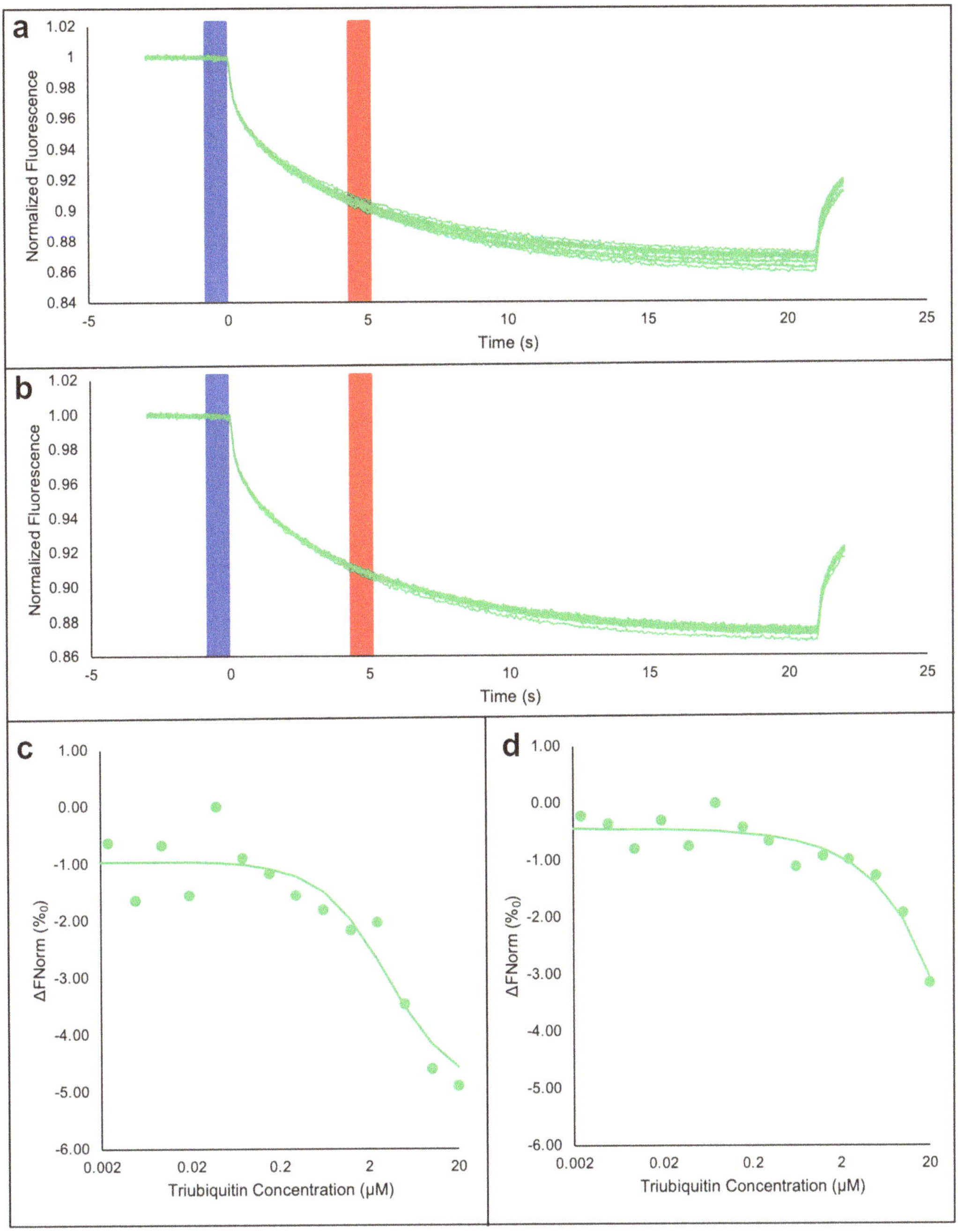

Fig. 2 Representative MST traces of linear triubiquitin titrated against (**a**) WT hTNIP1$^{417-509}$ or (**b**) D472N hTNIP1$^{417-509}$ with the cold zone set to time 0 s and the hot zone set to time 5 s. Zones represent data samples utilized in calculating the change in normalized fluorescence (ΔFNorm) with the cold zone being a point prior to laser activation and the hot zone being a point while the laser is active. Note the difference in the spread of traces in (**a**) indicative of differential migration due to binding versus more overlaid traces in (**b**). Panel (**c**) shows binding affinity derived from averaged duplicate MST traces; as shown in (**a**) it yields sigmoidal curve showing saturation and a K_D of 3.85 μM consistent with the related ubiquitin-binding region in NEMO [49, 50] prepared through different protocols. Panel (**d**) shows binding affinity derived from averaged quadruplicate MST traces; as shown in (**b**), with the poorer binding because of D472N, saturation was not achieved at the triubiquitin concentrations used. Predictive affinity software estimates a K_D of ~25 μM from these data. Citation: Created with BioRender.com

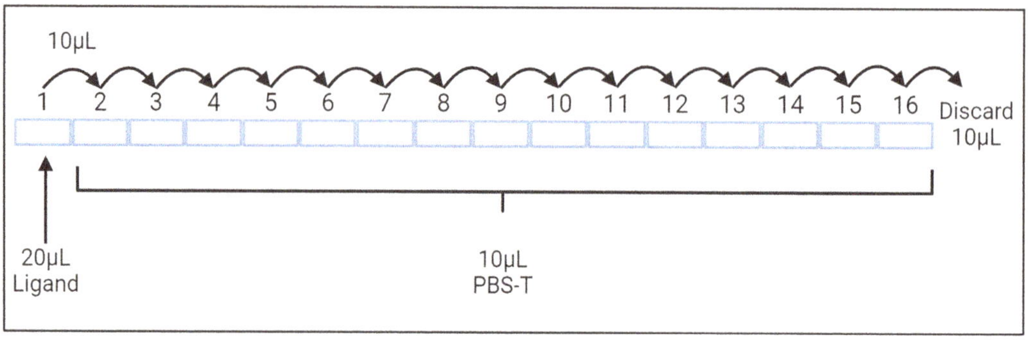

Fig. 3 Schematic of how a 16-step twofold serial dilution is established. It is paramount that the beginning ligand is in PBS-T to avoid minute buffer differences throughout the dilutions. From well 1 to well 16, there is a 65,536-fold dilution. Adapted from His-Tag Labeling Kit RED-tris-NTA 2nd Generation user manual. Citation: Created with Biorender.com

4 Notes

1. Unless otherwise stated, all water used in the methods and materials is type 1, 18.2 MΩ (Nanopure) water.
2. The high glycerol content in this buffer will make it difficult to mix. This issue can be alleviated by adding the 1 M Tris buffer and dye first, dissolving the SDS as best as possible, then adding the glycerol. Heating in a 50 °C water bath will help to resuspend the buffer evenly. Then water can be added to bring the buffer to 90% final volume (no β-mercaptoethanol). Solubilization may be enhanced with gentle heating and pipetting.
3. We make 1.2 L to fill the blotting box and soak materials. After mixing, placing the buffer in a −20 °C freezer for approximately an hour will serve to chill the buffer.
4. We filter sterilize a second time after Tween 20 addition to remove particles; gently heating the Tween 20 reagent bottle may also help by redissolving particles.
5. The secondary antibody and ladder, supplied by Li-Cor, are optimized for use with the Odyssey CLx and Image Studio. Please refer to manufacturer guidelines for use and optimization tips (https://www.licor.com/bio/image-studio/resources). Other detection software can be used but must be optimized through trial and error for best resolution and quantitation of bands.
6. Per manufacturer recommendations, one of these pellets vastly exceeds the protection required in this volume of buffer. That is to say, this buffer tends to be ~10-15 mL, and a single tablet is rated to inhibit proteases in 50 mL of buffer. Working more conservatively, one can create working stocks of the tablets in 2 mL of 100 mM phosphate buffer, pH 7.0 yielding a 25×

stock. This stock may be stored at 4 °C for 2 weeks or −20 °C for 12 weeks per manufacturer recommendation.

7. The presence of nucleic acids liberated from the bacterial pellet can lead to viscous lysate. The heat lysis tends to reduce viscosity somewhat but should be supplemented with sonication, needle shearing, or nuclease treatment; ordered protein production will innately require one or more of these methods due to the inability to perform heat lysis. We find it best to supplement the clarified supernatant with 8.33 units of Pierce nuclease per mL for 20 min at room temperature before application to the column.
8. Another point of optimization is in the elution step. With the above detailed EB1-EB4 buffers, our constructs elute in late EB3 or EB4 fractions; we sometimes extend EB4 to collect eight 5-mL fractions to maximize purified protein collection. Imidazole concentrations may be optimized based on elution profile or protein purity, as analyzed by gel.
9. Theoretically, addition of calcium chloride is not necessary for thrombin cleavage but rather for the conversion of prothrombin to active thrombin. While the manufacturer claims that no divalent ions are necessary for thrombin cleavage, calcium chloride was added based on the thrombin cleavage buffer in the Thrombin CleanCleave Kit (Sigma-Aldrich).
10. For triubiquitin, a 40 μM concentration is chosen because later addition of tagged protein will dilute this to 20 μM. Thus, the top ligand concentration of this assay is ~10× the anticipated K_D and should therefore provide distinct saturated and unsaturated plateaus in a sigmoidal binding curve.
11. Premium capillaries may be used to prevent sample adsorption; however, the Tween 20 in PBS-T takes care of adsorption in nearly all cases.
12. Given that small volumes do not fill a well in a plate, the plate must be worked with at an angle (about 60°). The MST incubations can be prepared in another vessel (i.e., PCR tubes, 1.5 mL tubes, etc.), but the caps on most other means of separating samples prove to be a hindrance throughout the process.
13. We use this from an abundance of caution regarding potential photobleaching of the tag. However, NanoTemper brochures claim that their tag is highly resistant to photobleaching.
14. While these IPTG values work for inducing our proteins of interest, proteins can be quite individual in this regard. We find a lower IPTG induction concentration can lead to a higher percentage of sample protein remaining intact, albeit an overall lower protein yield. Conversely, we use the higher IPTG

concentration on our ordered, globular protein because it is not prone to breakdown. Bacterial stress is no longer a concern in this case, and the amount of IPTG maximizes protein production.

15. Normalizing loading volumes of Laemmli samples of bacteria to culture OD_{600} can help one compare results during protocol optimization as well as across subsequent steps in purification. Normalizing to a load volume of 15 μL for an OD_{600} of 0.75 (1 cm path length) should give clearly visible bands. For western blots, 3.75 μL for an OD_{600} of 0.75 may be a good starting point.
16. Our typical ladder recipe is 7.5 μL at 1× Laemmli in final 20 μL loading volume (7.5 μL 2×, 5 μL 1×).
17. For best results, perform SDS-PAGE at 4 °C or keep the gel box in wet ice during the run. This will help to prevent heat-induced changes in band migration which creates a "smile." If using a fluorescent ladder, a dark room is best.
18. Due to our order of assembly, we put the gel on mirrored. While gel-imaging software can typically take mirroring into account, it is essential to keep track of the orientation through each step. To this end, we also use a pencil to mark the nitrocellulose paper with an asymmetrical letter, such as "P," oriented with the gel after the transfer.
19. It is best to perform transfer at 4 °C. We typically perform a quick transfer protocol, i.e., 100 V for 1 h, but an overnight transfer at a lower voltage like 30 V may be better for some proteins. One can stain the gel after transfer with Coomassie to assess completion of the transfer.
20. If an ultracentrifuge is available, one can spin at over 100,000× *g* for better results.
21. In our experience, Coomassie staining of samples from the first fractions collecting in EB4 contained detectable amounts of protein the same size as the protein of interest. While it could be assumed that this is uncleaved sample protein, a subsequent His-detecting Western blot suggested that the polyhistidine tag had been removed. However, the protein in these fractions was deemed not to be bioactive via lack of detectable PPI in MST. Investigators are encouraged to use multiple quality control checks on protein recovered from on-column thrombin cleavage, particularly in high imidazole elutions, before its inclusion in experimental assays.
22. For reliable MST results, assays should be run in at least duplicate. The amounts throughout this protocol are for a single run and should be scaled up as needed.

23. The dilution factor from our typical stock concentrations of protein to 200 nM simultaneously serves as a buffer swap into PBS-T.
24. To be cautious of freeze-thaw cycle effects on the dye, we typically aliquot the initial 25 μL available in the manufacturer's dye solution tube into 5 tubes containing 5 μL each and freeze these at −80 °C.
25. While this protein was purified via a His-tag nickel column chromatography system, both the ligand and the target having His-tags interfere with the ionically binding fluorescent tag, which can be transferred between polyhistidine tags and ultimately give odd results.
26. The incubation may run for longer (we have never allowed it to sit longer than 5 h); however, one should refrigerate the reaction after 1 h to ensure that the ordered ligand protein remains stable.
27. The values for excitation and MST power were determined by initially allowing the software to automatically select the optimal conditions. Following that assay, these values were continuously used so that all data was within the established precedent. Likewise, adherence to 23 °C is to ensure all data was collected under the same conditions.
28. The Monolith X machine performs its assay using the solution in the center of the capillary. Thus, if you smudge this section (or an air bubble is in this section), it will skew your results. A Kimwipe can be used to wipe the sides of a capillary that has accidentally been smudged, and a bubble may be disregarded if placed out of the line of detection.
29. This software will produce "warnings" about your data if the instrument detects adsorption in capillaries or if there is a greater than 20% standard deviation of fluorescence intensity among capillaries. Within reason, these warnings may be overridden to produce a K_D from the data. Otherwise, the capillaries in which the issue was detected may be removed from the dataset as outliers, also providing the user with a K_D.

Acknowledgments

Protocols described here for assessment of TNIP1-partner interactions were supported in part by a DoD CDMRP grant W81XWH-20-1-0081 (BJA). We thank Dr. Andrew Wiemer for access to the Li-Cor Odyssey CLx. MLS and LEC received graduate assistantship support from the Department of Pharmaceutical Sciences. Support for the University of Connecticut core facility NanoTemper Monolith for MST is provided by S10OD028574.

References

1. Siva Sankar D, Dengjel J (2021) Protein complexes and neighborhoods driving autophagy. Autophagy 17(10):2689–2705. https://doi.org/10.1080/15548627.2020.1847461
2. Ohsumi Y (2014) Historical landmarks of autophagy research. Cell Res 24(1):9–23. https://doi.org/10.1038/cr.2013.169
3. Feng Y, He D, Yao Z, Klionsky DJ (2014) The machinery of macroautophagy. Cell Res 24(1): 24–41. https://doi.org/10.1038/cr.2013.168
4. Kirkin V, Rogov VV (2019) A diversity of selective autophagy receptors determines the specificity of the autophagy pathway. Mol Cell 76(2):268–285. https://doi.org/10.1016/j.molcel.2019.09.005
5. Wesch N, Kirkin V, Rogov VV (2020) Atg8-family proteins-structural features and molecular interactions in autophagy and beyond. Cell 9(9):2008. https://doi.org/10.3390/cells9092008
6. Zaffagnini G, Martens S (2016) Mechanisms of selective autophagy. J Mol Biol 428(9 Pt A):1714–1724. https://doi.org/10.1016/j.jmb.2016.02.004
7. Gubas A, Dikic I (2022) A guide to the regulation of selective autophagy receptors. FEBS J 289(1):75–89. https://doi.org/10.1111/febs.15824
8. Rao VS, Srinivas K, Sujini GN, Kumar GN (2014) Protein-protein interaction detection: methods and analysis. Int J Proteomics 2014: 147648. https://doi.org/10.1155/2014/147648
9. Syu GD, Dunn J, Zhu H (2020) Developments and applications of functional protein microarrays. Mol Cell Proteomics 19(6): 916–927. https://doi.org/10.1074/mcp.R120.001936
10. Wu Z, Liao Q, Liu B (2020) A comprehensive review and evaluation of computational methods for identifying protein complexes from protein-protein interaction networks. Brief Bioinform 21(5):1531–1548. https://doi.org/10.1093/bib/bbz085
11. Hayes S, Malacrida B, Kiely M, Kiely PA (2016) Studying protein-protein interactions: progress, pitfalls and solutions. Biochem Soc Trans 44(4):994–1004. https://doi.org/10.1042/bst20160092
12. Syafrizayanti BC, Hoheisel JD, Kastelic D (2014) Methods for analyzing and quantifying protein-protein interaction. Expert Rev Proteomics 11(1):107–120. https://doi.org/10.1586/14789450.2014.875857
13. Ngounou Wetie AG et al (2014) Protein-protein interactions: switch from classical methods to proteomics and bioinformatics-based approaches. Cell Mol Life Sci 71(2): 205–228. https://doi.org/10.1007/s00018-013-1333-1
14. Piersimoni L, Kastritis PL, Arlt C, Sinz A (2022) Cross-linking mass spectrometry for investigating protein conformations and protein-protein interactionsa method for all seasons. Chem Rev 122(8):7500–7531. https://doi.org/10.1021/acs.chemrev.1c00786
15. Le Guerroué F et al (2023) TNIP1 inhibits selective autophagy via bipartite interaction with LC3/GABARAP and TAX1BP1. Mol Cell 83(6):927–941.e8. https://doi.org/10.1016/j.molcel.2023.02.023
16. Zhou J et al (2023) TBK1 phosphorylation activates LIR-dependent degradation of the inflammation repressor TNIP1. J Cell Biol 222(2):e202108144. https://doi.org/10.1083/jcb.202108144
17. G'Sell RT, Gaffney PM, Powell DW (2015) A20-Binding inhibitor of NF-κB activation 1 is a physiologic inhibitor of NF-κB: a molecular switch for inflammation and autoimmunity. Arthritis Rheumatol 67(9):2292–2302. https://doi.org/10.1002/art.39245
18. Shamilov R, Aneskievich BJ (2018) TNIP1 in autoimmune diseases: regulation of toll-like receptor signaling. J Immunol Res 2018: 3491269. https://doi.org/10.1155/2018/3491269
19. Magnez R, Bailly C, Thuru X (2022) Microscale thermophoresis as a tool to study protein interactions and their implication in human diseases. Int J Mol Sci 23(14):7672. https://doi.org/10.3390/ijms23147672
20. Wienken CJ et al (2010) Protein-binding assays in biological liquids using microscale thermophoresis. Nat Commun 1:100. https://doi.org/10.1038/ncomms1093
21. El Deeb S et al (2022) Microscale thermophoresis as a powerful growing analytical technique for the investigation of biomolecular interaction and the determination of binding parameters. Methods Appl Fluoresc 10(4): 10.1088/2050-6120/ac82a6
22. Asmari M, Ratih R, Alhazmi HA, El Deeb S (2018) Thermophoresis for characterizing biomolecular interaction. Methods 146:107–119. https://doi.org/10.1016/j.ymeth.2018.02.003

23. Carman LE, Samulevich ML, Aneskievich BJ (2023) Repressive control of keratinocyte cytoplasmic inflammatory signaling. Int J Mol Sci 24(15):11943. https://doi.org/10.3390/ijms241511943
24. Brady MP, Korte EA, Caster DJ, Powell DW (2020) TNIP1/ABIN1 and lupus nephritis: review. Lupus Sci Med 7(1):e000437. https://doi.org/10.1136/lupus-2020-000437
25. Verstrepen L, Carpentier I, Verhelst K, Beyaert R (2009) ABINs: A20 binding inhibitors of NF-kappa B and apoptosis signaling. Biochem Pharmacol 78(2):105–114. https://doi.org/10.1016/j.bcp.2009.02.009
26. Shamilov R, Vinogradova O, Aneskievich BJ (2020) The anti-inflammatory protein TNIP1 is intrinsically disordered with structural flexibility contributed by its AHD1-UBAN domain. Biomol Ther 10(11):1531. https://doi.org/10.3390/biom10111531
27. Ramirez VP, Gurevich I, Aneskievich BJ (2012) Emerging roles for TNIP1 in regulating post-receptor signaling. Cytokine Growth Factor Rev 23(3):109–118. https://doi.org/10.1016/j.cytogfr.2012.04.002
28. Verstrepen L et al (2010) Expression, biological activities and mechanisms of action of A20 (TNFAIP3). Biochem Pharmacol 80(12):2009–2020. https://doi.org/10.1016/j.bcp.2010.06.044
29. Morris OM, Torpey JH, Isaacson RL (2021) Intrinsically disordered proteins: modes of binding with emphasis on disordered domains. Open Biol 11(10):210222. https://doi.org/10.1098/rsob.210222
30. Mollica L et al (2016) Binding mechanisms of intrinsically disordered proteins: theory, simulation, and experiment. Front Mol Biosci 3:52. https://doi.org/10.3389/fmolb.2016.00052
31. Hong JY, Lin SC, Kuo BJ, Lo YC (2021) Structural and biochemical basis for higher-order assembly between A20-binding inhibitor of NF-κB 1 (ABIN1) and M1-linked Ubiquitins. J Mol Biol 433(18):167116. https://doi.org/10.1016/j.jmb.2021.167116
32. Herhaus L et al (2019) Molecular recognition of M1-linked ubiquitin chains by native and phosphorylated UBAN domains. J Mol Biol 431(17):3146–3156. https://doi.org/10.1016/j.jmb.2019.06.012
33. Mirza N, Sowa AS, Lautz K, Kufer TA (2019) NLRP10 affects the stability of Abin-1 to control inflammatory responses. J Immunol 202(1):218–227. https://doi.org/10.4049/jimmunol.1800334
34. Shinkawa Y et al (2022) ABIN1 is a signal-induced autophagy receptor that attenuates NF-κB activation by recognizing linear ubiquitin chains. FEBS Lett 596(9):1147–1164. https://doi.org/10.1002/1873-3468.14323
35. Lin SM et al (2017) Structural insights into linear tri-ubiquitin recognition by A20-binding inhibitor of NF-κB, ABIN-2. Structure 25(1):66–78. https://doi.org/10.1016/j.str.2016.11.005
36. Chen R (2012) Bacterial expression systems for recombinant protein production: E. Coli and beyond. Biotechnol Adv 30(5):1102–1107. https://doi.org/10.1016/j.biotechadv.2011.09.013
37. Aguilar Lucero D, Cantoia A, Ceccarelli EA, Rosano GL (2021) Starting a new recombinant protein production project in Escherichia coli. Methods Enzymol 659:3–18. https://doi.org/10.1016/bs.mie.2021.08.019
38. Umlauf BJ, McGuire MJ, Brown KC (2015) Introduction of plasmid encoding for rare tRNAs reduces amplification bias in phage display biopanning. BioTechniques 58(2):81–84. https://doi.org/10.2144/000114256
39. Oldfield CJ, Dunker AK (2014) Intrinsically disordered proteins and intrinsically disordered protein regions. Annu Rev Biochem 83:553–584. https://doi.org/10.1146/annurev-biochem-072711-164947
40. Cruz JA et al (2017) Interleukin-17 signaling triggers degradation of the constitutive NF-κB inhibitor ABIN-1. Immunohorizons 1(7):133–141. https://doi.org/10.4049/immunohorizons.1700035
41. Shamilov R, Staid MJ, Aneskievich BJ (2020) In Silico and in vitro considerations of keratinocyte nuclear receptor protein structural order for improving experimental analysis. Methods Mol Biol 2109:93–111. https://doi.org/10.1007/7651_2019_240
42. Li F et al (2020) Procleave: predicting protease-specific substrate cleavage sites by combining sequence and structural information. Genomics Proteomics Bioinformatics 18(1):52–64. https://doi.org/10.1016/j.gpb.2019.08.002
43. Wilkins MR et al (1999) Protein identification and analysis tools in the ExPASy server. Methods Mol Biol 112:531–552. https://doi.org/10.1385/1-59259-584-7:531
44. Graether SP (2018) Troubleshooting guide to expressing intrinsically disordered proteins for use in NMR experiments. Front Mol Biosci 5:118. https://doi.org/10.3389/fmolb.2018.00118

45. Pesarrodona M, Latorre I, Salvatella X (2022) Intrinsically disordered proteins (IDP): purification under denaturing conditions. Methods Mol Biol 2406:359–370. https://doi.org/10.1007/978-1-0716-1859-2_21
46. Hwang PM, Pan JS, Sykes BD (2012) A PagP fusion protein system for the expression of intrinsically disordered proteins in Escherichia coli. Protein Expr Purif 85(1):148–151. https://doi.org/10.1016/j.pep.2012.07.007
47. Goda N et al (2015) An optimized N(pro)-based method for the expression and purification of intrinsically disordered proteins for an NMR study. Intrinsically Disord Proteins 3(1): e1011004. https://doi.org/10.1080/21690707.2015.1011004
48. Bornhorst JA, Falke JJ (2000) Purification of proteins using polyhistidine affinity tags. Methods Enzymol 326:245–254. https://doi.org/10.1016/s0076-6879(00)26058-8
49. Hadian K et al (2011) NF-κB essential modulator (NEMO) interaction with linear and lys-63 ubiquitin chains contributes to NF-κB activation. J Biol Chem 286(29): 26107–26117. https://doi.org/10.1074/jbc.M111.233163
50. Vincendeau M, Krappmann D, Hadian K (2015) In vitro detection of NEMO-ubiquitin binding using DELFIA and microscale thermophoresis assays. Methods Mol Biol 1280: 311–320. https://doi.org/10.1007/978-1-4939-2422-6_18
51. Duff MR Jr, Grubbs J, Howell EE (2011) Isothermal titration calorimetry for measuring macromolecule-ligand affinity. J Vis Exp 55: 2796. https://doi.org/10.3791/2796
52. Dutta AK, Rösgen J, Rajarathnam K (2015) Using isothermal titration calorimetry to determine thermodynamic parameters of protein-glycosaminoglycan interactions. Methods Mol Biol 1229:315–324. https://doi.org/10.1007/978-1-4939-1714-3_25
53. Rajarathnam K, Rösgen J (2014) Isothermal titration calorimetry of membrane proteins – progress and challenges. Biochim Biophys Acta 1838(1 Pt A):69–77. https://doi.org/10.1016/j.bbamem.2013.05.023
54. Rose KJ, Strand FL (1990) ACTH/MSH (4-10) peptide increases postnatal metabolic activity of EDL muscle following early prenatal administration. Peptides 11(1):177–179. https://doi.org/10.1016/0196-9781(90)90128-r

Methods in Molecular Biology (2025) 2879: 83–92
DOI 10.1007/7651_2024_555

Published online: 13 July 2024

Proteomic Methods to Study Autophagy in Skin Exposed to Pollutants

Mohammad Fayyad-Kazan, Sandrine Bourgoin-Voillard, Walid Rachidi, and Michel Seve

Abstract

Autophagy refers to the natural cellular process by which cells degrade and recycle their own damaged or dysfunctional cellular components. It is an essential mechanism for maintaining cellular homeostasis removing toxic substances and providing energy during times of stress or nutrient deprivation. When autophagy is dysregulated or impaired, it can have detrimental effects on cell function and overall health. Studying autophagy in skin exposed to pollutants can provide valuable insights into the cellular mechanisms underlying pollutant-induced skin damage. Proteomic methods, which involve the large-scale analysis of proteins, can be employed to investigate the changes in protein expression associated with biological processes including autophagy. Here, we thus describe a method where LC-MS/MS was applied to identify the deregulated proteins in pollutant exposed-skin. Using bioinformatics and statistical analysis, we extracted the qualitative and quantitative information for proteins involved in autophagy. These deregulated proteins were then validated by immunohistochemistry (IHC). These methods help to understand how the pollutants affect the autophagy process.

Key words Autophagy, Diseases, Proteomics, Skin

Abbreviations

B[a]P	Benzo[a]Pyrene
DIA	Data Independent Acquisition
DMSO	Dimethyl sulfoxide
DTT	Dithiothreitol
FBS	Fetal bovine serum
IAA	Indole-3-acetic acid
IMDM	Iscove-modified Dulbecco media
IHC	Immunohistochemistry
LC-MS	Liquid chromatography-mass spectrometry
PAH	Polycyclic aromatic hydrocarbons
SDS	Sodium dodecyl sulfate
TCA	Trichloroacetic acid
UV	Ultraviolet

1 Introduction

Autophagy refers to a natural cellular process involved in the degradation and recycling of cellular components, including damaged proteins and organelles. It has an essential role in maintaining cellular homeostasis and ensuring overall cell health [1]. Autophagy can be triggered in response to various cellular stresses, such as nutrient deprivation, hypoxia, and infection and has been extensively studied in various tissues, including the skin [1].

Pollutants, such as air pollution, cigarette smoke, or ultraviolet (UV) radiation, can induce oxidative stress, trigger inflammatory responses, and cause damage to skin cells [2, 3]. Autophagic processes, in the skin, can be modulated in response to pollutant exposure. Under certain conditions, pollutants can disrupt autophagic processes, impairing the skin's ability to remove damaged cellular components effectively. This can result in the accumulation of toxic substances and dysfunctional organelles within skin cells, contributing to cellular damage and promoting skin aging [4–6]. On the other hand, autophagy can also be upregulated as a protective mechanism in response to pollutant-induced stress. Enhanced autophagy helps to clear damaged proteins and organelles, reducing oxidative stress and promoting cell survival. By eliminating harmful substances and repairing cellular damage, autophagy can contribute to maintain skin health and function in the face of pollutant exposure [4–6]. Researchers are actively investigating the intricate relationship between autophagy and skin exposed to pollutants to better understand how these processes interact and how they can be modulated for therapeutic purposes. Developing strategies to enhance autophagy in the skin may offer potential avenues for protecting against the detrimental effects of pollutants and promoting skin health [7]. However, it's important to note that more research is needed to fully understand the specific mechanisms and therapeutic implications of autophagy modulation in the context of pollutant-exposed skin.

Proteomic analysis is a powerful technique used to study the proteome, which refers to the whole proteins expressed by a cell, tissue, or organism. It involves the identification and quantification of proteins in a sample, providing insights into their abundance, post-translational modifications, interactions, and functional roles [8]. We have recently identified deregulated autophagic process in skin cells exposed to UV and the polycyclic aromatic hydrocarbons (PAH) mixtures including benzo[a]pyrene considered as carcinogenic [9, 10].

This was done by a label-free proteomic quantitative analysis that enabled differential protein expression analysis. The proteins extracted from exposed or control tissue were digested by trypsin, and the resulting peptides were purified before separation by reverse-phase nanochromatography and analyzed by mass spectrometry (*see* **Note 1**). The proteins involved in autophagy were identified and quantified by bioinformatics and gene ontology tools. This method unraveled deregulated expression of different proteins involved in the autophagic process (*see* **Note 2**). Further validation of deregulated autophagy was carried out using immunohistochemistry analysis to monitor the localization of two molecular markers: LC3-II, specifically associated with autophagosomes [9], and SQSTM1/P62, usually incorporated in the growing autophagosome as a cargo receptor where its degradation is a marker for active autophagy [9] (*see* **Note 3**).

2 Materials

2.1 For Human Skin Explant Samples Culture and Exposition

2.1.1 Reagents

1. Media consisting of two-thirds Iscove-modified Dulbecco media (IMDM), complete media, and one-third keratinocyte media with P/S and 5% FBS (Thermo Fisher Scientific).
2. UV irradiation (UVA 8 J/cm^2 and UVB 10 mJ/cm^2).
3. Benzo[a]pyrene (B[a]P) pollutant (Sigma, B1760).
4. Liquid nitrogen for storing samples at −80 °C for proteomics analysis.
5. Formalin for immunohistochemistry analysis.

2.1.2 Equipment

1. Vilber Lourmat stimulator (BIO-SUN) was used for irradiation.
2. UVA irradiation source (365 nm) composed of T-20.L-365 tubes (no UVB, no UVC emissions), mercury vapor tubes, low pressure and hot cathodes.
3. Stimulator fitted out with a UVB irradiation source (312 nm) composed of T-15.M-312 tubes (no UVA, no UVC emission).
4. A radiometer linked to an energy-programmable microprocessor allowing for the adjustment of the duration and the energy received by the skin explants.

2.2 For Processing of Proteomics' Samples

2.2.1 Reagents

1. 0.1 M Tris–HCl buffer (pH 8.5) containing 2% SDS and 0.1 M DTT.
2. Acetone/TCA (ratio 9:1).
3. 0.1 M Tris–HCl buffer (pH 8.5) containing 8 M urea and 2% DMSO to dissolve precipitated proteins and wash filters of Amicon Ultra.
4. ProteoExtract® Subcellular Proteome Extraction Kit, Millipore.
5. Filters were washed with 8 M urea in 0.1 M Tris (pH 8.5).

6. 10 mM DTT (dithiothreitol) (Euromedex, Souffelweyersheimas) as reducing reagent.
7. 50 mM IAA (Sigma) as alkylating reagent.
8. Trypsin at 1:50 ratio (Promega, Madison).
9. 0.5 M NaCl.
10. Acetonitrile (HPLC-MS quality).
11. H_2O (HPLC-MS quality).
12. 5% acetonitrile.
13. 10% trifluoroacetic acid at 0.1% final concentration.
14. Peptides were eluted with 1.2 mL of 70% acetonitrile, dried, and dissolved in 200 μl of 5% acetonitrile acidified with 0.1% formic acid.

2.2.2 Equipment

1. Amicon Ultra 0.5 10 kDa filters (Millipore, Billerica).
2. 500 mg SepPak tC18 sample extraction columns (Waters, Milford) for desalting peptides.

2.3 For LC-MS/MS Data Acquisition and Analysis

2.3.1 Reagents

1. Peptide digests (126 ng per run).
2. 3% to 40% buffer B (buffer A: 0.1% formic acid; buffer B: 95% acetonitrile, 0.1% formic acid).

2.3.2 Equipment

1. NanoACQUITY UPLC Symmetry C18 Trap Column, 180 μm × 20 mm (particle diameter 5 μm, pore size 100 Å).
2. ACQUITY UPLC Peptide BEH C18 nanoACQUITY Column 75 μm × 250 mm (particle diameter 1.7 μm, pore size 130 Å) (Both Waters, Milford).
3. Run gradient was performed by Eksigent Ultra Plus nano-LC 2D HPLC (Sciex, Framingham) system with a gradient from 3% to 40% buffer B (buffer A: 0.1% formic acid; buffer B: 95% acetonitrile, 0.1% formic acid).
4. Eksigent system was coupled to a TripleTOF® 5600 (Sciex, Framingham) mass spectrometer interfaced to a nanospray III source, which parameters were set as follows: IS at 2500 V, Curtain gas at 30 psi, Gas Sprayer 1 at 1 psi, and Interface Heater Temperature at 150 °C.
5. DDA spectra processing and database searching were performed with ProteinPilot (v4.5 beta, Sciex, Framingham) using the Paragon algorithm.

2.4 For Immunohistochemistry

1. Formalin solution (Sigma, HT50-1-1) was used for sample fixation.

2. Primary antibodies against LC3B (Abcam, ab192890) or SQSTM1 (Abcam, ab109012).
3. Secondary antibody conjugated with Alexa Fluor 488 dye (Life Technologies, A11008).

3 Methods

3.1 Exposure of Human Skin Explant Samples to Pollutants (UV + Benzo[a]pyrene)

1. Three abdominal skin explants (phototype I-II) were obtained from patients following the approved consent form.
2. Explants were cultured in media consisting of two-thirds of Iscove-modified Dulbecco media (IMDM), complete media, and one-third of keratinocyte media with P/S and 5%FBS (Thermo Fisher Scientific).
3. The explants were untreated, exposed to only UV irradiation (UVA 8 J/cm^2 and UVB 10 mJ/cm^2) [11], or exposed to benzo[a]pyrene pollutant (Sigma, B1760) at 100 nM in combination with UV irradiation.
4. The irradiation was carried out using a Vilber Lourmat stimulator (BIO-SUN). This was fitted out with a UVA irradiation source (365 nm) composed of T-20.L-365 tubes (no UVB, no UVC emissions), mercury vapor tubes, low pressure, and hot cathodes. Then, the stimulator was fitted out with a UVB irradiation source (312 nm) composed of T-15.M-312 tubes (no UVA, no UVC emission).
5. The radiometer was linked to an energy programmable microprocessor allowing the adjustment of the duration and the energy received by the skin explants.
6. These UV irradiations were sufficient to induce reproducible alterations in the epidermis and dermis.
7. The treatment schedule was on days 1, 2, 5, and 6. On day 7, skin explants were collected, and their epidermis was isolated from dermis. For proteomics analysis, epidermis samples were dilacerated and homogenized in liquid nitrogen and stored at −80 °C. For immunohistochemistry analysis, samples were fixed in formalin solution.

3.2 Processing of Proteomics' Samples

3.2.1 Preanalytical Sample Preparation

1. Homogenates were incubated in 0.1 M Tris–HCl pH 8.5 buffer containing 2% SDS and 0.1 M DTT during two periods of 5 min at 95 °C each, in between which freezing was applied for 15 min at −80 °C.
2. Samples were sonicated for 1 min, per 3 s pulse, each pulse followed by ice cooling.
3. The resulting lysates were centrifuged (15,000 × g, 10 min) and supernatants were collected.

4. Proteins were extracted sequentially to preferentially isolate cytosolic proteins, nuclear proteins, and membrane proteins from isolated epidermis (from skin explant) following a commercially available protein extraction kit (ProteoExtract® Subcellular Proteome Extraction Kit, Millipore) (*see* **Note 4**).
5. Proteins were precipitated in acetone/TCA (ratio 9:1) at −80 °C for 2 h.
6. Precipitated proteins were collected by centrifugation (20,000 × g, 20 min, 4 °C), and pellets were washed twice in an acetone/methanol/ acetic acid (ratio: 8:1:0.02).
7. Precipitated proteins were dissolved in 0.1 M Tris–HCl buffer containing 8 M urea and 2% DMSO.
8. The protein content was estimated by BCA assay (Pierce; Thermo Scientific, Rockford, IL),
9. The protein extract was digested according to the FASP procedure as described by [12], with slight modifications. The protein extracts (500 μg) were loaded onto Amicon Ultra 0.5 10 kDa filters (Millipore, Billerica), and filters were washed with 8 M urea in 0.1 M Tris (pH 8.5) to facilitate the removal of protein-bound small molecules (*see* **Note 5**).
10. Proteins were reduced with 10 mM DTT (dithiothreitol) (Euromedex, Souffelweyersheim) for 60 min and alkylated with 50 mM IAA for 30 min.
11. Urea buffer was exchanged by ultrafiltration for 50 mM ammonium bicarbonate. Cleaned proteins were digested overnight on the filter with trypsin at 1:50 ratio (Promega, Madison), and the resulting peptides were released from the filter by centrifugation (14,000 × g, 20 min), followed by a first wash of the filter in 0.5 M NaCl and second wash of the filter by 50% acetonitrile.
12. All three fractions were pooled, and peptides were diluted at 5% acetonitrile and acidified by 0.1% trifluoroacetic acid at 0.1% final concentration. Peptides were desalted using 500 mg Sep-Pak tC18 sample extraction columns (Waters, Milford), eluted with 1.2 mL of 70% acetonitrile, dried, and dissolved in 200 μL of 5% acetonitrile acidified with 0.1% formic acid.

3.2.2 LC-MS/MS Data Acquisition and Analysis

1. Peptide digests (126 ng per run) were loaded onto a nanoACQUITY UPLC Symmetry C18 Trap Column, 180 μm × 20 mm (particle diameter 5 μm, pore size 100 Å) in trap and elute mode with ACQUITY UPLC Peptide BEH C18 nanoACQUITY Column 75 μm × 250 mm (particle diameter 1.7 μm, pore size 130 Å) (Both Waters, Milford).
2. Run gradient was performed by Eksigent Ultra Plus nano-LC 2D HPLC (Sciex, Framingham) system over 90 min with a

gradient from 3% to 40% buffer B (buffer A: 0.1% formic acid; buffer B: 95% acetonitrile, 0.1% formic acid) at a flow rate of 300 nL/min. Eksigent system was coupled to a TripleTOF® 5600 (Sciex, Framingham) mass spectrometer interfaced to a nanospray III source, which parameters were set as follows: IS at 2500 V, Curtain gas at 30 psi, Gas Sprayer 1 at 1 psi, and Interface Heater Temperature at 150 °C.

3. Acquisition parameters were as follows: for DDA (Data Dependent Acquisition) mode, one 250 ms MS scan (>30 K resolution).
4. Following each survey MS1 scan, MS/MS spectra for the 30 most abundant parent ions (*m/z* range 350–1250) were acquired (high sensitivity mode, >15 K resolution). For DIA (Data Independent Acquisition) mode, one 150 ms MS scan was acquired (>30 K resolution), followed by 35 fixed SWATH windows each with a 75 ms accumulation time and a 350–1250 *m/z* range.
5. MS/MS SWATH scans (high sensitivity mode, >15 K resolution) were set at 26 amu window; Q1 isolation windows were covering the entire mass range.
6. DDA spectra processing and database searching were performed with ProteinPilot (v4.5 beta, Sciex, Framingham) using the Paragon algorithm. Search parameters were as follows—sample type: identification; cys-alkylation: iodoacetamide; digestion: trypsin; instrument: TripleTOF 5600; special factors: urea denaturation; ID focus: biological modifications.
7. The database was downloaded from Uniprot (June 2014), filtering for reviewed human proteins only (20194 entries).
8. The resulting group file was loaded into Peakview® (v2.0, Sciex, Framingham), and peaks from SWATH runs were extracted with a peptide confidence threshold of 99% and a false discovery rate < 1%. Label-free quantification was performed using Marker View (v1.2.1, Sciex, Framingham).
9. Selection of the proper peak was performed using the automated assistance of PeakView.
10. Absolute signals of peptides and proteins were calculated by summing the extracted area of all unique fragment ions.

3.2.3 Bioinformatic Analysis

1. Among the identified proteins (*see* **Note 6**), we selected for further analysis the proteins that passed both Student's t-test ($p < 0.05$) and whose fold changes were ≥2 (upregulated proteins) or ≤0.5 (downregulated proteins).
2. The enrichment analysis for gene ontology terms (biological processes, molecular functions, or cell components) was performed with DAVID functional annotation analysis

(Bioinformatics Resource v6.8, https://david.ncifcrf.gov/tools.jsp) by setting the human genome as background [13, 14].

3. Enrichment analysis works by (1) building associations between identified proteins from MS assay whose expression is considered as being modulated because of the differential conditions, and terms describing their biological, cellular, or molecular functions (GO annotations, Gene Ontology), metabolic pathways to which they belong; (2) grouping proteins associated with the same descriptive terms; and (3) looking for the significance of these protein groups/descriptive terms associations through statistical test.

4. Enrichment analysis uses Fisher's exact test, and multiple testing procedure is Bonferroni correction, as it is the most conservative one. Results are considered significant for a p-value lesser than or equal to 0.05.

3.3 Immunohistochemistry Analysis

1. On day 7, following the different treatments stated above, samples were fixed in formalin solution (Sigma, HT50-1-1) for 48 to 72 h.

2. After fixation, the explants were embedded in paraffin, and slices of 5 μm of thickness were studied.

3. The slices were stained with primary antibodies against LC3B (Abcam, ab192890) or SQSTM1 (Abcam, ab109012), followed by a secondary antibody conjugated with Alexa Fluor 488 dye (Life technologies, A11008).

4. A total of 17 to 20 pictures per condition were acquired by fluorescence microscopy, and the intensity of staining in the epidermis was analyzed with Visilog software.

5. Wilcoxon's test was used to analyze the results, and a p-value lower than 0.05 indicated a statistically significant difference.

4 Notes

1. It should be noted that no PTM (such as phosphorylation, acetylation, glycosylation) was considered in this study. But, it may be added by proceeding with adequate enrichment strategies prior to LC-MS analysis.

2. This procedure enabled identification of target proteins whose modulations are related to their level of expression. However, other proteins that are modulated at posttranslational level could not be detected by this method.

3. In this procedure, IHC was applied to validate the deregulated expression of the identified autophagy-related protein. The

choice of this technique is derived from the fact that during autophagy, the cytoplasmic LC3-I protein becomes lipidated and conjugated to phosphatidylethanolamine (PE) to form LC3-II, which in turn is localized on the surface of nascent autophagosomes. Therefore, quantifying LC3-II-positive puncta via immunohistochemistry was the best approach for assessing cellular autophagic activity. However, other techniques such as Western blot or ELISA could be applied in order to validate the deregulated expression of other membrane-free proteins.

4. Verification of subcellular fractionation: it is recommended to perform Western blot on specific markers.
5. A minimum of 500 μg protein extracts is recommended for loading onto Amicon Ultra 0.5 10 kDa filters (Millipore, Billerica).
6. While numerous proteins were identified in this study (1487 proteins in cytosolic fraction, 1387 proteins in the cell membrane fraction, 1516 proteins in the nuclear fraction), the whole proteome is not totally achieved.

References

1. Ichimiya T, Yamakawa T, Hirano T, Yokoyama Y, Hayashi Y, Hirayama D, Wagatsuma K, Itoi T, Nakase H (2020) Autophagy and autophagy-related diseases: a review. Int J Mol Sci 21:1–21
2. Baudouin C, Charveron M, Tarroux R, Gall Y (2002) Environmental pollutants and skin cancer. Cell Biol Toxicol 18:341–348
3. Drakaki E, Dessinioti C, Antoniou CV (2014) Air pollution and the skin. Front Environ Sci 2:11
4. Wang M, Charareh P, Lei X, Zhong JL (2019) Autophagy: multiple mechanisms to protect skin from ultraviolet radiation-driven photoaging. Oxidative Med Cell Longev 2019:1
5. Jeong D, Qomaladewi NP, Lee J, Park SH, Cho JY (2020) The role of autophagy in skin fibroblasts, keratinocytes, melanocytes, and epidermal stem cells. J Invest Dermatol 140:1691–1697
6. Sample A, He YY (2017) Autophagy in UV damage response. Photochem Photobiol 93:943–955
7. Umar SA, Shahid NH, Nazir LA, Tanveer MA, Divya G, Archoo S, Raghu SR, Tasduq SA (2021) Pharmacological activation of autophagy restores cellular homeostasis in ultraviolet-(B)-induced skin photodamage. Front Oncol 11
8. Aslam B, Basit M, Nisar MA, Khurshid M, Rasool MH (2017) Proteomics: technologies and their applications. J Chromatogr Sci 55:182–196
9. Fayyad-Kazan M, Kobaisi F, Nasrallah A, Matarrese P, Fitoussi R, Bourgoin-Voillard S, Seve M, Rachidi W (2023) Effect of ultraviolet radiation and Benzo[a]pyrene co-exposure on skin biology: autophagy as a potential target. Int J Mol Sci 24
10. IARC Working Group (2010) Evaluation of Carcinogenic Risks to Humans: Some non-heterocyclic polycyclic aromatic hydrocarbons and some related exposures. IARC Monogr Eval Carcinog Risks Hum 92:1
11. Mildner M, Weninger W, Trautinger F, Ban J, Tschachler E (1999) UVA and UVB radiation differentially regulate vascular endothelial growth factor expression in keratinocyte-derived cell lines and in human keratinocytes. Photochem Photobiol 70:674–679
12. Wiśniewski JR, Zougman A, Nagaraj N, Mann M (2009) Universal sample preparation

method for proteome analysis. Nat Methods 2009 6:5 6:359–362
13. Huang DW, Sherman BT, Lempicki RA (2008) Systematic and integrative analysis of large gene lists using DAVID bioinformatics resources. Nat Protoc 2009 4:1 4:44–57
14. Huang DW, Sherman BT, Lempicki RA (2009) Bioinformatics enrichment tools: paths toward the comprehensive functional analysis of large gene lists. Nucleic Acids Res 37:1–13

Methods in Molecular Biology (2025) 2879: 93–111
DOI 10.1007/7651_2024_562

Published online: 10 August 2024

The Chorioallantoic Membrane (CAM) Assay for the Analysis of Starvation-Induced Autophagy

Nazlı Şevval Menemenli, Özün Özcan, H. Hazal Hüsnügil, Aliye Ezgi Güleç Taşkıran, Göksu Oral, Aytekin Akyol, and Sreeparna Banerjee

Abstract

During avian development, the chorioallantoic membrane (CAM) is generated around 4 days after fertilization following the fusion of the allantois and the chorion. The CAM develops rapidly over the next several days and gets heavily vascularized and therefore has been explored widely as a tool for the study of angiogenesis. Additionally, being immunodeficient, the CAM can be used for tumor growth of human origin and its metastasis. Of note, the CAM assay is minimally invasive for the chicken embryo and lacks innervation, which gives this in vivo model a low ethical burden. Here, we describe the protocol for the generation of microtumors from human colorectal cancer cell lines on the CAM, incubated in a nutrient-deficient medium for the activation of autophagy. We show that pre-inoculation markers of autophagy induced through nutrient deficiency are retained in the microtumors generated on the CAM.

Key words Autophagy, Chorioallantoic membrane, Colorectal cancer, LC3, Nutrient deficiency, p62

1 Introduction

Macroautophagy (referred to as autophagy) is a cellular process whereby cells under nutrient stress degrade or recycle organelles such as mitochondria or long-living proteins in order to survive. During autophagy, cellular materials are enclosed in double membrane structures called autophagosomes [1]. The biogenesis of autophagosomes involves several closely regulated steps such as initiation, followed by nucleation of the phagophore with the recruitment of phospholipids and autophagy-related proteins, expansion of the double membrane with the continual incorporation of phospholipids into the membrane structure, and its closure to form the autophagosome. The autophagosome then fuses with a lysosome for the enzymatic digestion of the cargo and the recycling of macromolecules [2]. Cancer cells often activate autophagy at later stages as a stress response mechanism to survive low nutrient availability, whereas at the early stages, the degradation of organelles such as mitochondria by mitophagy has been shown to have tumor suppressive properties. Therefore, targeting autophagy through various activators or inhibitors has been considered as a therapeutic option [3].

Various techniques have been considered in the evaluation of autophagy. The formation of the double membrane autophagosome can be shown by electron microscopy. Microtubule-associated protein 1A/1B-light chain 3 (LC3-I) is a cytosolic protein that gets conjugated to phosphatidylethanolamine (PE) to form LC3-PE conjugate (LC3-II). LC3-II runs at a slightly lower molecular weight than LC3-I, which can be shown by Western blot. Additionally, punctate structures can be detected with the expression of GFP-tagged LC3 in cells undergoing autophagy, among various other techniques [4]. Autophagic flux is usually evaluated through the degradation of the cargo protein p62 (encoded by the gene *SQSTM1*). In cells with active autophagic flux, p62 levels should be lower as the protein gets degraded [4].

The demonstration of autophagy and autophagic flux in vivo is especially challenging. We have developed a protocol to evaluate autophagy and autophagic flux induced through nutrient deficiency in microtumors generated on the chorioallantoic membrane (CAM). The CAM is generated from the fusion of the allantois and the chorion. The allantois appears as an extension from the ventral wall of the endodermal hind gut. It pushes out of the body of the embryo into extraembryonic coelom, a cavity that contains the chorion. The CAM is a highly vascularized structure that is connected to the embryonic circulation by two allantoic arteries and one allantoic vein and serves as a respiratory exchange surface for the embryo [5]. As the CAM is immune deficient until developmental day 18, it serves as a perfect receptacle for the growth of human cancer cells. Tumor xenografts can be generated in as less as 3–5 days after the implantation of human cancer cells and processed for paraffin sectioning or protein isolation. These microtumors were shown to closely resemble murine xenografts and retained human tumor characteristics [6].

We have previously reported that human colorectal cancer cells undergoing autophagy through nutrient depletion can form microtumors on the CAM that are similar in volume as the tumors developed from cells grown in complete medium [7, 8]. Here, we describe the protocol for the generation of microtumors from nutrient-depleted and control cells on the CAM and also report that the microtumors retain autophagic markers such as LC3 lipidation and autophagic flux, demonstrated via a decrease in the levels of the cargo protein p62.

2 Materials

Materials necessary for the propagation and cryopreservation of Caco-2, HCT-116, and LoVo cells are given. All cell lines were authenticated by short tandem repeat (STR) analysis prior to the start of experiments and were found to resemble the original cell lines by over 95%. All cell lines were routinely tested for mycoplasma contamination. All disposable plasticware used in this study were supplied by Sarstedt (Germany) or Jet Biofil (China).

2.1 Reagents and Materials for Cell Culture

1. Fetal bovine serum (FBS) (Capricorn Scientific, Germany, cat no: FBS-HI-11A, or PAN Biotech, Aidenbach, Germany, cat no: P30-3031).
2. 100 U–100 mg Penicillin-streptomycin (Sartorius, Germany, cat no: 03-031-1B, or PAN Biotech, Aidenbach, Germany, cat no: P06-07100).
3. 200 mM L-glutamine (Sartorius, Germany, cat no: 03-020-1B).
4. Non-essential amino acids (100×) (Sartorius, Germany, cat no: 01-340-1B).
5. 100 mM sodium pyruvate (Sartorius, Germany, cat no: 03-042-1B).
6. 200 g/l (+) D-glucose solution (Gibco, USA, cat no: A24940-01).
7. Plasmocin™ (InvivoGen, France, cat no: ant-mpt).
8. Culture medium for Caco-2 cells: Eagle's minimum essential medium (EMEM), without L-glutamine, with Earl's salt (Sartorius, Germany, cat no:01-025-1A) supplemented with 20% FBS, 1 mM sodium pyruvate, 1× non-essential amino acid, 2 mM L-glutamine, 100 U/mL penicillin–100 μg/mL streptomycin, 2.5 mg/mL Plasmocin.
9. Culture medium for HCT-116 and LoVo cells: RPMI 1640 medium without glucose (Sartorius, Germany, cat no: 01–103-1A) supplemented with 10% FBS, 2 mM L-glutamine, 100 U/mL penicillin–100 μg/mL streptomycin, 2.5 mg/mL Plasmocin.
10. Nutrient-deficient medium for Caco-2 cells: Dulbecco's modified Eagle's medium (DMEM), without glucose, L-glutamine or sodium pyruvate (Sartorius, Germany, cat no: 01-057-1A) supplemented with 0.1 g/l glucose, 0.2 mM L-glutamine, 1% FBS, 1 mM sodium pyruvate, 1× non-essential amino acids, 100 U/mL penicillin–100 μg/mL streptomycin.
11. Nutrient-deficient medium for HCT-116 and LoVo cells: RPMI 1640 medium, without glucose or L-glutamine

(Sartorius, Germany, cat no:01–101-1A) supplemented with 0.1 g/l glucose, 0.2 mM L-glutamine, 1% FBS, 100 U/mL penicillin–100 μg/mL streptomycin.

2.1.1 Subculturing Materials

1. DPBS (Dulbecco's phosphate buffered saline, without Ca^{2+} and Mg^{2+}) (Sartorius, Germany, cat no:02-023-1A).
2. Trypsin-EDTA (0.5%) in DPBS (10×) (Capricorn Scientific, Germany, cat no: TRY-1B-10). Prepare 1× trypsin-EDTA solution by diluting with DPBS.
3. Complete growth medium.
4. 37 °C water bath.
5. Incubator supplemented with 5% CO_2 at 37 °C.

2.1.2 Cryopreservation Materials

1. Dimethyl sulfoxide (DMSO) (SERVA, Germany, cat no: 20385.02).
2. Mr. Frosty™ Freezing Container (Thermo Fisher Scientific, USA, cat no: 5100-0001).
3. Liquid nitrogen.

2.2 Protein Isolation and Western Blot Materials

2.2.1 Protein Isolation

1. DPBS (Dulbecco's phosphate-buffered saline, without Ca^{2+} and Mg^{2+}) (Sartorius, Germany, cat no:02-023-1A).
2. M-PER Mammalian Protein Extraction Reagent (Thermo Fisher Scientific, USA, cat no: 78501).
3. Mini Protease Inhibitor Cocktail, cOmplete™ (Roche Diagnostic GmbH, Germany, cat no: 11836170001), and phosphatase inhibitor tablets, PhosSTOP™ (Roche Diagnostic GmbH, cat no: 04906837001). Prepare 10× stocks by dissolving the tablets in 1 mL DPBS.
4. Prepare the cell lysis buffer by mixing M-PER Reagent with 1× protease and phosphatase inhibitors.
5. Sample loading dye (6×): 12% sodium dodecyl sulfate (SDS) (PanReac AppliChem GmbH, Germany, cat no: A2572), 30% β-mercaptoethanol (Amresco, cat no: M131), 30% glycerol (AppliChem, cat no: A2926/1000), 0.012% bromophenol blue (Sigma Aldrich, USA, cat no: B0126), 375 mM Tris–HCl pH 6.8.
6. Pierce™ Coomassie (Bradford) Protein Assay Kit (Thermo Fisher Scientific, USA, cat no:23200).
7. Cooling centrifuge.
8. 95 °C heat block.

2.2.2 SDS Gel Electrophoresis, Protein Transfer and Immunoblotting

1. 4% Stacking gel: For 1.5 mm spacer, mix 3.1 mL dH_2O, 650 μL 30% (w/v) acrylamide/bis solution (37.5:1) (SERVA, Germany, cat no: 10688.01), 1.25 mL stacking buffer (0.5 M Tris–HCl, pH 6.8), 5 μL TEMED (SERVA, Germany, cat no:

Table 1
Primary antibodies used for Western blot

Primary antibodies	Vendor and catalog number	Conditions for use
LC3A/B (D3U4C) XP®, (14/16 kD)	Cell Signalling Technology, cat no: 12741	1:1000 in 5% BSA in TBS-T
SQSTM1/p62 (D-3) (62 kD)	Santa Cruz Biotechnology, USA, cat no: sc-28359	1:1000 in 5% BSA in TBS-T
GAPDH (FL-335) (37 kD)	Santa Cruz Biotechnology, sc-25778	1:4000 in 5% skim milk or BSA in TBS-T

35925.01), 50 μL 10% ammonium persulfate (APS) (Pierce, cat no: 17874).

2. 12% Separating gel: For 1.5 mm spacer, mix 3.4 mL dH_2O, 4 mL 30% (w/v) acrylamide/bis solution (37.5:1) (SERVA, Germany, cat no: 10688.01), 2.5 mL separating buffer (1.5 M Tris–HCl, pH 8.8), 10 μL TEMED (SERVA, Germany, cat no: 35925.01), 100 μL 10% APS.
3. PageRuler™ Plus Prestained Protein Ladder (Thermo Fisher Scientific, USA, cat no: 26619).
4. SDS-PAGE running buffer (10×): 25 mM Trizma® base (Sigma Aldrich, USA, cat no: T1503), 192 mM glycine (SERVA, Germany, cat no: 23390.04), 0.1% SDS (PanReac AppliChem GmbH, Germany, cat no: A2572). Store the running buffer at room temperature. Prepare 1X SDS-PAGE running buffer by diluting 100 mL of 10X SDS PAGE running buffer with 900 mL of dH_2O.
5. Transfer buffer (10×): 250 mM Trizma® base (Sigma Aldrich, USA, cat no: T1503), 1.92 M glycine (SERVA, Germany, cat no: 23390.04). Adjust pH to 8.3 in 1 L dH_2O. Store the transfer buffer (10×) at 4 °C. For 1× transfer buffer, mix 100 mL 10× transfer buffer and 200 mL methanol. Complete the volume to 1 L with dH_2O and pre-chill at 4 °C.
6. Mini-PROTEAN® Tetra Vertical Electrophoresis Cell for Mini Precast Gels (Bio-Rad, USA, cat no: 1658005) for SDS gel electrophoresis.
7. Polyvinylidene difluoride (PVDF) membrane with a pore size of 0.2 μm (Roche Diagnostic GmbH, Germany, cat no: 03010040001).
8. Mini Trans-Blot® Module (Bio-Rad, cat no: 1703935).
9. PowerPac Basic Power Supply (Bio-Rad, USA, cat no: 164-5050).

10. Rotilabo® Blotting paper, 0.35 mm thick (Carl Roth GmbH, Karlsruhe, Germany, cat no: CL67.1).
11. Tris buffered saline (TBS, 20X): Mix 50 mM Trizma® base (Sigma Aldrich, USA, cat no: T1503), 300 mM NaCl (Tekkim Kimya, cat no: TK.170540), 4 mM KCl (Merck Millipore, USA, cat no: 104935). Adjust pH to 7.4 in 1 L dH_2O. Sterilize by autoclaving and store at room temperature.

    For 1X TBS with 0.1% Tween 20 (0.1% TBS-T), add 1 mL Tween 20 (VWR International, USA, cat no: 97062-332) to 1X TBS solution prepared with 50 mL 20X TBS and 950 mL autoclaved distilled dH_2O.
12. Blocking solution: 5% skim milk (Sigma-Aldrich, USA, cat no: 70166) in 1X TBS-T or 5% BSA (Albumin Bovine Fraction V, pH 7.0, standard grade, lyophilized, SERVA, Germany, cat no: 11943) in 1X TBS-T. Store at 4 °C.
13. Antibody diluent solutions: 5% skim milk or 5% BSA in 1X TBS-T. Store at 4 °C.
14. Primary and secondary antibodies (*see* Table 1).
15. Chemiluminescent Clarity™ ECL Substrate (250 mL Clarity Western Peroxide Reagent and 250 mL Clarity Western Luminol/Enhancer Reagent) (Bio-Rad, USA, cat no: 1705061) for detection of horseradish peroxidase (HRP).
16. ChemiDoc MP Imaging System (Bio-Rad, USA) for chemiluminescent detection.
17. Membrane stripping buffer: Dissolve 15 g glycine (SERVA, Germany, cat no: 23390.04), 1 g SDS (PanReac AppliChem GmbH, Germany, cat no: A2572), and 10 mL Tween 20 (VWR International, USA, cat no: 97062-332) in 800 mL dH_2O. Adjust pH to 2.2, then complete volume up to 1 L with dH_2O.

2.3 Chorioallantoic Membrane (CAM) Assay Materials

1. Fertilized, specified pathogen-free (SPF) Leghorn eggs (obtained from Tavukçuluk Research Institute, Ankara, Türkiye or VALO Biomedia, Osterholz-Scharmbeck, Germany).
2. Matrigel® Basement Membrane Matrix, LDEV-free (Corning, USA, cat no: 354234).
3. Trypan blue solution 5 mg/mL in saline (Biological Industries/Sartorius, Germany, cat no:03-102-1).
4. Counting chamber (Hemocytometer) 0.1 mm depth, (Marienfeld Superior, Germany, cat no: 0640030).
5. Cover glass (Microcult cat no: 217230, 22 × 22mm).
6. Tweezers, curved fine end forceps, curved end scissors, spoon micro-spatula.
7. Eosin (Sigma, cat no: 212954).

8. Paraformaldehyde (PFA) (PanReac AppliChem GmbH, Germany, cat no: A3813). Prepare 36% (w/v) stock solution in DPBS. Dilute to 4% PFA in DPBS before use.

2.4 Materials for CAM Microtumor Sectioning and Immunohistochemistry

1. Paraffin (cat no: VO-5-1002, Vogel MedTec GmbH, Fernwald, Germany).
2. Tissue embedding and processing cassettes with lid (Histo-Plus™, Turkey, cat no: HEM-CA2-W250 or Carl Roth, cat no: XY29.1, Karlsruhe, Germany).
3. Xylene (cat no. XTR-200L-97-1; SAV Liquid Production GmbH, Flintsbach, Germany).
4. Ethyl alcohol (cat no: 100983, Millipore, Burlington, MA, USA) at 96% and 100% (v/v in water).
6. Microscopy slides (StarFrost adhesive, cat no: 11250, Engelbrecht GmbH, Edermuende, Germany).
7. Hematoxylin (Cytopath Hematoxylin Gill III, cat no: CP826, Diapath, Martinengo, Italy), eosin (cat no: C0356, Diapath, Martinengo, Italy).
8. Ki67 (Dako/Agilent, Santa Clara, CA, USA, cat no: M7240, 1: 100 dilution).
9. Ventana BenchMark Ultra automated instrument (Ventana Medical Systems, Oro Valley, AZ, USA).
10. UltraView Universal DAB Detection Kit (cat no: 05269806001, Roche, Switzerland).

2.5 Protein Isolation from CAM Microtumors

1. Cell recovery solution (Corning, USA, cat no: 354253).
2. Ultrasonic bath (Bandelin Sonorex, 31200038343.02) for protein isolation.
3. DPBS (Dulbecco's phosphate buffered saline, without Ca^{2+} and Mg^{2+}) (Sartorius, Germany, cat no: 02-023-1A).
4. M-PER (Thermo Fisher Scientific, USA, cat no: 78501) containing Mini Protease Inhibitor Cocktail, cOmplete™ (Roche Diagnostic GmbH, Germany, cat no: 11836170001), and phosphatase inhibitor tablets, PhosSTOP™ (Roche Diagnostic GmbH, cat no: 04906837001).

2.6 Statistical Analyses

Results are expressed as the mean ± standard error of mean (SEM). Data analyses and graphing were carried out with GraphPad (Prism version 9.00 for Windows, GraphPad Software, La Jolla, California, USA). Statistical analysis was carried out using Mann–Whitney U test or unpaired t-test. $p < 0.05$ was considered as statistically significant.

3 Methods

3.1 Cell Culture

3.1.1 Cell Thawing

1. Add 9 mL of complete growth medium for each cell line to a sterile 15 mL falcon tube.
2. Place the cryovial containing the frozen cells into a pre-warmed water bath at 37 °C.
3. Swirl the cryovial in the 37 °C water bath. Thawing should be rapid and continued until a small portion of ice is left in the vial.
4. Dry the outside of the vial with a paper towel first and then clean with 70% ethanol.
5. Transfer the vial into a sterile laminar flow hood.
6. Transfer the entire thawed cell suspension (approximately 1 mL) from the cryovial into the falcon tube containing complete growth medium (9 mL).
7. Centrifuge the cells at room temperature for 5 min at 3000 rpm.
8. Aspirate the supernatant aseptically and gently without disturbing the visible cell pellet.
9. Add 5 mL of complete medium for each cell line onto the cell pellet and gently resuspend the cell pellet by pipetting.
10. Transfer the cell suspension into a sterile 25 cm^2 culture flask and place in a humidified incubator supplemented with 5% CO_2 at 37 °C.

3.1.2 Cell Maintenance and Passaging

Propagate the cells in a 25 cm^2 culture flask by changing the growth medium every 2 days. When the cell confluency reaches 60–70%, sub-culture the cells into a fresh 25 cm^2 culture flask by following the steps below.

1. Aspirate the medium from the 25 cm^2 culture flask and add DPBS in a volume such that the entire flask surface is covered.
2. Remove the DPBS and add 500 μL of 1X trypsin-EDTA solution to the flask.
3. Incubate the cells at 37 °C until all cells are detached from the flask surface, usually 5–10 min.
4. Check the cells under an inverted microscope to make sure that the cell layer is dispersed.
5. Add 4 mL of complete growth medium and gently pipette the cells to obtain a homogenized cell suspension.
6. Subculture the cell suspension to a fresh 25 cm^2 culture flask according to the dividing ratio/doubling time of the cell line.
7. Add Plasmocin at 2.5 μg/mL concentration routinely to the culture media of cells to prevent mycoplasma contamination.

8. Incubate cell culture in a humidified incubator supplemented with 5% CO_2 at 37 °C and change the medium every 2 days.

3.1.3 Cell Freezing

1. Cells at 60–70% confluency should be processed using the same method until **step 4** described in Subheading 3.1.2. Take an aliquot from the sample for routine mycoplasma testing.
2. Centrifuge the cell suspension at room temperature for 5 min at 3000 rpm and aspirate the growth medium.
3. Thoroughly resuspend the cell pellets in 5 mL complete growth medium.
4. Transfer the 950 μL from this cell suspension to a cryovial, add 50 μL DMSO (final concentration 5%), and pipette the mixture gently 4–5 times.
5. Put the cryovials immediately in Mr. Frosty™ Freezing Container and store at −80 °C overnight to be transferred to the vapor phase of liquid nitrogen the next day.

3.1.4 Induction of Nutrient Deficiency

1. Seed approximately 4×10^6 cells to a fresh 75 cm^2 tissue culture flask.
2. Incubate the cells in a humidified incubator supplemented with 5% CO_2 at 37 °C to allow for attachment overnight.
3. The next day, discard the medium and wash the cells with DPBS.
4. Add 8–10 mL of nutrient-deficient medium for each cell line (Subheading 2.1).
5. As a control, add 8–10 mL of the complete growth medium to the cells in a separate 75 cm^2 tissue culture flask.
6. After 48 h of nutrient deficiency, process the cells for counting by trypan blue exclusion assay to prepare the cells for the CAM assay.

3.2 Chorioallantoic Membrane (CAM) Assay

3.2.1 Sterilization of Materials to Be Used for CAM Experiments

1. At the start and end of the CAM assay, the egg incubator should undergo a decontamination cycle to minimize the risk of fungal contamination. The use of a dedicated egg incubator with 100% air and set at 37 °C is essential. Ensure that the egg incubator is located far from the mammalian cell culture facility to prevent any cross contamination.
2. Clean the surface of the incubator and the shelves with Acto-sept AF (Acto Pharma, cat no: 05.2242.5) to prevent bacterial, mycobacterial, yeast, and viral contamination.
3. Start the decontamination cycle according to the manufacturer's instructions by placing approximately 300 mL autoclaved distilled water in a tray. This cycle involves heating the incubator to 180 °C for 11 h.

4. Autoclave all materials including tweezers, curved fine end forceps, spoon micro-spatula, and curved-end scissors for at least 30 min just before the CAM experiment.

3.2.2 Egg Delivery and Handling

1. Fertilized Leghorn eggs are collected under refrigeration at approximately 12 °C.
2. Allow eggs to equilibrate to room temperature (*see* **Note 1**).
3. Clean the eggs by gently wiping any visible dirt from the eggshells with distilled water (*see* **Note 2**).
4. Place the eggs on egg holders in the incubator at 37 °C with the more rounded pole oriented upward for optimal gas exchange for the developing embryo. This day is considered embryonic developmental day (EDD) 1.
5. Turn the eggs by 5–10° three times a day until egg opening. This prevents the CAM from sticking to the eggshell.

3.2.3 Egg Opening on EDD 8

1. Examine each egg with a flashlight to determine the exact location of the air sac. Place a piece of sterile silk tape (Durapore 3 M) on the round pole of the egg where the air sac is located (*see* **Note 3**).
2. Introduce a small hole into the eggshell in the middle of the air sac using sterile, sharp-tipped forceps.
3. Open a window with a diameter of 1–1.5 cm into the eggshell using sharp, curved scissors, starting from the initial hole (*see* **Note 4**).
4. Place a drop of sterile DPBS onto the eggshell membrane. Pinch the membrane with sharp forceps, allowing the DPBS to flow into the space between the membrane and the CAM.
5. Remove the membrane without damaging the CAM or blood vessels. Strive to remove as much of the membrane as possible to avoid interference with CAM development.
6. Seal the window with silk tape to keep the egg contents from drying out and place the eggs back in the incubator.

3.2.4 Trypan Blue Exclusion Assay

The inoculations are carried out on EDD 9. All treatments on the cell lines should be completed on the day of the inoculation.

1. Aspirate the medium from the cells and wash with DPBS.
2. Remove the DPBS and add 500 μL of 1X trypsin-EDTA solution to the flask.
3. Incubate the cells in a 37 °C incubator until all cells are detached from the flask surface.
4. Add 4 mL of complete growth medium and gently pipette the cells to obtain a homogeneous cell suspension.

5. Centrifuge the cells at room temperature for 5 min at 3000 rpm.
6. Resuspend the cell pellets in 8 mL complete growth medium thoroughly.
7. Mix 20 μL of 0.4% trypan blue and 20 μL of cell suspension in a microcentrifuge tube (1:1 dilution of 0.4% trypan blue and cell suspension) and incubate for 3 min (*see* **Note 5**).
8. Apply 10 μL of the trypan blue/cell mixture to a hemocytometer. Place the hemocytometer on the stage of a binocular microscope and focus on the cells.
9. Count the unstained (viable) cells in the hemocytometer to obtain the total number of viable cells per milliliter of aliquot. Multiply the total number of viable cells by 2 (the dilution factor for trypan blue).
10. Calculate the percentage of viable cells as follows:
    Number of viable cells $\times$ 10^4 = cells/mL culture.

3.2.5 Preparation of Cell Pellets in Matrigel

1. Pool cells from 2–3 75 cm^2 flasks and count the number of viable cells by trypan blue staining. Distribute the needed volume of cells to obtain 1 $\times$ 10^6 cells per egg (at least 8–10 eggs per experimental group) into 1.5 mL microcentrifuge tubes.
2. Centrifuge at 300× *g* for 5 min in a precooled centrifuge (4 °C) and discard the supernatant completely. Separate an aliquot (approximately 10%) of the cells prior to centrifugation, pellet the cells, and store at −80 °C. These pellets will be used for protein isolation to determine the pre-inoculation levels of autophagy markers by Western blot.
3. Prepare a 1:1 mixture of ice-cold Matrigel and complete medium (total volume depends on the size of the pellet; approximately 40–45 μL per pellet).
4. Resuspend the cell pellets in the 1:1 dilution of ice-cold Matrigel and the complete medium mixture.
5. Pipette the cell suspension/Matrigel mixture on the smooth surface of a 10 cm cell culture dish at a vertical angle to form small droplets. Avoid bubbles and other disruptions on the droplets.
6. Incubate the droplets for 10 min at room temperature and then for 1 h at 37 ° C until the Matrigel solidifies.

3.2.6 Implantation of the Cells onto the CAM on EDD9

1. Place a small drop of sterile DPBS near the cell suspension/Matrigel mixture and allow it to soak briefly; this will make collecting the mixture from the dish surface easier.

2. Remove the mixture from the surface of the dish with the flattened side of a spoon micro-spatula, keeping it as horizontal as possible.
3. Place all of the cell suspension on a pre-identified well-vascularized spot on the CAM.
4. Seal the window with silk tape to keep the egg contents from drying out and place the eggs back in the incubator.

3.2.7 CAM Microtumor Harvesting for Protein Isolation

1. Keep the inoculated eggs in the incubator for an additional 5 days (EDD 14).
2. Discard any eggs showing signs of death of the embryo (*see* **Note 6**).
3. On EDD 14, remove the silk tape, gently cut the shell with sharp sterile scissors until the opening is 2–3 cm wide, and examine the inside of the egg to locate the microtumor.
4. With the help of sterilized curved fine end forceps, gently hold the CAM tissue near the microtumor. Harvest the microtumor by cutting the surrounding CAM tissue and the blood vessels with sterilized sharp curved scissors.
5. Place a few drops of DPBS on the microtumor to remove any remaining blood and then place the microtumor onto a filter paper with the help of sterilized tweezers to let it dry.
6. Sacrifice the embryos by quickly cutting the umbilical cord and the neck of the embryo with a sharp pair of scissors to cause minimum distress. Dispose the embryo and the egg as biological waste for incineration or decontamination by autoclaving.
7. Take a picture of the microtumors with a 10 cm ruler under a dissecting microscope and measure the height, width, and length of the microtumor for the calculation of microtumor volume (Fig. 1).
8. Place the microtumors into sterile microcentrifuge tubes and store them at −80 °C until use for protein isolation.
9. Calculate the microtumor volume with the formula below [9]:

$$\text{Tumor volume} = \text{tumor length} \times \text{width} \times \text{height} \times 0.52.$$

3.2.8 CAM Microtumor Harvesting for Paraffin Embedding

1. Follow the first four steps as described in Subheading 3.2.7.
2. Label the embedding cassettes with the name of the cell line used to generate the microtumor, and the treatment applied, if any. A code for each microtumor is useful.
3. Place a few drops of 4% paraformaldehyde (PFA) solution onto the microtumors.

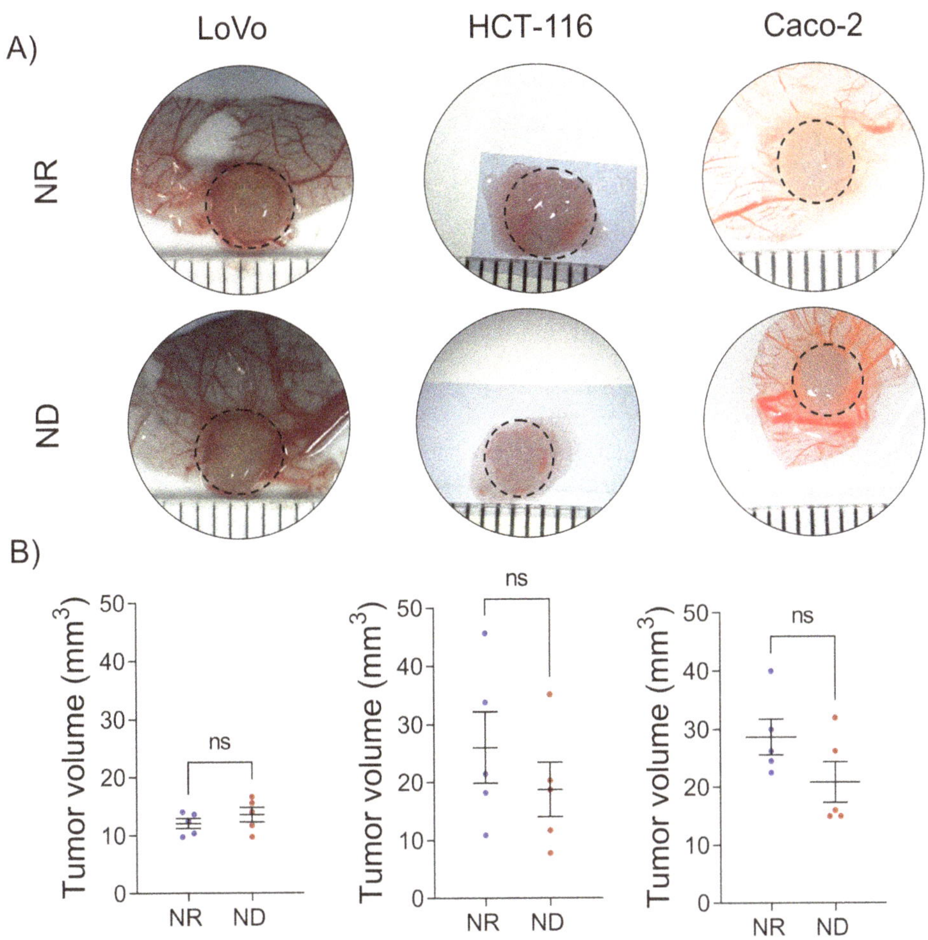

Fig. 1 Determination of microtumor volume. (**a**) Microtumors were formed on the CAM from LoVo, HCT-116, and Caco-2 cells pre-incubated in the nutrient-deficient (ND) or complete medium (nutrient rich, NR) for 48 h. (**b**) The microtumor volumes were measured and compared. No significant difference was observed. Five microtumors isolated from 8 to 10 eggs inoculated for each experimental group (NR or ND) in each cell line were used. The lines on the ruler correspond to 1 mm. Mann–Whitney U test was used for statistical analysis, *ns* not significant

4. After 30–60 s, harvest the microtumor by cutting the surrounding CAM tissue and the vessels with the help of sterilized curved fine end scissors.
5. Place the microtumor on a cellulose acetate filter paper to allow it to dry for a few minutes.
6. Place a drop of 0.2% eosin solution onto the microtumors on cellulose acetate filter paper. This will help identify the microtumor after embedding.
7. The parts of the filter paper outside of the margins of the microtumor can be cut out and removed.

8. Gently wrap the eosin stained microtumors in a 3×3 cm tissue paper pre-wet with 4% PFA, and place them into tissue embedding and processing cassettes.
9. Place the cassettes in 4% PFA for overnight fixation at room temperature. Do not exceed the duration of fixation.

3.2.9 Sectioning and Staining of CAM Microtumors

1. Follow protocol until **step 9** in Subheading 3.2.8.
2. After overnight fixation of the tumors in 4% PFA, process the cassettes using routine tissue processing and paraffin-embedding protocols.
3. Obtain 1–3 μm thick sections from paraffin-embedded microtumors and gently place on glass slides.
4. Deparaffinize the sections by incubation at 75 °C for 40 min.
5. Wash 2× with xylene for 4 min each.
6. Rehydrate the sections in decreasing concentrations of 100% isopropanol and 80% isopropanol for 2 min 30 s each.
7. Stain the sections with hematoxylin and eosin for histological assessment of the samples.
8. Stain additional sections from the same tumors for Ki67 using routine protocols on a Ventana BenchMark Ultra automated instrument (Ventana Medical Systems, Inc., Oro Valley, AZ, USA) with the UltraView Universal DAB Detection Kit (Roche, Basel, Switzerland) according to the manufacturer's recommendations (Fig. 2).
9. Dehydrate the sections using 96% ethanol for 1 min, 2× 100% ethanol for 2 min, followed by 2X Xylene for 4 min each.
10. Mount the coverslips using Permount medium.

3.2.10 Protein Isolation from CAM Microtumors and Pro-inoculation Cells

1. To remove the Matrigel from the harvested microtumors, place 200 μL recovery solution onto each pellet in a 1.5 mL microcentrifuge tube. Microtumors should be cut into pieces with sterile scissors to ensure that the recovery solution reaches the inside of the tumors. The volume of the recovery solution should be scaled up proportionately if the microtumors are pooled.
2. Gently mix the microtumors with the recovery solution by inverting the tube up and down approximately 10× over 30 min and incubate on ice for 60–90 min.
3. After the Matrigel is completely dissolved, centrifuge the tubes at 300× *g* for 5 min at 4 °C and discard the supernatant.
4. Wash the pellet with cold DPBS and centrifuge at 500× *g* for 5 min at 4 °C. Repeat two more times.

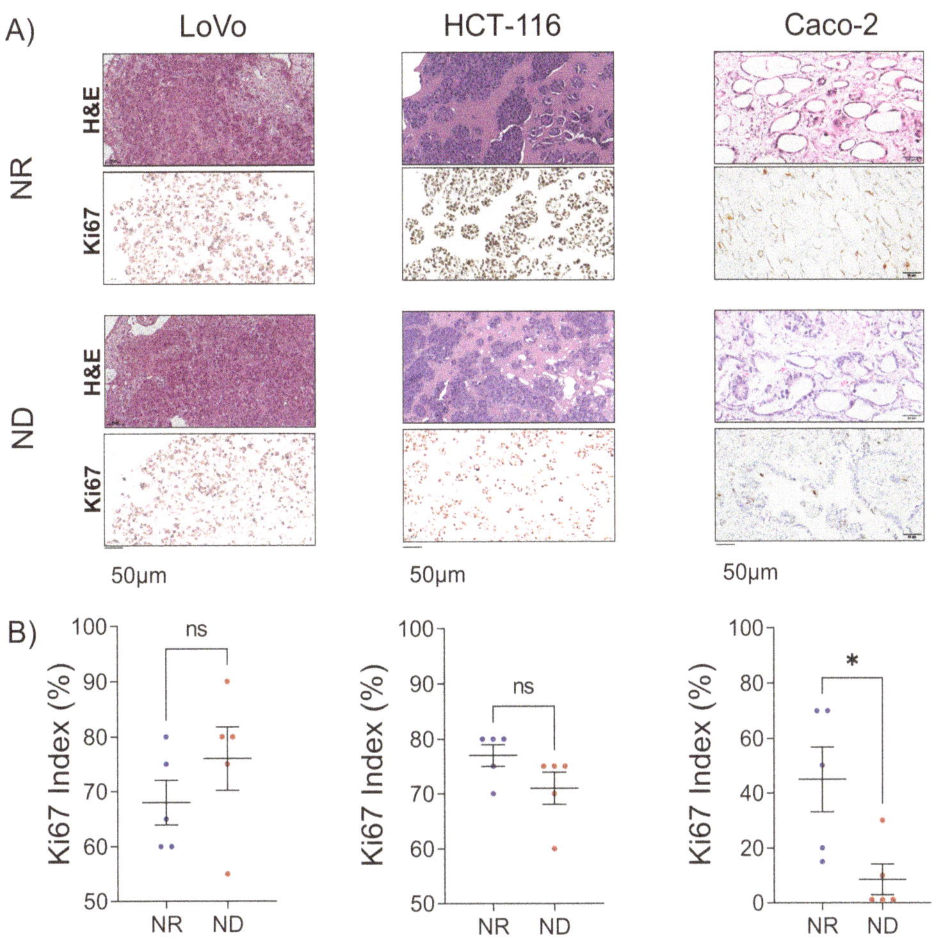

Fig. 2 Staining of the microtumor sections with hematoxylin and eosin for morphology and with Ki67 for proliferation. (**a**) Microtumors formed on the CAM from LoVo, HCT-116, and Caco-2 cells pre-incubated in the nutrient-deficient (ND) or complete medium (nutrient rich, NR) for 48 h ($n = 5$) were fixed, paraffin embedded, and sectioned. The sections were processed for hematoxylin and eosin and Ki67 staining. No difference in cellular morphology was observed with nutrient deficiency. (**b**) Ki67 staining suggested no significant change in proliferation in LoVo and HCT-116 cells while Caco-2 cells showed a significant decrease in Ki67 index with nutrient deficiency. Five microtumors isolated from 8 to 10 eggs inoculated for each experimental group (NR or ND) were used. Mann–Whitney U test was used for statistical analysis, $^*p < 0.05$

5. Process the pellet immediately for total protein isolation as described below or store at −80 °C until use.
6. For protein isolation, aspirate the DPBS gently and resuspend the pellets in lysis buffer, prepared with M-PER containing 1X protease and phosphatase inhibitors. The volume of the lysis buffer will depend on the pellet volume, usually 150 μL.
7. Vortex the pellets for 30 s and incubate for 10 min on ice; repeat this process 3×.

8. Sonicate the pellet for 10 s and then incubate on ice for 10 s. Repeat this step 6×.
9. Centrifuge the tubes at 14,000× *g* for 10 min at 4 °C. Collect the supernatant into fresh microcentrifuge tubes and store at −80 °C until use.
10. For protein isolation from pre-inoculation cells, resuspend the cell pellets in lysis buffer prepared with M-PER containing 1X protease and phosphatase inhibitors. The volume of the lysis buffer can change according to the pellet volume; approximately 40-45 μL per pellet may be sufficient.
11. Vortex the pellets for 30 s and incubate for 10 min on ice; repeat this process 3×.
12. Centrifuge the tubes at 14,000× *g* for 10 min at 4 °C. Collect the supernatant into fresh microcentrifuge tubes and store at −80 °C until use.
13. To determine the protein amount of the lysates isolated from pre-inoculation cells or from microtumors, mix 1 μL of the protein solution (1:5 dilution) with 250 μL of Coomassie Protein Assay Reagent in a 96-well microplate. Prepare at least three technical replicates for each protein sample.
14. Measure the protein absorbance at 595 nm in a MultiSkan GO Microplate Spectrophotometer.
15. Quantify the protein amount from a standard curve generated with bovine serum albumin (BSA) at concentrations from 25 to 2000 μg/mL mixed with Coomassie Protein Assay Reagent.

3.2.11 Western Blot for Autophagy Detection in Lysates from Pre-inoculation Cells and Microtumors

1. Mix the protein isolates (10–30 μg) from the microtumors with 6X Laemmli Buffer and heat for 7 min at 95 °C in a heat block.
2. Load the proteins into the wells of 12% SDS-polyacrylamide gel.
3. Run the samples at 60 V until all the proteins have moved into the separating gel, and then separate protein samples at 90 V for approximately 2 h.
4. Proceed with a wet transfer of the proteins to a polyvinylidene fluoride (PVDF) membrane by using the sandwich model at 115 V for 90 min with continuous cooling.
5. Block the membrane in 5% skimmed milk or 5% BSA in TBS-T buffer at room temperature for 1 h on an orbital shaker.
6. Incubate the membrane overnight on an orbital shaker at 4 °C with the primary antibody diluted in 5% skim milk or BSA in 0.1% TBS-T according to the manufacturer's instructions.
7. The next day, wash the membranes 3× with TBS-T for 10 min each on a rocker.

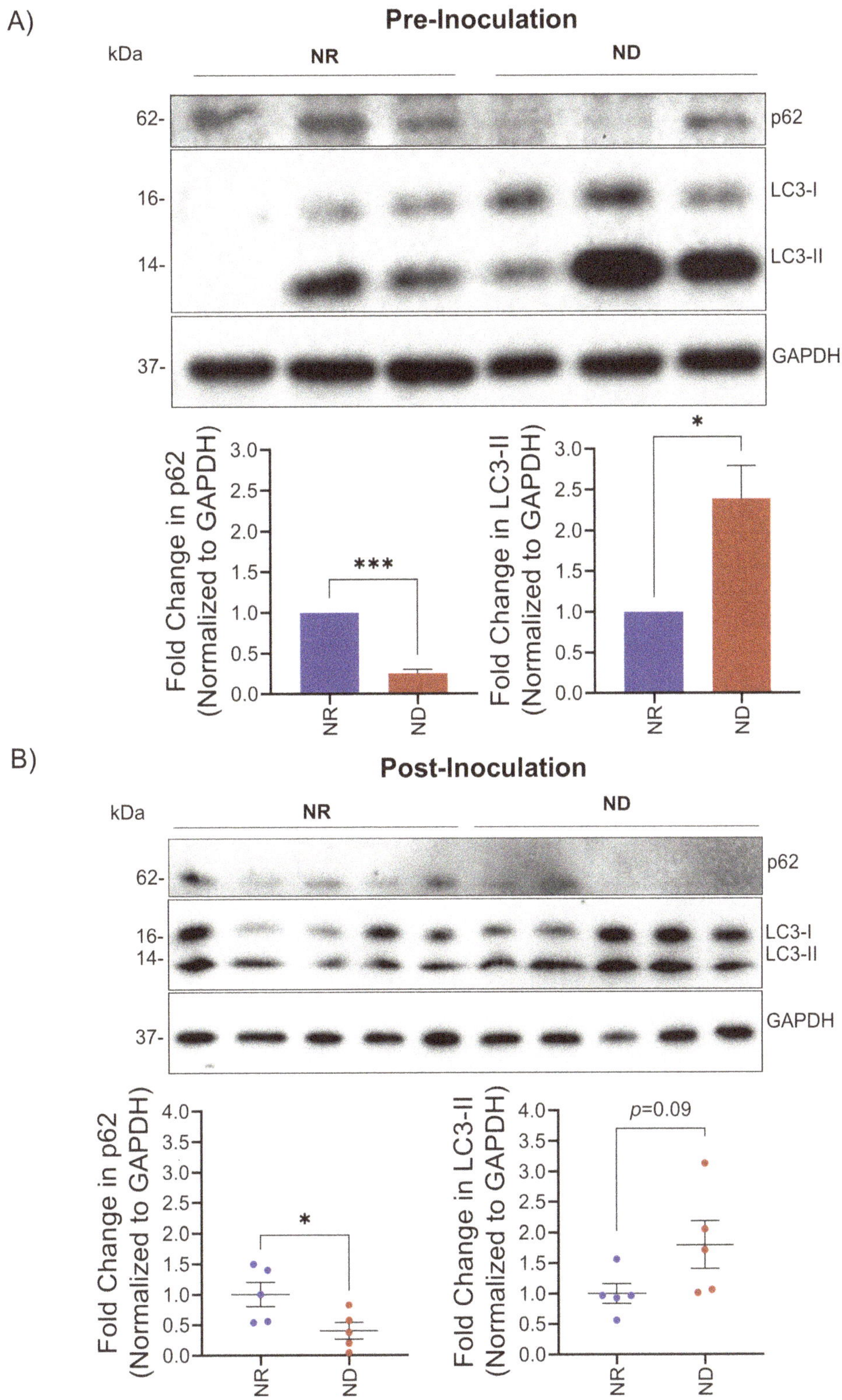

Fig. 3 Western blot showing the activation of autophagy with nutrient deficiency in CAM microtumors. HCT-116 cells were incubated in the nutrient-deficient (ND) or complete medium (nutrient rich, NR). (**a**) Proteins were isolated from pre-inoculation samples; data from three biological replicates is shown. A

8. Incubate the membranes with an HRP-conjugated secondary antibody for 1 h at room temperature on an orbital shaker.
9. After washing the membranes 3× with TBS-T for 10 min each, incubate the membranes with the ECL substrate in the dark and at room temperature for approximately 1 min followed by visualization.
10. Visualize the protein band intensities in ChemiDoc MP Imaging System (Bio-Rad) with an exposure time until the most saturated intensity is obtained.
11. Obtain the adjusted band intensities for protein of interest with ImageLab software and normalize the adjusted band intensities to that of a loading control for densitometric analysis (Fig. 3).
12. If further incubation with a primary antibody is required, incubate the membranes for 2–4 min at 70–80 °C in a pre-heated mild stripping buffer.
13. Before adding a new primary antibody, wash the membranes 3× with TBS-T for 10 min each.

4 Notes

1. The sterilized incubator for the eggs should be set to a constant temperature of 37 °C, 100% air and a relative humidity of 60–70%. Use autoclaved, sterile water to maintain the humidity inside the incubator. The use of a dedicated egg incubator is essential. Ensure that the egg incubator is located far from the mammalian cell culture facility to prevent any cross contamination.
2. Do not use alcohol for cleaning the eggshells, as it can damage the eggshell membrane and compromise embryo viability.
3. If any of the eggshells crack while cutting an opening, cover the crack with a piece of silk tape before continuing the incision.
4. Make sure that the cut for opening the shell avoids the region where the embryo is attached to the eggshell.

Fig. 3 (continued) statistically significant increase in LC3-II levels and a decrease in p62 levels, signifying autophagy induction and autophagic flux, respectively, were observed with nutrient deficiency. (**b**) Western blot showing that the activation of autophagy was retained in the CAM microtumors ($n = 5$). The cells were incubated in the ND or NR medium for 48 h prior to inoculation. Five microtumors isolated from 8 to 10 eggs inoculated for each experimental group (NR or ND) were used for protein isolation followed by Western blot. An increase in LC3-II and a decrease in p62 protein levels suggest that autophagic signaling was maintained in the microtumors. Unpaired t-test was used for statistical analyses ($^{*}p < 0.05$, $^{***}p < 0.001$)

5. For accurate cell viability assessment before egg inoculation using the trypan blue exclusion assay, count the cells within 3–5 min of mixing with trypan blue. Longer incubation times can cause cell death and lead to underestimating viability.
6. Inspect the egg contents to evaluate blood vessel development, embryo visibility, and overall viability prior to ending the experiment. Discard any eggs where the embryos are not properly developed or appear dead.

Acknowledgments

The authors acknowledge funding received from TUBITAK (Project no. 118Z116, 122Z320) and METU BAP (Project no. GAP-108-2022-10893) to SB. Chuanpit Hampel, Kerstin Huebner, Regine Schneider-Stock, and Katharina Erlenbach-Wuensch from the Institute of Pathology, University Hospital Erlangen, Friedrich Alexander University, Erlangen-Nürnberg, Germany, are gratefully acknowledged for the training and data provided for the CAM assay. COST action CA17118 is acknowledged for funding a Short-Term Scientific Mission fellowship to HHH.

References

1. Metur SP, Lei Y, Zhang Z, Klionsky DJ (2023) Regulation of autophagy gene expression and its implications in cancer. J Cell Sci 136:jcs260631
2. Ravanan P, Srikumar IF, Talwar P (2017) Autophagy: the spotlight for cellular stress responses. Life Sci. https://doi.org/10.1016/j.lfs.2017.08.029
3. Ma W, Lu Y, Jin X et al (2024) Targeting selective autophagy and beyond: from underlying mechanisms to potential therapies. J Adv Res. https://doi.org/10.1016/j.jare.2024.05.009
4. Klionsky DJ, Abdel-Aziz AK, Abdelfatah S et al (2021) Guidelines for the use and interpretation of assays for monitoring autophagy (4th edition) 1. Autophagy 17(1):1–382
5. Ribatti D (2017) The chick embryo chorioallantoic membrane (CAM) assay. Reprod Toxicol 70. https://doi.org/10.1016/j.reprotox.2016.11.004
6. Ranjan RA, Muenzner JK, Kunze P et al (2023) The chorioallantoic membrane xenograft assay as a reliable model for investigating the biology of breast cancer. Cancers (Basel) 15. https://doi.org/10.3390/cancers15061704
7. Pastorino GA, Sheraj I, Huebner K et al (2024) A partial epithelial-mesenchymal transition signature for highly aggressive colorectal cancer cells that survive under nutrient restriction. J Pathol 262:347–361. https://doi.org/10.1002/path.6240
8. Hüsnügil HH, Güleç Taşkıran AE, Güderer I et al (2024) Lysosomal alkalinization in nutrient restricted cancer cells activates cytoskeletal rearrangement to enhance partial epithelial to mesenchymal transition. Transl Oncol 41:101860. https://doi.org/10.1016/j.tranon.2023.101860
9. Böhm J, Muenzner JK, Caliskan A et al (2019) Loss of enhancer of zeste homologue 2 (EZH2) at tumor invasion front is correlated with higher aggressiveness in colorectal cancer cells. J Cancer Res Clin Oncol 145:2227–2240. https://doi.org/10.1007/s00432-019-02977-1

Methods in Molecular Biology (2025) 2879: 113–122
DOI 10.1007/7651_2023_502

Published online: 07 November 2023

Keeping an Eye Out for Autophagy in the Cornea: Sample Preparation for Single-Cell RNA-Sequencing

Han Peng, Nihal Kaplan, Min Liu, Huimin Jiang, and Robert M. Lavker

Abstract

Single-cell RNA-sequencing (scRNA-seq) is a powerful technique that can barcode individual cells and thus used to obtain a gene expression profile for every individual cell within a tissue. This makes scRNA-seq an excellent method for characterizing rare cell populations such as stem cells. We describe how scRNA-seq can be utilized to examine limbal epithelial stem cell population as well as investigate the contribution of autophagy to the function of limbal epithelial stem cells. To accomplish this, we used the Beclin1 heterozygous (Beclin1 het) mouse, a well-established model of autophagy deficiency. We provide a protocol that we developed for the isolation of viable, single-cell suspensions of limbal/corneal tissues, as well as the analysis of scRNA-seq data.

Key words Beclin1 heterozygous mouse, Cell cluster identification, Dissection of cornea, Enzyme digestion, Single-cell suspension

1 Introduction

Autophagy is a process by which cells break down and recycle their own cellular components, including damaged proteins and organelles [1]. Thus, autophagy plays an important role in maintaining cellular homeostasis under resting settings and promotes survival during stress conditions [2–8]. The functions of autophagy in the maintenance of homeostasis and in stress response are particularly critical for the health of stem cells, which are normally relatively quiescent [9–12] and therefore cannot eliminate damaged cellular components by proliferation. In addition, autophagy plays an important role in various physiological conditions, and abnormalities of autophagy have been associated with many pathological conditions, including cancer, neurodegenerative diseases, and inflammatory disorders [13–24]. In corneal/limbal epithelia, dysregulation of autophagy has been associated with the development of several pathological conditions, including dry eye syndrome and disrupted corneal epithelial wound healing [13–22]. For example,

Han Peng and Nihal Kaplan contributed equally with all other contributors.

transient stimulation of autophagy by treatment of lacritin, a tear protein that is selectively deficient in dry eye, restores corneal epithelial cell homeostasis and promotes wound healing [22].

We have demonstrated that the EphA2/PLD1 pathway coordinately regulated end-stage autophagy and ciliogenesis, which contributes to the maintenance of corneal epithelial homeostasis [7]. This suggests a positive role for autophagy in corneal tissue maintenance. We have also demonstrated that miRs-103/107 are preferentially expressed in the limbal epithelium [25] where the corneal epithelial stem cells reside [12]. Depletion of miRs-103/107 in the stem cell-enriched limbal epithelial cells in vitro and in vivo resulted in an inhibition of end-stage autophagy [6]. Inhibition of autophagy in vitro negatively affects epithelial proliferative capacity [6], a hallmark of epithelial stem cells. Inhibition of autophagy in limbal epithelium in vivo results in an abnormal response of stem cells to wound healing [6], suggesting that the ability of limbal epithelial stem cells to positively respond to wounding may be compromised. Taken together, these findings reveal that autophagy is essential for the maintenance of homeostasis and proper function of limbal epithelial stem cells. However, knowledge of the molecular signaling which underlies autophagy functions in limbal epithelial stem cells is incomplete.

Limbal epithelial stem cells are a rare cell population [9–12], making their characterization challenging. In addition, the limbal/corneal epithelial basal layer is heterogeneous [10–12, 26], further complicating the ability to isolate a discrete stem cell population. Single-cell RNA-sequencing (scRNA-seq) techniques can barcode individual cells and thus provide a gene expression profile for each cell in the limbal epithelial population [8]. Therefore, we utilized scRNA-seq to characterize the limbal epithelial stem cell population and interrogate the contribution of autophagy to the function of limbal epithelial stem cells using the Beclin1 heterozygous (Beclin1 het) mouse, a well-established mouse model of autophagy deficiency. To perform scRNA-seq, we developed a protocol specifically designed for the isolation of viable, single-cell suspensions of limbal/corneal epithelial cells.

2 Materials

1. Mice: C57BL6 wild-type mouse and Beclin1 heterozygous mouse.
2. Ca2+/Mg2+ free PBS (Sigma).
3. Collagenase A solution:

   20 mg collagenase A (Millipore Sigma Cat# 10103578001) is dissolved in DMEM containing 10% FBS,

50 μg/mL gentamicin, and 1.25 μg/mL amphotericin B, and the solution is filtered through a 0.22 μm syringe filter.

4. DMEM containing 10% FBS.
5. 0.25% trypsin in 1 mM EDTA-PBS
6. Sterile Ca2+-/Mg2+-free PBS containing 0.04% BSA,
   0.04 g BSA is added to 100 mL Ca2+-/Mg2+-free PBS and the solution is filtered through a 0.22 micron syringe filter.
7. FLOWMI CELL STRAINERS (SP Bell-Art)
   https://www.belart.com/flowmi-cell-strainers-for-1000-l-pipette-tips.html

3 Methods

3.1 Dissecting the Mouse Cornea

1. After euthanasia, mouse eyeballs were proptosed from the socket and a curved forceps was used to remove the eyeballs (Fig. 1).
2. The eyeballs were rinsed with PBS, and dissection of the corneal rim was conducted under a stereo microscope. Initially, eyeballs were penetrated with a pair of sharp forceps to release pressure and cut about 1 mm behind the limbus toward the back of the eyeball (red line in Fig. 2) using ophthalmological surgical scissors (Fig. 2). This yields a corneal button.
3. Cornea buttons were dissected and placed in PBS with the epithelium facing downward and corneal endothelium upward. The iris was scraped off using the round handle side of a forceps. In this way, a limbal/corneal rim containing primarily epithelium, stroma, and endothelium is obtained (Fig. 3).

3.2 Mouse Cornea Single-Cell Isolation by Collagenase A/Trypsin

1. Two corneal rims with limbal and limbal-conjunctival junction (iris scraped off) from a single mouse were digested in a *37 °C incubator for 1 h* with *1 mL of 2 mg/mL collagenase A* (Millipore Sigma Cat# 10103578001) in DMEM containing 10% FBS, 50 μg/mL gentamicin, and 1.25 μg/mL amphotericin B in a 12 well culture dish. Briefly, corneal rims were placed in 1 mL collagenase A in a well of a 12 well plate. Using retractable microdissection scissors, rims were chopped into small pieces to increase the surface area for digestion. The digestion mixture was incubated at *37 °C* and swirled every 15 min to increase exposure to the enzyme. After 30 min of incubation and at the end of the digestion, the digestion mixture was perturbed by pipetting up and down for 5 times with a 1 mL pipette tip. At this stage, attention should be paid to ensure that all cornea pieces are removed from the pipette tip.
2. After a 1 h incubation, the small, digested limbal/corneal clusters were centrifuged at 300 × g for 5 min and resuspended

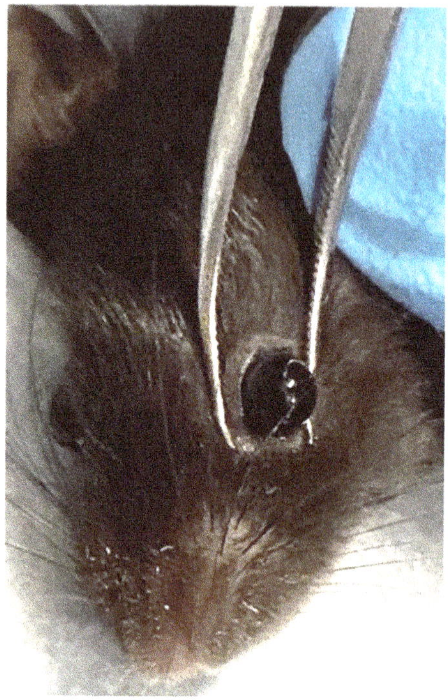

Fig. 1 Enucleation of eyeball

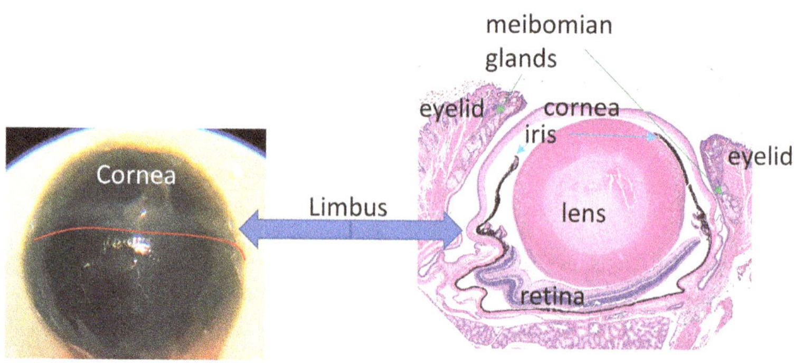

Fig. 2 Dissecting corneal rim

in DMEM containing 10% FBS to rinse off the collagenase. The clusters were centrifuged again and resuspended in 10 mL Ca2+-/Mg2+-free PBS 1× to remove any serum containing media.

3. After the PBS rinse and repeated centrifugation, clusters were further digested with 1 mL of *0.25% trypsin* in 1 mM EDTA-PBS at *37 °C for 10 min* to yield viable single cells. (Gently pipette up and down to disassociate cells *5 min* after *trypsin* treatment and then continue for another 5 min.).

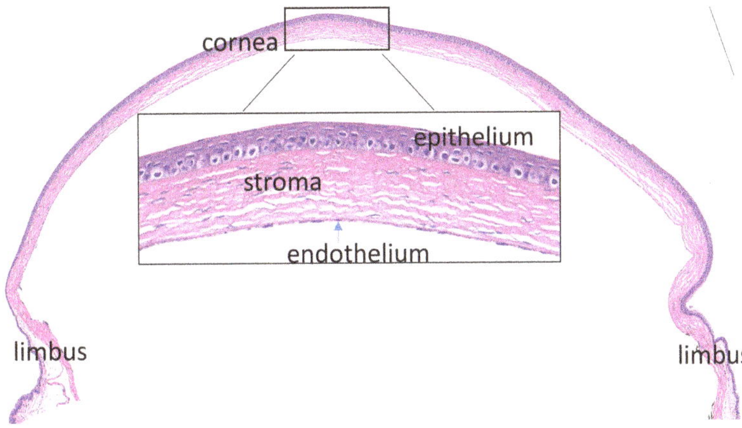

Fig. 3 Corneal/limbal rim after removing retina, lens, and iris

4. Add 5 mL DMEM+10% FCS media to deactivate the trypsin. Centrifuge at 300 × g for 5 min to remove the trypsin from the cells.
5. Centrifuge and resuspend cells 2× with sterile Ca2+-/Mg2+-free PBS containing 0.04% BSA.
6. After a final centrifugation to pellet the cells, cells are resuspended in 60 μL of PBS w/o Ca-Mg containing 0.04% BSA and filtered into an Eppendorf tube using FLOWMI CELL STRAINERS (SP Bell-Art).

3.3 Yield of Single-Cell Suspensions and Quality Check

At the end of the procedure, the yield and viability were examined under an inverted microscope using trypan blue staining. Two corneal buttons should yield 50–100 K cells with at least 80% viability. Finally, the single-cell suspension should immediately be sent out for single-cell RNA-sequencing using a 10× Genomics platform.

3.4 Analysis of Single-Cell RNA-Sequencing Data from the Corneal Samples

Raw sequencing data were processed to generate matrix data using Cell Ranger following the instructions from 10× Genomics. The matrix data were further processed to yield clustering using Seurat following the Vignettes for Seurat (https://satijalab.org/seurat/articles/integration_introduction.html). Using the genes that were specifically detected in each cluster, we identified cell types present in our datasets. By analyzing the differentially expressed genes in cells from wild type and Beclin1 het (autophagy deficient) mouse cornea, we determined how autophagy affects limbal/corneal epithelial biology.

1. Identify cell types using cell type markers that are conserved in both wild type and Beclin1 het cells.

   In the cells from wild type and Beclin1 het corneas, single-cell RNA-seq yielded 10 clusters [8]. Three clusters were identified as corneal stromal keratocytes based on the exclusive expression of high levels of vimentin (*Vim*), lumican (*Lum*), and keratocan (*Kera*), which are well established genes associated with mesenchymal tissues [8, 27]. Interestingly, within the keratocyte cluster, a small subcluster of cells was classified as corneal stromal stem cells (CSSCs) due to high levels of *Fhl1* and *Cd90* (Thy1), which are considered CSSC markers [28].

The remaining clusters expressed high levels of epithelial markers such as *Pax6*, adherens, and tight junctional genes, as well as keratin genes [29–31]. Among the epithelial clusters, a stem/early TA cell population was identified based on (i) high expression of Lrig1, a marker for slow cycling limbal epithelial stem cells [32]; (ii) relatively low expression of the proliferation marker,Mki67, suggesting a quiescent phenotype; and (iii) low levels of connexins, which is a characteristic of stem cell-enriched tissues [25, 33–35]. Early TA cells are phenotypically and functionally similar to the stem cells [36]. This concept is supported by our scRNA-seq analysis since the stem/early TA cells were clustered into one group. Grouping stem cells and early TA cells together helps explain many observations concerning proliferation and repair. Having identified a discrete population of stem/early TA cells, we investigated novel regulators that were potentially involved in limbal epithelial stem cell homeostasis. We identified a group of genes which were highly expressed in the stem/early TA cell population. Interestingly, the association of these genes with stem cell physiology had not previously been appreciated. For example, one of the highest expressed genes was thioredoxin-interacting protein (*Txnip*). TXNIP positively regulates the p53-mediated antioxidant pathway and thus is critical in the maintenance of the stem cell response to oxidative stress [37]. Immunostaining confirmed that TXNIP was exclusively detected in a subset of limbal basal cells [8], which mimicked the spatial distribution of limbal epithelial stem/early TA cells that are believed to be scattered in the limbal basal layer. This observation suggests that TXNIP may also have an antioxidant effect in limbal epithelial stem cells. The heterogeneous expression of TXNIP in limbal epithelial basal cells would be most difficult to detect using bulk transcriptomic approaches; however, it was clearly revealed by our single-cell RNA-seq analysis.

2. Define cell-type specific responses to autophagy deficiency by analysis of differentially expressed genes (DEGs)

   Seurat FindMarkers function generated DEGs in each cluster between wild-type and Beclin1 het corneas. These DEGs were subjected to gene-annotation enrichment analysis (GO by DAVID [38, 39]). Not surprisingly, GO analysis of DEGs in the stem/early TA cell cluster indicated an enrichment for cell fate commitment and regulation of pluripotency of stem cells. Conversely, GO analysis of the mature TA cell cluster indicated that DEGs were enriched for cell division and cell cycle properties. Among them, PDZ-binding kinase (PBK) was identified as a novel regulator of quiescence in stem/early TA cells [8]. PBK induced cell cycle arrest in corneal epithelial cells specifically at the G2/M phase. The gene that showed the highest increase in the mature TA cell cluster of Beclin1 het mice was the transcription factor ATF3, which negatively controls the proliferation of limbal epithelial cells [8]. These results suggested that autophagy regulated epithelial proliferation via increased PBK and H2AX expression and decreased ATF3 expression, resulting in stem/TA cell activation.

3.5 Conclusion

Single-cell RNA-sequencing provides an opportunity to examine in detail the transcriptional profile of mesenchymal cells, as well as limbal and corneal epithelial cells at the single-cell level. One significant advancement is the use of single-cell RNA-sequencing, which enables the direct assessment of rare cell types like stem/early TA cells and mature TA cells. These cell populations play critical roles in maintaining limbal and corneal homeostasis.

The translational relevance of scRNA-seq is clearly demonstrated since it enables a comparative investigation of discrete cell populations derived from the limbus/cornea of wild-type and Beclin1 het (autophagy-deficient) mice [8]. The scRNA-seq approach seems to be highly suitable for studying corneal abnormalities associated with excessive inflammation (e.g., dry eye diseases, alkali burn, sulfur mustard injury). By utilizing scRNA-seq techniques, it should be relatively straightforward to examine changes in inflammation in the limbal/corneal region of both wild-type and autophagy-deficient mouse models as well as gene profiling the infiltrated immune cells. scRNA-seq is also an ideal approach to investigate corneal diseases that involve dysregulation of stem cells (e.g., diabetic keratopathies and aniridia). For example, in diabetic patients, the expression of putative stem cell markers is reduced in limbal epithelial stem cells, suggesting that this cell population may be compromised [40]. By utilizing scRNA-seq techniques described in this study, a comparison between the stem/early TA cell and mature TA cell populations from both wild-type and mouse models of diabetes [41] could be conducted. Insights gained from these studies should not only enhance our understanding of how autophagy affects inflammation in the limbus/cornea in response to stress but also form a basis for developing novel treatments for excessive inflammation by modulation of autophagy.

4 Notes

1. Collagenase A and trypsin must be prewarmed at 37 °C.
2. Tissues should not be kept in trypsin more than 25 min.
3. In each step after centrifugations during rinsing, pipetting should be extremely gentle, no direct contact with cells should be made, and no bubbles should form.
4. Run the whole procedure as quickly as possible.

Acknowledgments

This research was supported by the National Institutes of Health Grants EY028560 and EY019463 (to R.M.L.), and EY032922 (to H.P.); a Dermatology Foundation research grant and Career Development Award (to H.P.); an Eversight research grant (to H. P.). Sequencing was performed by NUseq Core at Center for Genetic Medicine.

References

1. Klionsky DJ et al (2016) Guidelines for the use and interpretation of assays for monitoring autophagy (3rd edition). Autophagy 12:1–222. https://doi.org/10.1080/15548627.2015.1100356
2. Eskelinen EL, Saftig P (2009) Autophagy: a lysosomal degradation pathway with a central role in health and disease. Biochim Biophys Acta 1793:664–673
3. Martin LM et al (2019) Autophagy in corneal health and disease: a concise review. Ocul Surf 17:186–197. https://doi.org/10.1016/j.jtos.2019.01.008
4. Peng H, Park JK, Lavker RM (2017) Autophagy and macropinocytosis: keeping an eye on the corneal/limbal epithelia. Invest Ophthalmol Vis Sci 58:416–423
5. Peng H, Park JK, Lavker RM (2017) Eyeing autophagy and macropinocytosis in the corneal/limbal epithelia. Autophagy:1–3. https://doi.org/10.1080/15548627.2017.1287658
6. Park JK et al (2016) MicroRNAs-103/107 coordinately regulate macropinocytosis and autophagy. J Cell Biol 215:667–685. https://doi.org/10.1083/jcb.201604032
7. Kaplan N et al (2021) Ciliogenesis and autophagy are coordinately regulated by EphA2 in the cornea to maintain proper epithelial architecture. Ocul Surf 21:193–205. https://doi.org/10.1016/j.jtos.2021.06.006
8. Kaplan N et al (2019) Single-cell RNA transcriptome helps define the limbal/corneal epithelial stem/early transit amplifying cells and how autophagy affects this population. Invest Ophthalmol Vis Sci 60:3570–3583. https://doi.org/10.1167/iovs.19-27656
9. Sun TT, Tseng SC, Lavker RM (2010) Location of corneal epithelial stem cells. Nature 463:E10–E11; discussion E11. https://doi.org/10.1038/nature08805
10. Sun TT, Lavker RM (2004) Corneal epithelial stem cells: past, present, and future. J Investig Dermatol Symp Proc 9:202–207. https://doi.org/10.1111/j.1087-0024.2004.09311.x
11. Lavker RM, Tseng SC, Sun TT (2004) Corneal epithelial stem cells at the limbus: looking at some old problems from a new angle. Exp Eye Res 78:433–446
12. Cotsarelis G, Cheng SZ, Dong G, Sun TT, Lavker RM (1989) Existence of slow-cycling limbal epithelial basal cells that can be preferentially stimulated to proliferate: implications on epithelial stem cells. Cell 57:201–209
13. Kempuraj D, Mohan RR (2022) Autophagy in extracellular matrix and wound healing modulation in the Cornea. Biomedicines 10. https://doi.org/10.3390/biomedicines10020339

14. Deretic V, Saitoh T, Akira S (2013) Autophagy in infection, inflammation and immunity. Nat Rev Immunol 13:722–737. https://doi.org/10.1038/nri3532
15. Liu Z et al (2020) Autophagy activation protects ocular surface from inflammation in a dry eye model in vitro. Int J Mol Sci 21. https://doi.org/10.3390/ijms21238966
16. Benischke AS et al (2017) Activation of mitophagy leads to decline in Mfn2 and loss of mitochondrial mass in Fuchs endothelial corneal dystrophy. Sci Rep 7:6656. https://doi.org/10.1038/s41598-017-06523-2
17. Choi SI et al (2012) Impaired autophagy and delayed autophagic clearance of transforming growth factor beta-induced protein (TGFBI) in granular corneal dystrophy type 2. Autophagy 8:1782–1797. https://doi.org/10.4161/auto.22067
18. Liu Z et al (2020) Trehalose induces autophagy against inflammation by activating TFEB signaling pathway in human corneal epithelial cells exposed to hyperosmotic stress. Invest Ophthalmol Vis Sci 61:26. https://doi.org/10.1167/iovs.61.10.26
19. Ma S et al (2019) Corneal autophagy and ocular surface inflammation: a new perspective in dry eye. Exp Eye Res 184:126–134. https://doi.org/10.1016/j.exer.2019.04.023
20. Wei C et al (2022) Enhanced autophagy alleviated corneal allograft rejection via inhibiting NLRP3 inflammasome activity. Am J Transplant. https://doi.org/10.1111/ajt.16968
21. Stuard WL, Titone R, Robertson DM (2022) IGFBP-3 functions as a molecular switch that mediates mitochondrial and metabolic homeostasis. FASEB J 36:e22062. https://doi.org/10.1096/fj.202100710RR
22. Karnati R, Talla V, Peterson K, Laurie GW (2016) Lacritin and other autophagy associated proteins in ocular surface health. Exp Eye Res 144:4–13. https://doi.org/10.1016/j.exer.2015.08.015
23. Nixon RA (2013) The role of autophagy in neurodegenerative disease. Nat Med 19:983–997. https://doi.org/10.1038/nm.3232
24. Mulcahy Levy JM, Thorburn A (2020) Autophagy in cancer: moving from understanding mechanism to improving therapy responses in patients. Cell Death Differ 27:843–857. https://doi.org/10.1038/s41418-019-0474-7
25. Peng H et al (2015) microRNA-103/107 family regulates multiple epithelial stem cell characteristics. Stem Cells. https://doi.org/10.1002/stem.1962
26. Lehrer MS, Sun TT, Lavker RM (1998) Strategies of epithelial repair: modulation of stem cell and transit amplifying cell proliferation. J Cell Sci 111(Pt 19):2867–2875
27. Syed-Picard FN et al (2015) Dental pulp stem cells: a new cellular resource for corneal stromal regeneration. Stem Cells Transl Med 4:276–285. https://doi.org/10.5966/sctm.2014-0115
28. Funderburgh ML, Du Y, Mann MM, SundarRaj N, Funderburgh JL (2005) PAX6 expression identifies progenitor cells for corneal keratocytes. FASEB J 19:1371–1373. https://doi.org/10.1096/fj.04-2770fje
29. Rubsam M et al (2018) Adherens junctions and desmosomes coordinate mechanics and signaling to orchestrate tissue morphogenesis and function: an evolutionary perspective. Cold Spring Harb Perspect Biol 10. https://doi.org/10.1101/cshperspect.a029207
30. Sun TT et al (1983) Keratin classes: molecular markers for different types of epithelial differentiation. J Invest Dermatol 81:109s–115s
31. Tsukita S, Tanaka H, Tamura A (2019) The Claudins: from tight junctions to biological systems. Trends Biochem Sci 44:141–152. https://doi.org/10.1016/j.tibs.2018.09.008
32. Parfitt GJ et al (2015) Immunofluorescence tomography of mouse ocular surface epithelial stem cells and their niche microenvironment. Invest Ophthalmol Vis Sci 56:7338–7344. https://doi.org/10.1167/iovs.15-18038
33. Chen Z, Evans WH, Pflugfelder SC, Li DQ (2006) Gap junction protein connexin 43 serves as a negative marker for a stem cell-containing population of human limbal epithelial cells. Stem Cells 24:1265–1273
34. Matic M, Evans WH, Brink PR, Simon M (2002) Epidermal stem cells do not communicate through gap junctions. J Invest Dermatol 118:110–116
35. Matic M et al (1997) Stem cells of the corneal epithelium lack connexins and metabolite transfer capacity. Differentiation 61:251–260
36. Potten CS, Loeffler M (1990) Stem cells: attributes, cycles, spirals, pitfalls and uncertainties. Lessons for and from the crypt. Development 110:1001–1020. https://doi.org/10.1242/dev.110.4.1001
37. Jung H et al (2013) TXNIP maintains the hematopoietic cell pool by switching the function of p53 under oxidative stress. Cell Metab 18:75–85. https://doi.org/10.1016/j.cmet.2013.06.002
38. da Huang W, Sherman BT, Lempicki RA (2009) Bioinformatics enrichment tools: paths toward the comprehensive functional

analysis of large gene lists. Nucleic Acids Res 37:1–13. https://doi.org/10.1093/nar/gkn923

39. da Huang W, Sherman BT, Lempicki RA (2009) Systematic and integrative analysis of large gene lists using DAVID bioinformatics resources. Nat Protoc 4:44–57. https://doi.org/10.1038/nprot.2008.211

40. Saghizadeh M et al (2011) Alterations of epithelial stem cell marker patterns in human diabetic corneas and effects of c-met gene therapy. Mol Vis 17:2177–2190

41. Ljubimov AV (2017) Diabetic complications in the cornea. Vis Res 139:138–152. https://doi.org/10.1016/j.visres.2017.03.002

Methods in Molecular Biology (2025) 2879: 123–138
DOI 10.1007/7651_2024_523

Published online: 19 March 2024

Evaluation of Autophagy in Conjunctival Fibroblasts

Parvaneh Mehrbod, Paola Brun, Umberto Rosani, Andrea Leonardi, and Saeid Ghavami

Abstract

Vernal keratoconjunctivitis (VKC) is a serious eye allergy characterized by poorly understood pathogenic mechanisms and a lack of effective treatments. Autophagy, a process involved in both triggering and suppressing immune and inflammatory responses, plays a role in VKC's pathophysiology. Understanding autophagy's involvement in VKC could lead to new treatment possibilities, such as utilizing specific topical substances to induce or inhibit autophagy and prevent severe complications of this eye condition. In our current protocol, we present a robust methodology established in our laboratory for studying autophagy in primary conjunctival fibroblasts. We assess autophagy through techniques like immunocytochemistry, immunoblotting, and qPCR.

Key words Autophagy flux, Chloroquine, Vernal keratoconjunctivitis

1 Introduction

Macroautophagy, also known as autophagy, is a meticulously controlled catabolic process preserved throughout evolution, functioning at baseline levels in the majority of cells to maintain cellular equilibrium by breaking down surplus proteins and eliminating impaired organelles through lysosomal degradation [1–4]. Yet, it can swiftly escalate in reaction to various metabolic stresses such as nutrient deficiency, oxygen deprivation, or heightened metabolic demands associated with rapid cell growth, all aimed at preserving cellular balance [5–8]. Recent advancements reveal a nuanced and multifaceted relationship between the autophagy pathway and immunity [9, 10].

The autophagy pathway plays a role in both triggering and dampening immune and inflammatory reactions. Conversely, chemokines and cytokines can either stimulate or hinder the autophagy pathway in mammalian cells [11–14]. MyD88 and TRIF (Toll-like receptor adaptors) along with HGB1 disrupt the interaction

Authors Parvaneh Mehrbod and Paola Brun have co-first authorship

Authors Andrea Leonardi and Saeid Ghavami have co-senior authorship

between BECLIN-1 and BCL2 by binding to BECLIN-1 and displacing BCL2 [15, 16]. Additional evidence regarding the regulation of autophagy and immune signaling molecules was presented through genome-wide siRNA screening [17]. In this analysis, 219 genes were identified that inhibit basal autophagy, operating independently of the mammalian target of rapamycin complex 1 (*mTORC1*). These genes encompassed cytokines (*CLCF1, LIF, IGF1, FGF2*), a chemokine (*SDF1*, also recognized as *CXCL12*), and cellular signaling molecules influenced by cytokines (*STAT3*) [17]. Considering the broad role of autophagy in maintaining cellular balance and modulating immune and inflammatory signaling, alterations in autophagy levels induced by cytokines in immune cells could have a significant impact on immunity and inflammation [9].

Studies have demonstrated the essential role of autophagy in the secretion of bronchial epithelial mucus in a Th2 model of asthma [18]. Autophagy might also participate in airway remodeling [19], presenting a potentially innovative focus in treating steroid-resistant severe asthma [20], indicating its involvement in allergic conditions [14, 18].

Inflammation and autophagy are pivotal biological processes engaged in various physiological and pathological scenarios. Pro-inflammatory cytokines exert autophagy-mediated effects across diverse inflammatory conditions. The interplay between autophagy and inflammation represents a fundamental mechanism regulating the function of both the innate and adaptive immune systems [21]. Autophagy has the capacity to both regulate and be influenced by a broad spectrum of pro-inflammatory cytokines [22, 23]. These interactions have the potential to either ameliorate or exacerbate inflammatory states. Consequently, the interplay between autophagy and inflammatory cytokines holds significant importance and serves as a primary focus of research in modulating autophagy and immunity [22].

Vernal keratoconjunctivitis (VKC) is a serious and uncommon chronic eye allergy that predominantly impacts children and teenagers residing in warm regions, significantly disrupting their daily lives and overall well-being [24]. The condition is marked by severe ocular symptoms including itching, sensitivity to light, a feeling of foreign objects in the eye, redness of the conjunctiva, and excessive mucus production. These symptoms are often aggravated during or after outdoor activities. Additionally, patients may develop large papillae on the upper eyelid or inflammation and nodules near the cornea. This discomfort not only significantly impacts the quality of life of patients and their caregivers but also poses risks of complications such as cataracts and glaucoma due to long-term corticosteroid treatment [25].

VKC has been recognized as an allergic reaction mediated by both IgE and T cells, wherein the CD4+ T helper (Th)2 inflammatory pattern, prompted by IL-4, plays a significant role. This pattern involves the secretion of IL-4, IL-5, and IL-13 by cells and has been shown to influence different levels of corneal involvement and tissue restructuring [26–28].

Similar to other atopic conditions like atopic dermatitis and asthma, the precise trigger for VKC often remains unknown in the majority of cases. A range of nonspecific factors, including warm climates, sunlight, dry air, oxidative stress, and potential panallergens, can worsen or sustain ocular inflammation. In VKC patients, there is an increase in various cytokines (including Th2-type and Th1-type), chemokines, growth factors, and enzymes [29, 30], suggesting VKC is associated with unique patterns of inflammation in the conjunctiva.

Studies have demonstrated that the activities and characteristics of conjunctival fibroblasts and epithelial cells can be altered in laboratory settings by various mediators implicated in ocular allergic reactions, including histamine, IL-1, IL-4, and TNF-α [30–35]. These cytokines also play a role in triggering autophagy, underscoring the interconnectedness between autophagy, inflammatory agents, and the development of VKC.

Histamine, a biogenic amine with diverse biological roles [36], triggers an elevation in the ratio of LC3II/LC3I and stimulates the production of intracellular reactive oxygen species (ROS), consequently inducing autophagy that is stress-induced within the endoplasmic reticulum (ER) [37]. TNF-α, a multifaceted cytokine known for its modulation of pro-inflammatory reactions, is implicated in interfering with the process of autophagy [38]. IL-1β binds to IL-1 receptor 1, initiating inflammatory cascades by stimulating the production of IL-1α and IL-23. This, in turn, prompts the formation of autophagosomes, which may induce autophagy as a component of a negative feedback mechanism to mitigate the inflammatory response [39]. IL-4, a crucial Th2 cytokine, serves as a key effector in the immune response. Harris et al. demonstrated that unlike pro-inflammatory cytokines, the secretion of anti-inflammatory cytokines like IL-4 inhibits the induction of autophagy by activating mTOR [40].

Hence, in this chapter, we introduced a novel methodology for assessing autophagy in conjunctivitis fibroblasts in response to inflammatory mediators. We employed established techniques including immunohistochemistry, qPCR, and Western Blotting (WB).

2 Materials

2.1 Conjunctival Biopsies

1. Optimal cutting temperature (OCT) compound (Histo-Line Laboratories, Pantigliate, MI, Italy).

2.2 Immunohistochemistry (IHC)

1. Superfrost plus slides (Thermo scientific, Waltham, MA, USA).
2. Formaldehyde (VWR International, Milan, Italy).
3. TRIZMA (2-Amino-2-(hydroxymethyl)-1,3-propanediol, Tris–HCl (hydroxyamino methane) maleate buffer 0.05 M, pH 7,6 (Sigma, St. Louis, MO, USA).
4. Normal horse serum (Vector Laboratories Inc, Burlingame, CA, USA – S 2000).
5. Normal rabbit serum (Vector Laboratories Inc, Burlingame, CA, USA – S 5000).
6. Normal goat serum (Vector Laboratories Inc, Burlingame, CA, USA – S 1000).
7. Secondary antibody anti-mouse made in horse (Vector Laboratories Inc, Burlingame, CA, USA – BA 2000).
8. Secondary antibodies anti-rabbit made in goat (Vector Laboratories Inc, Burlingame, CA, USA – BA 1000).
9. Secondary antibodies anti-goat made in rabbit (Vector Laboratories Inc, Burlingame, CA, USA – BA 5000)
10. Rabbit, goat, or mouse non-specific immunoglobulins (Santa Cruz Biotechnologies, Santa Cruz, CA, USA).
11. ABC kit HRP Elite (Vectstain, Vector Laboratories Inc, Burlingame, CA, USA – PK6100).
12. Mayer hematoxylin (AppliChem, Darmstadt, Germany).

2.3 Cell Cultures and Treatments

1. Type I collagenase (Worthington Biochemical, Lakewood, NJ, USA).
2. Wong-Kilbourne derived Chang (ChWK) conjunctival epithelial cells (American Type Culture Collection (ATCC), Manassas, VA, USA).
3. Human monocytes U937 (Thermo Scientific (Wilmington, DE, USA).
4. Dulbecco's Modified Eagle's Medium (DMEM, Gibco; Grand Island, NY, USA).
5. RPMI 1640 medium (Gibco, Grand Island, NY, USA).
6. 10% Fetal Bovine Serum (FBS) (Gibco, Grand Island, NY, USA).
7. Penicillin/streptomycin (P/S, Euroclone, Milan, Italy).
8. Phorbol-12-myristate-134 acetate (PMA, Sigma, St. Louis, MO, USA).
9. Lipopolysaccharide (LPS, Sigma, St. Louis, MO, USA).
10. Histamine (Sigma, St. Louis, MO, USA) and IL-1β, IL-4, and TNF-α (Peprotech, London, UK).

11. Insulin, Transferrin, Selenium supplement (ITS, GIBCO, Grand Island, NY, USA).
12. Chloroquine (CQ, Sigma, St. Louis, MO, USA).

2.4 Single and Double Staining Immunofluorescence

1. Coverslips (Thermo scientific, Waltham, MA, USA).
2. Human primary conjunctival fibroblasts obtained as described above.
3. TNF-α (Peprotech, London, UK).
4. DMEM medium (Gibco, Grand Island, NY, USA).
5. Insulin, Transferrin, Selenium supplement (ITS, GIBCO, Grand Island, NY, USA).
6. Formaldehyde (VWR International, Milan, Italy).
7. PBS, Triton X-100, and BSA (Sigma, St. Louis, MO, USA).
8. Anti-goat IgG conjugated with Alexa Fluor 488 (Invitrogen, Thermo scientific, Waltham, MA, USA).
9. Anti-rabbit IgG conjugated with Alexa Fluor 647 (Invitrogen, Thermo scientific, Waltham, MA, USA).
10. Hoechst (Sigma, St. Louis, MO, USA).
11. Glycerol and DABCO (Sigma, St. Louis, MO, USA).

2.5 RNA Isolation and qPCR Analysis

1. TRIzol (Life Technologies, Carlsbad, CA, USA).
2. Oligo(dT) (Life Technologies, Carlsbad, CA, USA).
3. SuperScript II Reverse Transcriptase 200 U (Life Technologies, Carlsbad, CA, USA).
4. SYBR Green I dye (Roche, Mannheim, Germany).

2.6 Western Blotting

1. Triton X-100, deoxycholic acid, EDTA, Tween 20 and PBS (Sigma, St. Louis, MO, USA).
2. Bradford reagent (Serva, Heidelberg, Germany).
3. Acrylamide solution (AppliChem, Darmstadt, Germany).
4. Tris–HCl pH 6.8, Glycerol, sodium dodecyl sulfate (SDS), β-mercaptoethanol and bromophenol blue, glycine and Methanol (Sigma, St. Louis, MO, USA).
5. Dry milk (AppliChem, Darmstadt, Germany).
6. Horseradish Peroxidase (HRP)-conjugated anti-mouse IgG (GE Healthcare, Buckinghamshire, UK).
7. Horseradish Peroxidase (HRP)-conjugated anti-rabbit IgG and anti-goat IgG (R&D Systems, Minneapolis, MN, USA).
8. ECL Plus (GE Healthcare, Buckinghamshire, UK).
9. Anti-β-actin antibody (Sigma, St. Louis, MO, USA).

3 Methods

3.1 Conjunctival Biopsies

1. Prepare conjunctival tissues from active VKC patients and healthy age-matched control subjects who underwent eyelid surgery (*see* **Notes 1** and **2**).
2. Snap freeze upper tarsal conjunctival biopsies with optimal cutting temperature (OCT) compound (Histo-Line Laboratories, Pantigliate, MI, Italy) in liquid nitrogen and maintain at −80 °C until use for the immunohistochemistry (data not shown) and qPCR analysis (*see* **Note 3**).

3.2 Immunohistochemistry (IHC)

1. Stain two sections from each biopsy sample applying immunohistochemical methods with a panel of antibodies specific for some autophagic components (Table 1).
2. Prepare serial 5-μm-thick cryosections and fix them in 4% buffered formaldehyde and incubate with 0.3% hydrogen peroxide for 30 min. (*see* **Note 4**).
3. Block the samples with appropriate serum (1:20) in TRIZMA (2-amino-2-(hydroxymethyl)-1,3-propanediol, Tris–HCl (hydroxyamino methane, Sigma, St. Louis, MO, USA), and maleate 0.05 M, pH 7.6.
4. Incubate the sections with adequate dilution of the primary antibody in TRIZMA maleate for 1 h at room temperature in a humid chamber.
5. Demonstrate antibody binding using secondary antibodies anti-mouse (Vector, BA 2000), anti-rabbit (Vector BA 1000), or anti-goat (Vector, BA 5000) followed by ABC kit HRP Elite, PK6100 (Vectstain), and diaminobenzidine substrate (brown color) (Sigma, St. Louis, MO, USA).

Table 1
Autophagy markers

Target	Supplier	Catalogue number	Source	Methods
MAP1LC3A	Abcam	ab52628	Rabbit	IHC
LC3B	Abcam	ab64782	Rabbit	IHC
LC3B	Proteintech	18725-1-AP	Rabbit	IF
LC3B	Sigma	L7543	Rabbit	WB
Beclin-1	Cell signalling Tech.	#3495P	Rabbit	IHC, WB
Cathepsin D	Santa Cruz Biotech.	Sc-6486	Goat	IHC, WB, IF
P62	Millipore	MABC32	Mouse	WB
LAMP 1	Sino Biological Inc	11215-R107	Rabbit	IHC, IF

6. Counterstain the slides with hematoxylin and then rinse with tap water.
7. Perform negative controls using normal non-specific goat, mouse, or rabbit immunoglobulins (Santa Cruz Biotechnologies, Santa Cruz, CA, USA).
8. Observe with a light microscope (Leica Axioplan, Wetzlar, Germany) at 400× magnification.
9. Evaluate the positive reaction for all the studied antigens in the epithelium and sub-epithelial stroma of conjunctival tissues using a 0–3 score by two independent investigators (*see* **Note 5**).

3.3 Cell Cultures and Treatments

3.3.1 Methods per Cell Cultures and Inflammatory Conditioned Medium (CM) Production

1. Obtain primary conjunctival fibroblasts after digestion of VKC tarsal biopsies with 20 U/mL type I collagenase (Worthington Biochemical) at 37 °C for 12 h.
2. Prepare Wong-Kilbourne-derived Chang (ChWK) conjunctival epithelial cells from the American Type Culture Collection (ATCC; Manassas, VA, USA) as control cell lines.
3. Obtain human monocytes U937 from Thermo Scientific (Wilmington, DE, USA) for the production of the CM.
4. Grow the conjunctival cells under the standard cell culture practices in complete Dulbecco's modified Eagle's medium (DMEM) supplemented with 10% fetal bovine serum (FBS), 1% (v/v) penicillin/streptomycin (P/S), 2 mM glutamine at 37 °C, in a 5% humidified CO_2 atmosphere.
5. Culture U937 cells in suspension in complete RPMI 1640 medium (Gibco; Grand Island, NY) and sub-culture twice weekly by dilution using a seeding density of 10^6 cells/mL.
6. Differentiate U937 monocytes to macrophages by addition of 20 ng/mL of phorbol-12-myristate-134 acetate (PMA) into the cell culture medium for 48 h and subsequently incubate for 1 h with 1 μg/mL lipopolysaccharide (LPS).
7. Examine differentiation of monocytes to macrophages under an inverted phase-contrast microscope and also assess the mRNA expression of CD68 (the macrophage differentiation marker) by quantitative real-time PCR (qPCR).
8. Wash the cells and culture in complete DMEM for 24 h to produce the inflammatory conditioned medium (CM).
9. Subsequently withdraw, centrifuge, filter, and store at −80 °C.

3.3.2 Methods for Conjunctival Cells Treatments

1. Seed VKC conjunctival fibroblasts in 6-well culture dishes and treat for 24 h with the U937 culture medium (CM) (*see* **Note 6**).

2. Prior to each treatment, starve the cell cultures with DMEM in the presence of 1% insulin, transferrin, and selenium supplement (ITS) for 24 h, to make them quiescent (*see* **Note 7**).
3. Replace the medium by complete DMEM medium and detach the cells at 4, 10, and 24 h and freeze at −80 °C until lysis for protein and mRNA extraction to analyze the expression of autophagic markers (*see* **Note 8**).
4. In another series of experiments, seed both conjunctival cell types in 6-well culture dishes and treat with 10 ng/mL of histamine, 1 ng/mL of human recombinant IL-1β, 10 ng/mL IL-4, or 10 ng/mL TNF-α (*see* **Note 9**).
5. In some experiments, expose TNF-α-treated fibroblasts in 1% ITS to 100 μM CQ for 4, 10, and 24 h and subsequently analyze LC3B expression by Western Blotting (*see* **Note 10**).

3.4 Single and Double Staining Immunofluorescence

1. Seed conjunctival fibroblasts at a concentration of 5000 cell/cm^2 coverslips at the bottom of 24-well culture dishes and cultivate in the appropriate medium with 10% FBS.
2. Allow the cells to grow to 50% confluency and treat them with 10 ng/mL TNF-α in DMEM with 1% ITS for 4, 10 and 24 h.
3. Fix the cells in 4% formaldehyde and quench with 100 mM glycine in PBS and permeabilize in 0.25% Triton X-100.
4. Block the cells with 3% bovine serum albumin (BSA) in phosphate buffered saline (PBS) for 1 h at −20 °C.
5. Incubate the slides for the single reaction with the primary anti-LC3B antibody and for the double reaction with the rabbit anti-LAMP1 for 1 h and subsequently with the goat anti-Cathepsin D antibody overnight at 4 °C.
6. Wash the cells with PBS, then incubate them with the appropriate secondary antibodies (anti-goat IgG conjugated with Alexa Fluor 488; anti-rabbit IgG conjugated with Alexa Fluor 647) diluted at 1:200 in PBS.
7. Counterstain the nuclei with Hoechst (Sigma) for 5 min and mount the slides using 90% glycerol and 5% DABCO.
8. Examine the antibody reactivity with fluorescence microscope (Zeiss, Heidelberg, Germany) by two independent observers (*see* **Note 11**).

3.5 RNA Isolation and qPCR Analysis

1. Extract total RNA from conjunctival tissues, conjunctival cell cultures, or U937 cell cultures using TRIzol (Life Technologies, Carlsbad, CA, USA) according to the manufacturer's protocol.
2. Quantify the RNA concentration of each sample using NanoDrop 2000c Spectrophotometer (Thermo Scientific, Wilmington, DE, USA).

Table 2
Genes to investigate and primers to use

Gene (Accession number)	Gene name	Primer sequences
BECN1 (NM_003766.4)	Beclin-1	Fw 5′-ATGGAGAACCTCAGCCGAAG-3′ Rv 5′-CTGTAACTGTTCACTGT CATCCTCA-3′
BNIP3 (NM_004052.3)	BCL2 interacting protein 3	Fw 5′-TGCAGGGCTCCTGGGT -3′ Rv 5′-GGTGGGCTGTCACATGAG -3′
CTSD (NM_001909.4)	Cathepsin D	Fw 5′-CACAAGTACAACAGCGACA AGTC-3′ Rv 5′-TGGCTGCGATGAAGGTGA-3′
LAMP1 (NM_005561.3)	Lysosomal associated membrane protein 1	Fw 5′-CACCATCCAGGCGTACCTTT-3′ Rv 5′-TCACCGTCGTGTTGTCCTTC3′
MAP1LC3A (NM_032514.3)	Microtubule associated protein 1 light chain 3 alpha	Fw 5′-AGCGAGTTGGTCAAGAT CATCC-3′ Rv 5′-AAGCCGTCCTCGTCTTTCTC-3′
MAP1LC3B (NM_022818.4)	Microtubule associated protein 1 light chain 3 beta	Fw 5′-AGAACCACACCCAAAGTCCTC-3′ Rv 5′-TGCTGCTTTCCGTAACAACAC-3′
p62 (M88108.1)	p62	Fw 5′-CCCTGAGGAACAGATGGAGTC-3′ Rv 5′-CCTCTGGCGGGAGATGTG-3′
IL-6 (NM_000600.4)	Interleukin 6	Fw 5′-CACGCCTTGGACAGAATCCA-3′ Rv 5′-CCTCCAGCAACCAGGAATGT-3′
IL-1β (NM_000576.2)	Interleukin 1 beta	Fw 5′-GAATCTCCGACCACCA CTACAG-3′ Rv 5′-TGATCGTACAGGTGCATCGTG-3′
IL-4 (NM_000589.3)	Interleukin 4	FW 5′GGCTTGGTAAGCTGCACTGT-3′ Rv 5′-AATGGGTTGACCAAGGGTGC-3′
TNF-α (NM_000594.3)	Tumor necrosis factor alpha	Fw 5′-AAGCCTGTAGCCCATGTTGT-3′ Rv 5′-GGACCTGGGAGTAGATGAGGT-3′
PPIA (NM_021130.3)	Peptidylprolyl Isomerase A	Fw 5′-GGGCTTTAGGCTGTAGGTCAA-3′ Rv 5′-AACCAAAGCTAGGGAGAGGC-3′

3. Use the primers pairs (Table 2) (*see* **Note 12**).
4. Retro-transcribe 1 μg of total RNA into cDNA using oligo $(dT)_{12-18}$, 25 ng of primers, and 1 μL of SuperScript II Reverse Transcriptase 200 U (Life Technologies).
5. Perform the qPCR analysis with real-time PCR system (Rotor Gene RG-3000A, Corbett Research) using SYBR Green I dye (Roche, Mannheim, Germany) and combine sense and antisense primers at 300 nM final concentration.

6. Use the following cycling conditions: initial denaturation step of 10 min at 95 °C, 45 cycles of amplification consisting of 15 s at 95 °C, 30 s at 60 °C, and 30 s at 72 °C.
7. Evaluate genes expression with ΔΔCt method using peptidyl-prolyl isomerase A (PPIA) as reference gene (*see* **Note 13**).

3.6 Western Blotting

1. Analyze conjunctival fibroblasts and epithelial cells by Western Blotting for the presence of LC3B, Beclin-1, Cathepsin D, and p62 (Table 1).
2. Treat cell pellets with RIPA lysis buffer (1% v/v Triton X-100, 0.5% w/v deoxycholic acid, 10 mM EDTA in PBS, Sigma) supplemented with protease inhibitor cocktail for 45 min at 4 °C.
3. Remove the particulate material by centrifugation (12,000 g for 10 min, at 4 °C), collect the supernatants, and determine protein concentration using the Bradford Assay.
4. Add the lysates (20 μg) to the sample loading buffer (62.5 mM Tris–HCl pH 6.8, 10% v/v glycerol, 2% w/v sodium dodecyl sulfate, 5% v/v β-mercaptoethanol, and 0.1% w/v bromophenol blue) and denature at 98 °C for 5 min and separate by 10% sodium dodecyl sulfate-polyacrylamide gel electrophoresis (SDS-PAGE).
5. Transfer the proteins to the nitrocellulose membrane overnight at 4 °C with the constant current of 100 mA, in blotting buffer (25 mM Tris–HCl, 192 mM Glycine and 20% Methanol).
6. Block non-specific binding sites incubating the nitrocellulose membrane for 1 h at 23 °C in 5% w/v non-fat dry milk in 20 mM Tris–HCl [pH 7.6], 150 mM NaCl (TBS).
7. Incubate the nitrocellulose membrane with anti-LC3B, LC3A, Beclin-1, Cathepsin D, or p62 appropriately diluted antibodies in TBS, 0.1% Tween-20 (TBST) for 2 h at 23 °C.
8. Wash the membrane three times for 10 min in TBST (Tris-Buffered Saline and Tween 20, Sigma, St. Louis, MO, USA).
9. Incubate the membrane with HRP (Horseradish Peroxidase)-conjugated antibody (anti-mouse IgG, anti-rabbit IgG, or anti-goat IgG) diluted in TBST, for 1 h at 23 °C.
10. Reveal the antibody reaction by chemiluminescence using ECL Plus (GE Healthcare).
11. Incubate the blots sequentially at 23 °C for 2 h with monoclonal anti-β-actin antibody (Sigma) diluted at 1:4000 and process as described above.
12. Normalize the densitometry values of positive bands with the corresponding value of β-actin.

13. For the statistical analysis, compare the groups using Student's unpaired *t*-test. Compare the protein levels between VKC and control biopsies using the non-parametric Mann-Whitney U-test (*see* **Notes 14** and **15**).

3.7 Interpretation of Results

For autophagy evaluation, the autophagy proteins are first evaluated in conjunctivitis tissue using IHC. An example of immunohistochemical analysis for BECLIN in VKC biopsy is shown in Fig. 1.

Following that, the positive cells are cultured and modulated by diverse mediators involved in the ocular allergic reaction. Conducting qPCR, the results will show increase in the autophagy-related genes that guide us for the next step, IHC. Then, the colocalization of autophagosome and lysosome markers in IHC will guide us for another confirmation by WB. By switching to WB, LC3 lipidation, p62 degradation, and other related markers are evaluated. Application of CQ for confirmation of autophagy flux will be the final step. Figure 2 shows an example of WB analysis.

A schematic of the methods used is shown in Fig. 3.

4 Notes

1. None of the control subjects should use contact lenses or have inflammatory signs and symptoms or history of allergy.
2. Diagnosis of VKC should be based on the typical clinical history and evaluation of the signs and symptoms. All VKC patients must be free of topical mast cell stabilizers and/or anti-histamines for at least 3 days and free of topical

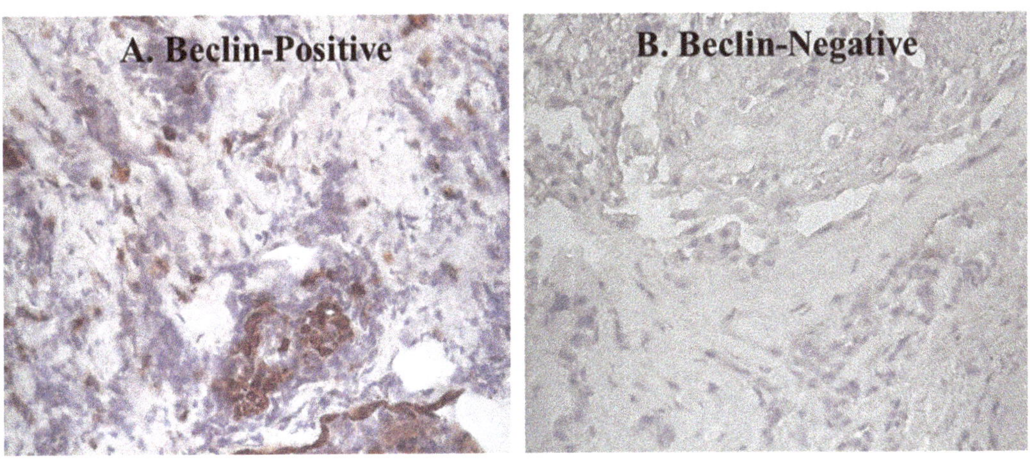

Fig. 1 Immunohistochemical analysis for Beclin in human vernal keratoconjunctivitis (VKC) biopsy. The biopsy samples were treated with the anti-Beclin antibody. (**a**) The positive reaction was obtained in the stroma of conjunctival tissue incubated with the anti-Beclin antibody for 1 h. (**b**) Negative control was performed using normal nonspecific rabbit immunoglobulins

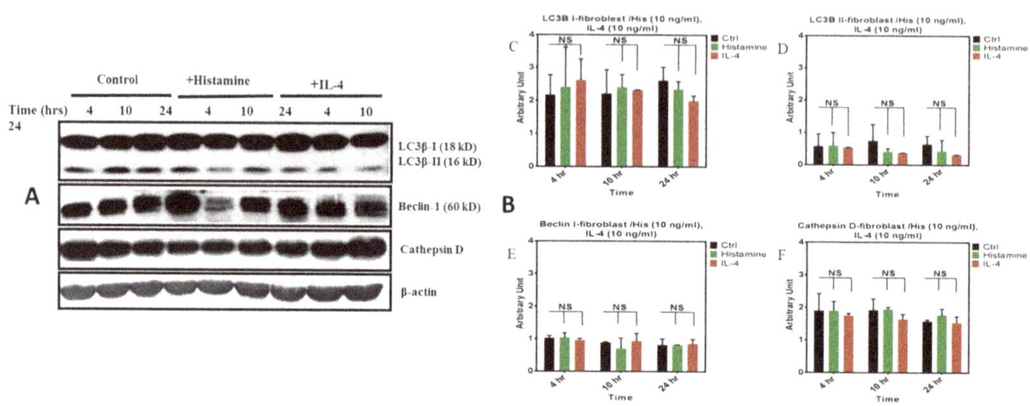

Fig. 2 Western Blot analysis of autophagic markers expression in conjunctival fibroblasts stimulated by Histamine and IL4. (**a**) Primary human conjunctival fibroblast cultures were stimulated by histamine or IL-4, and then the expression of LC3BI and II, Beclin, and Cathepsin D was evaluated by WB analysis at 4, 12, and 24 h after stimulation. (**b**) Densitometric values of positive bands were normalized with the corresponding value of β-actin. The results demonstrated that in vitro expression of LC3B, BECLIN, and Cathepsin D was not changed

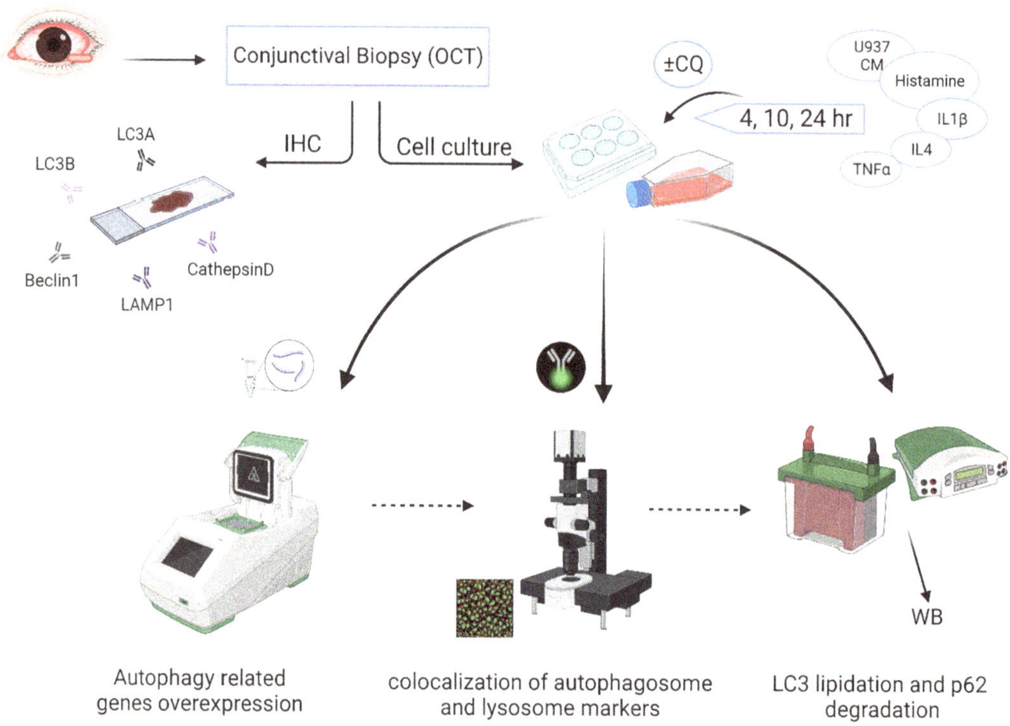

Fig. 3 The schematic of the methods used for this methodology. (Created with BioRender.com)

corticosteroids for at least 7 days before tissue samples collection. VKC patients may be positive to serum-specific IgE for at least one aeroallergen, such as bluegrass, mites, parietaria, compositae, and/or tree pollens.

3. A written informed consent must be obtained from all the subjects or their parents before obtaining tissue specimens. The research study should be approved by the Institutional Review Board and Local Ethical Committee and adhered to the tenets of the Declaration of Helsinki.
4. Incubation with 0.3% hydrogen peroxide for 30 min is necessary to inhibit endogenous peroxidase activity (IHC).
5. 0–3 score is as follow: (from 0 = absence of immunostaining to 3 = extensive intense immunostaining).
6. Autophagic proteins expression is analyzed in ChWK and conjunctival fibroblasts stimulated with activated U937 culture medium (CM) or with various pro-inflammatory molecules.
7. Making cells quiescent ensures that all the cells are synchronized in the same phase of the cell cycle and when stimulated their responses would be comparable.
8. The experiment is repeated three times to make sure of the accuracy of the results.
9. The preliminary unpublished data in this regard showed that these concentrations of the cytokines yielded the best results for the purposes of this study.
10. The reason to expose the cells to chloroquine (CQ) is that CQ treatment inhibits the autophagic flux [41] and is used as control treatment.
11. The immunofluorescence protocol has been performed according to our established method which has been described in our previous investigations [42, 43].
12. The primers pairs (Table 2) were designed by Primer3 from a DNA sequence program and synthesized.
13. The cDNA samples must be analyzed in triplicate.
14. For the statistical significance, differences are indicated at $P < 0.05$ (*), $P < 0.01$ (**) and $P < 0.001$ (***).
15. Immunoblotting has been performed according to our established protocol [44, 45].

Acknowledgments

SG and AL conceptualized and designed the study. PB performed the experiments. SG and PB and PM analyzed the data. PM and PB wrote the manuscript. UR, PM, PB, AL, and SG reviewed and edited the manuscript. AL funded and SG supervised the experiments.

Funding

Supported in part by grants MIUR 60A07-4701/13 and 60A06-1891/12 from the Italian.
Institute of Health.

References

1. Shintani T, Klionsky DJ (2004) Autophagy in health and disease: a double-edged sword. Science 306:990–995
2. Emami A, Shojaei S, da Silva Rosa SC, Aghaei M, Samiei E, Vosoughi AR et al (2019) Mechanisms of simvastatin myotoxicity: the role of autophagy flux inhibition. Eur J Pharmacol 862:172616
3. Golbabapour S, Bagheri-Lankarani K, Ghavami S, Geramizadeh B (2019) Autoimmune hepatitis and stellate cells; an insight into the role of autophagy. Curr Med Chem 27:6073–6095
4. Yeganeh B, Rezaei Moghadam A, Alizadeh J, Wiechec E, Alavian SM, Hashemi M et al (2015) Hepatitis B and C virus-induced hepatitis: apoptosis, autophagy, and unfolded protein response. World J Gastroenterol 21: 13225–13239
5. Ghavami S, Mutawe MM, Schaafsma D, Yeganeh B, Unruh H, Klonisch T, Halayko AJ (2012) Geranylgeranyl transferase 1 modulates autophagy and apoptosis in human airway smooth muscle. Am J Physiol Lung Cell Mol Physiol 302:L420–L428
6. Mehrbod P, Ande SR, Alizadeh J, Rahimizadeh S, Shariati A, Malek H et al (2019) The roles of apoptosis, autophagy and unfolded protein response in arbovirus, influenza virus, and HIV infections. Virulence 10: 376–413
7. Alizadeh J, da Silva Rosa SC, Weng X, Jacobs J, Lorzadeh S, Ravandi A et al (2023) Ceramides and ceramide synthases in cancer: focus on apoptosis and autophagy. Eur J Cell Biol 102: 151337
8. Alizadeh J, Kavoosi M, Singh N, Lorzadeh S, Ravandi A, Kidane B et al (2023) Regulation of autophagy via carbohydrate and lipid metabolism in cancer. Cancers (Basel) 15:2195
9. Levine B, Mizushima N, Virgin HW (2011) Autophagy in immunity and inflammation. Nature 469:323–335
10. Hajiahmadi S, Lorzadeh S, Iranpour R, Karima S, Rajabibazl M, Shahsavari Z, Ghavami S (2023) Temozolomide, simvastatin and acetylshikonin combination induces mitochondrial-dependent apoptosis in GBM cells, which is regulated by autophagy. Biology (Basel) 12:302
11. Alizadeh J, Glogowska A, Thliveris J, Kalantari F, Shojaei S, Hombach-Klonisch S, Klonisch T, Ghavami S (2018) Autophagy modulates transforming growth factor beta 1 induced epithelial to mesenchymal transition in non-small cell lung cancer cells. Biochim Biophys Acta 1865:749–768
12. Martelli A, Omrani M, Zarghooni M, Citi V, Brogi S, Calderone V et al (2022) New visions on natural products and cancer therapy: autophagy and related regulatory pathways. Cancers (Basel) 14:5839
13. Zeki AA, Ott S, Sandhu S, Ghavami S, Kenyon NJ (2014) The complex roles of endoplasmic reticulum stress and autophagy in modulating eotaxin-3 production and secretion from human airway epithelial cells. Am J Respir Crit Care Med 189:A5683
14. Zeki AA, Yeganeh B, Kenyon NJ, Post M, Ghavami S (2016) Autophagy in airway diseases: a new frontier in human asthma? Allergy 71:5–14

15. Kroemer G, Mariño G, Levine B (2010) Autophagy and the integrated stress response. Mol Cell 40:280–293
16. Tang D, Kang R, Livesey KM, Cheh CW, Farkas A, Loughran P et al (2010) Endogenous HMGB1 regulates autophagy. J Cell Biol 190: 881
17. Lipinski MM, Hoffman G, Ng A, Zhou W, Py BF, Hsu E et al (2010) A genome-wide siRNA screen reveals multiple mTORC1 independent signaling pathways regulating autophagy under normal nutritional conditions. Dev Cell 18: 1041–1052
18. Dickinson JD, Alevy Y, Malvin NP, Patel KK, Gunsten SP, Holtzman MJ, Stappenbeck TS, Brody SL (2016) IL13 activates autophagy to regulate secretion in airway epithelial cells. Autophagy 12:397–409
19. Ghavami S, Yeganeh B, Zeki AA, Shojaei S, Kenyon NJ, Ott S et al (2018) Autophagy and the unfolded protein response promote profibrotic effects of TGF-beta1 in human lung fibroblasts. Am J Physiol Lung Cell Mol Physiol 314:L493–L504
20. Ban GY, Pham DL, Trinh TH, Lee SI, Suh DH, Yang EM et al (2016) Autophagy mechanisms in sputum and peripheral blood cells of patients with severe asthma: a new therapeutic target. Clin Exp Allergy 46:48–59
21. Klionsky DJ (2007) Autophagy: from phenomenology to molecular understanding in less than a decade. Nat Rev Mol Cell Biol 8: 931–937
22. Ge Y, Huang M, Yao YM (2018) Autophagy and proinflammatory cytokines: interactions and clinical implications. Cytokine Growth Factor Rev 43:38
23. Netea-Maier RT, Plantinga TS, van de Veerdonk FL, Smit JW, Netea MG (2016) Modulation of inflammation by autophagy: consequences for human disease. Autophagy 12:245–260
24. Vichyanond P, Pacharn P, Pleyer U, Leonardi A (2014) Vernal keratoconjunctivitis: a severe allergic eye disease with remodeling changes. Pediatr Allergy Immunol 25:314–322
25. Leonardi A, Bonini S (2013) Is visual function affected in severe ocular allergies? Curr Opin Allergy Clin Immunol 13:558–562
26. Leonardi A (2002) Vernal keratoconjunctivitis: pathogenesis and treatment. Prog Retin Eye Res 21:319–339
27. Brun P, Tarricone E, Di Stefano A, Mattiuzzo E, Mehrbod P, Ghavami S, Leonardi A (2020) The regulatory activity of autophagy in conjunctival fibroblasts and its possible role in vernal keratoconjunctivitis. J Allergy Clin Immunol 146:1210–1213
28. Ghavami S, Hashemi M, de Serres FJ, Bajestani SN, Mehrabifar H, Leonardi A (2007) Trypsin inhibitory capacity in vernal keratoconjunctivitis. Invest Ophthalmol Vis Sci 48:264–269
29. Leonardi A (2013) Allergy and allergic mediators in tears. Exp Eye Res 117:106–117
30. Leonardi A, Di Stefano A, Motterle L, Zavan B, Abatangelo G, Brun P (2011) Transforming growth factor-beta/Smad – signalling pathway and conjunctival remodelling in vernal keratoconjunctivitis. Clin Exp Allergy 41:52–60
31. Leonardi A, DeFranchis G, De Paoli M, Fregona I, Plebani M, Secchi A (2002) Histamine-induced cytokine production and ICAM-1 expression in human conjunctival fibroblasts. Curr Eye Res 25:189–196
32. Leonardi A, Radice M, Fregona IA, Plebani M, Abatangelo G, Secchi AG (1999) Histamine effects on conjunctival fibroblasts from patients with vernal conjunctivitis. Exp Eye Res 68: 739–746
33. Leonardi A, Curnow SJ, Zhan H, Calder VL (2006) Multiple cytokines in human tear specimens in seasonal and chronic allergic eye disease and in conjunctival fibroblast cultures. Clin Exp Allergy 36:777–784
34. Leonardi A, Brun P, Tavolato M, Plebani M, Abatangelo G, Secchi AG (2003) Tumor necrosis factor-alpha (TNF-alpha) in seasonal allergic conjunctivitis and vernal keratoconjunctivitis. Eur J Ophthalmol 13:606–610
35. Kumagai N, Fukuda K, Fujitsu Y, Yamamoto K, Nishida T (2006) Role of structural cells of the cornea and conjunctiva in the pathogenesis of vernal keratoconjunctivitis. Prog Retin Eye Res 25:165–187
36. Lamas DJM, Nicoud MB, Sterle HA, Cremaschi GA, Medina VA (2015) Histamine: a potential cytoprotective agent to improve cancer therapy? Cell Death Dis 6:e2029
37. Jakhar R, Paul S, Bhardwaj M, Kang SC (2016) Astemizole–Histamine induces Beclin-1-independent autophagy by targeting p53-dependent crosstalk between autophagy and apoptosis. Cancer Lett 372:89–100
38. Lin NY, Stefanica A, Distler JHW (2013) Autophagy: a key pathway of TNF-induced inflammatory bone loss. Autophagy 9:1253–1255
39. Zhang M, Kenny SJ, Ge L, Xu K, Schekman R (2015) Translocation of interleukin-1β into a vesicle intermediate in autophagy-mediated secretion. elife 4:e11205

40. Harris J, De Haro SA, Master SS, Keane J, Roberts EA, Delgado M, Deretic V (2007) T helper 2 cytokines inhibit autophagic control of intracellular Mycobacterium tuberculosis. Immunity 27:505–517
41. Mauthe M, Orhon I, Rocchi C, Zhou X, Luhr M, Hijlkema KJ et al (2018) Chloroquine inhibits autophagic flux by decreasing autophagosome-lysosome fusion. Autophagy 14:1435–1455
42. Yeganeh B, Ghavami S, Kroeker AL, Mahood TH, Stelmack GL, Klonisch T, Coombs KM, Halayko AJ (2015) Suppression of influenza A virus replication in human lung epithelial cells by noncytotoxic concentrations bafilomycin A1. Am J Physiol Lung Cell Mol Physiol 308: L270–L286
43. Ghavami S, Sharma P, Yeganeh B, Ojo OO, Jha A, Mutawe MM et al (2014) Airway mesenchymal cell death by mevalonate cascade inhibition: integration of autophagy, unfolded protein response and apoptosis focusing on Bcl2 family proteins. Biochim Biophys Acta 1843:1259–1271
44. Alizadeh J, Zeki AA, Mirzaei N, Tewary S, Rezaei Moghadam A, Glogowska A et al (2017) Mevalonate cascade inhibition by simvastatin induces the intrinsic apoptosis pathway via depletion of isoprenoids in tumor cells. Sci Rep 7:44841
45. Sheikholeslami K, Ali Sher A, Lockman S, Kroft D, Ganjibakhsh M, Nejati-Koshki K, Shojaei S, Ghavami S, Rastegar M (2019) Simvastatin induces apoptosis in medulloblastoma brain tumor cells via mevalonate cascade prenylation substrates. Cancers (Basel) 11:1–19

Methods in Molecular Biology (2025) 2879: 139–150
DOI 10.1007/7651_2024_521

Published online: 28 February 2024

Real-Time Visualization of Cholesterol Trafficking in Human Granulosa Cells Using Confocal Live Cell Microscopy as a Tool to Study the Novel Role of Autophagy in Sex Steroid Synthesis

Yashar Esmaeilian and Ozgur Oktem

Abstract

Autophagy is an evolutionarily conserved process that aims to maintain the energy homeostasis of the cell by recycling long-lived proteins and organelles. We have very recently demonstrated that lipophagy, a special form of autophagy, mediates the association of the lipid droplets (LDs) with lysosomes to deliver the lipid cargo within the LDs to lysosomes for degradation in order to release free cholesterol required for steroid synthesis in human ovary and testis. In this chapter, we describe live cell confocal microscopy technique that allows us to monitor real-time cholesterol trafficking and the association of cholesterol-laden LDs with lysosome (lipophagy) in human granulosa cells.

Key words Autophagy, Cholesterol, Lipid droplet, Luteinized granulosa cell, Live cell imaging, Confocal microscopy, ImageJ

1 Introduction

All steroid hormones including sex steroids (androgens, estrogens, and progesterone) are synthesized from a common precursor molecule cholesterol in a complex biosynthetic pathway in mitochondria [1]. Most of the steroidogenic cholesterol is derived from the circulating low-density lipoproteins (LDLs) that are internalized by receptor-mediated endocytosis after binding to LDL receptors. The resultant endocytic vesicles fuse with lysosome, and engulfed LDL proteins are degraded by proteolysis and release the cholesterol esters, which are subsequently hydrolyzed to free cholesterol by lysosomal acid lipase (LAL) [1]. In steroidogenic cells, free cholesterol is either transported to mitochondria via a carrier protein StAR (steroidogenic acute regulatory protein) or stored in the lipid droplets (LDs following re-esterification by Acyl-Coenzyme A

Supplementary Information The online version contains supplementary material available at https://doi.org/10.1007/7651_2024_521.

cholesterol acyltransferase (ACAT) also known as sterol-O-acetyltransferase (SOAT1) [1]. Other alternative sources of cholesterol are de novo synthesis from acetyl-CoA by β-Hydroxy β-methylglutaryl-CoA (HMG-CoA) reductase and HDL (high-density lipoprotein) (Fig. 1). HDL is internalized by scavenger receptor class B type I (SR-BI) and hydrolyzed to free cholesterol through hormone-sensitive lipase (HSL). LDs synthesized in the ER can serve as a potential source of free cholesterol following their hydrolysis by HSL.

The intracellular trafficking of cholesterol for steroid hormone production has been the subject of extensive research for the last 50 years and has not been fully elucidated. The autophagic process is an evolutionarily conserved process that aims to maintain the energy homeostasis of the cell by recycling long-lived proteins and organelles [2]. It was also shown that autophagy also regulates lipid metabolism through lysosomal degradation of lipids, a special form of autophagic process called lipophagy [3]. In broader terms, autophagy, a highly regulated intracellular process, involves the controlled degradation and recycling of cellular components, enabling the removal of damaged organelles, clearance of intracellular pathogens, energy homeostasis, and cellular metabolism. Autophagy is induced by different factors such as oxidative stress [4], nutrient deprivation [5], hypoxia [6], and ER stress [7]. In terms of lipophagy, a specific form of autophagy, fasting or caloric restriction [8], exercise [9], and gonadotropin hormones (FSH, LH, and hCG) [10] induce lipophagy. The autophagy mechanism consists of several steps that are regulated through a network of genes. It typically begins with the initiation step, where specific signals or cellular stresses trigger the formation of a structure called the isolation membrane or phagophore. The phagophore then elongates and engulfs the targeted cellular cargo, including damaged organelles, protein aggregates, lipids, and other cellular components. Subsequently, the double-membraned vesicle, now known as an autophagosome, encapsulates the cargo. The autophagosome then fuses with a lysosome, resulting in the formation of an autolysosome. Within the autolysosome, the cargo is subjected to degradation by lysosomal enzymes [11, 12] (Fig. 2).

Previous studies have revealed the role of impaired lipophagy in liver fibrosis, cancer, and obesity [13–17]. In terms of the link between autophagy and hormone production, studies on *Drosophila* [18], mouse [19–21], and porcine [22] have demonstrated the role of lipophagy in sex steroid synthesis. We have very recently discovered that lipophagy, a selective form of autophagy, plays a pivotal role in sex steroid hormone biosynthesis by mediating lysosomal degradation of LDs to release free cholesterol (Fig. 1) for steroid hormone biosynthesis in human ovary and testis; progesterone synthesis through this mechanism is defective in women and is known to have luteal phase defect [10, 23].

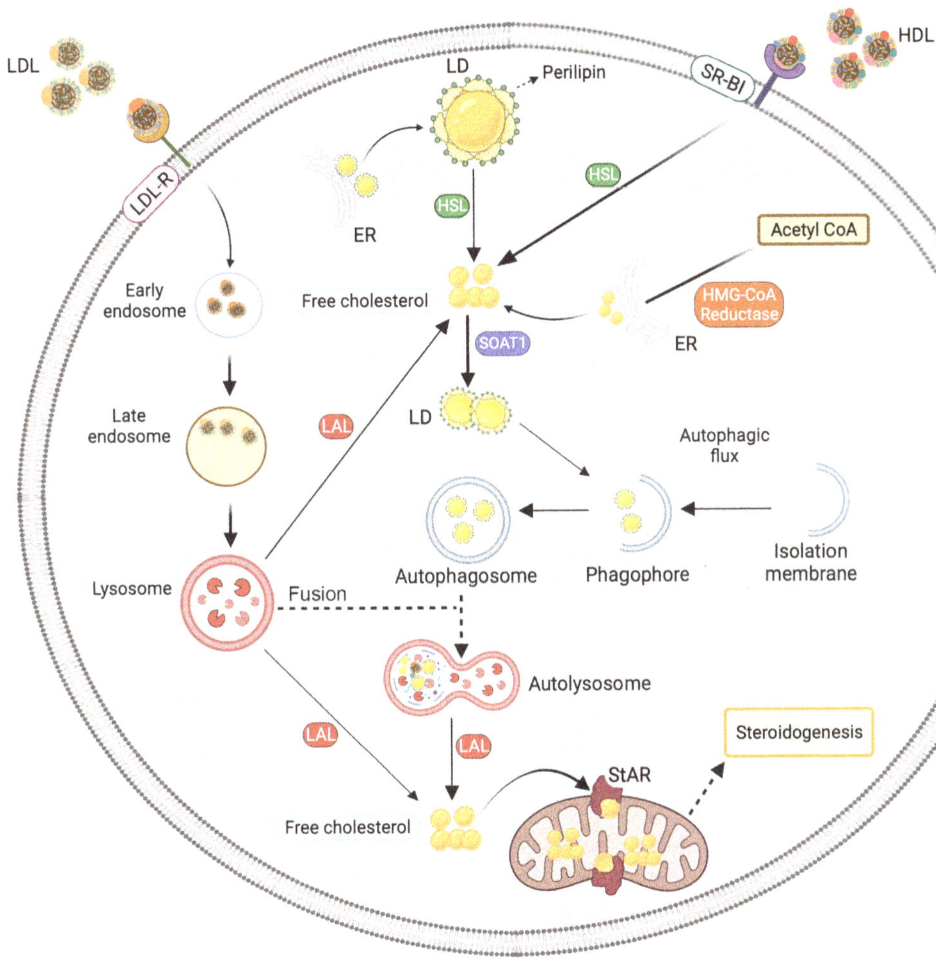

Fig. 1 Schematic diagram of the impact of lipophagy on sex steroid hormone biosynthesis in steroidogenic cells. Cholesterol for steroidogenesis is acquired from four resources. Circulating lipoproteins (LDL and HDL), de novo synthesis from acetyl coenzyme A via HMG-CoA reductase, and hydrolysis of LDs through HSL that are synthesized in ER. LDL-R mediates LDL uptake by endocytic mechanism that delivers them to the lysosomes, where the free cholesterol is released through LAL-hydrolyzing activity. HDL is taken up by the SR-BI receptor and subsequently hydrolyzed to release free cholesterol via HSL. Free cholesterol is either transported to the mitochondria through the carrier protein StAR for steroidogenesis or re-esterified by SOAT1 and stored in LDs. Autophagy mediates LDs trafficking and degradation by engulfing them into autophagosomes and delivering their lipid cargo into lysosomes for degradation (lipophagy) to release the free cholesterol required for steroidogenesis in mitochondria

This chapter provides the experimental methodologies that allowed us to monitor real-time cholesterol trafficking and its association with autophagic machinery in human primary granulosa cells using live confocal imaging techniques.

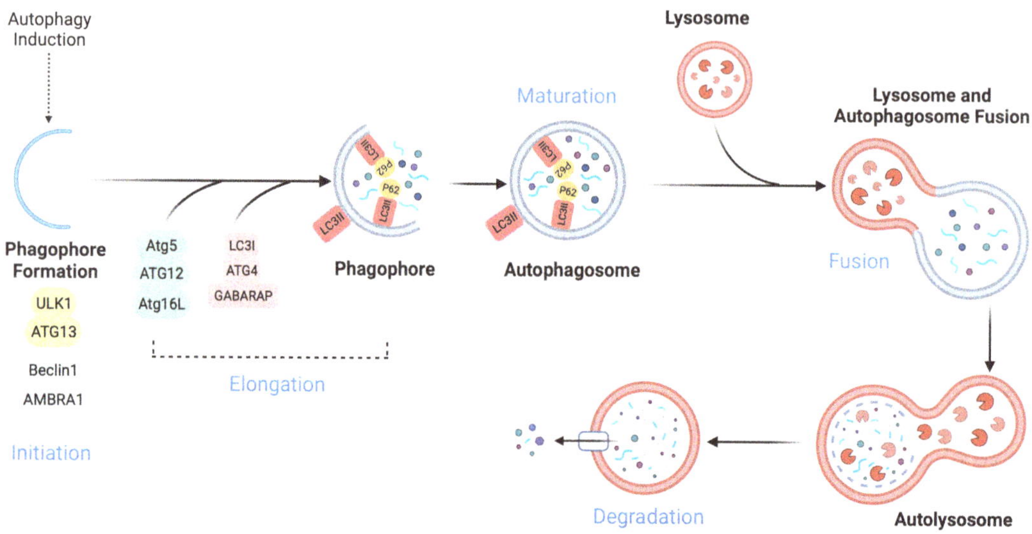

Fig. 2 Steps of the autophagy or lipophagy mechanism. Autophagy is induced by several factors mentioned above. The regulation of autophagy involves a set of proteins, with some of them depicted in the schematic diagram. These steps include the formation of the phagophore, which elongates and engulfs cellular cargo, leading to the creation of an autophagosome, or maturation step. This autophagosome subsequently fuses with a lysosome, forming an autolysosome where lysosomal enzymes break down the enclosed cargo

2 Materials

2.1 Luteinized Granulosa Cells (LGCs)

After the surge of luteinizing hormone in mid-cycle, ovulation occurs and the remaining follicle structure undergoes the luteinization process and forms corpus luteum. Corpus luteum is characterized by lipid accumulation and a high amount of progesterone (P_4) production required for preparation of endometrium for implantation and maintenance of pregnancy [24]. Mural luteinized granulosa cells (LGCs) line the follicle wall and recover during the oocyte retrieval procedure in IVF patients as we previously described [24–26].

2.1.1 LGCs of IVF Patients with Normal Luteal Function (NLF)

In patients with NLF, LGCs display normal reactions to ovarian stimulation, resulting in normal follicular development and oocyte retrieval numbers. They also express gonadotropin receptors and steroidogenic enzymes within physiologically expected ranges [26–28].

2.1.2 LGCs of IVF Patients with Defective Luteal Function (DLF)

The luteal function of the patients in whom ovulation is triggered with gonadotropin-releasing hormone (GnRH) analog is defective. The LGCs of these patients have reduced and unstable progesterone (P_4) production [25, 29–31].

2.2 Cell Culture

1. Dulbecco's Modified Eagle's Medium/Nutrient Mixture F-12 Ham (DMEM/F12), Sigma-Aldrich, D8437.
2. Fetal bovine serum (FBS), heat-inactivated (HI), Gibco, 10500-064.
3. Penicillin/streptomycin/amphotericin B solution (PSA), Millipore, 516104.
4. Nunc™ Glass Base Dish (27 mm), Thermo Scientific, 150682.

2.3 Treatment

1. Chloroquine diphosphate salt, Sigma-Aldrich, C6628.

   For preparing 100 mM stock solution, 50 mg chloroquine (CQ) was dissolved in 1 mL autoclaved water in darkness, sterilized by filtration through a 0.2 μm membrane, aliquoted, and stored at −20 °C.
2. OVITRELLE® recombinant human chorionic gonadotropin (hCG), 250 μg/0.5 mL, Merck.

2.4 Immunofluorescent (IF) Staining

1. Dimethyl sulfoxide (DMSO), Sigma, D2650.
2. Dulbecco's phosphate buffer (DPBS), with calcium with magnesium, Biowest, L0625.
3. Phosphate buffered saline (PBS) without calcium without magnesium, Biowest, L0616.
4. BODIPY 493/503 (4,4-difluoro-1,3,5,7,8-pentamethyl-4-bora-3a,4a-diaza-s-indacene), Invitrogen™, D3922.

   For preparing 5 mM stock solution, 1.3 mg BODIPY was dissolved in 1 mL DMSO in darkness, aliquoted, and stored at −20 °C.
5. LysoTracker Red DND-99, Invitrogen™, L7528.

   1 mM stock solution was diluted to 1 μM by adding 1 μL of stock solution to 1 mL PBS in darkness, aliquoted, and stored at −20 °C.
6. Hoechst 33342, Cell Signaling, 4082.

   For preparing 1 mg/mL stock solution, 1 mg Hoechst was dissolved in 1 mL PBS in darkness, aliquoted, and stored at 4 °C.

2.5 Live Cell Imaging

1. Confocal microscopy (Leica TCS SP8/DMI8, Germany) equipped with incubation chamber (37 °C and 5% CO_2).
2. Type F immersion liquid, Leica.

2.6 ImageJ Analysis

ImageJ software, National Institutes of Health (NIH).

3 Methods

3.1 Cell Seeding

1. 1.5×10^5 LGCs were cultured in glass base dishes (*see* **Note 1**).
2. 2 mL DMEM/F12 supplemented with 10% FBS and 1% PSA was used as a culture medium.
3. The cells were incubated overnight at 37 °C and 5% CO_2.
4. The following day, in the case of treatment, the cells were treated with CQ and hCG (*see* Subheading 3.3); otherwise, they were allocated for live cell IF staining as an untreated group (*see* Subheading 3.2).

3.2 Untreated Group

In the first set of experiments, LGCs from the patients with normal and defective luteal function were cultured without any treatment to compare cholesterol-laden LDs uptake and its association with lysosomes at basal state. For this purpose, on the following day of seeding (see Subheading 3.1, **step 4**), cells were transferred for live cell IF staining without any treatment (*see* Subheading 3.4).

3.3 Treated Group

The cells from both NLF and DLF groups were treated with the combination of hCG and CQ to analyze the role of autophagy in the association of cholesterol with lysosome [10]. hCG accelerates autophagic flux and CQ inhibits the final step of the lysosomal hydrolysis stage of autophagy [10, 15].

1. In the first step, the medium was aspirated (*see* **Note 2**).
2. 2 mL fresh medium containing 10 IU/mL hCG and 60 μM CQ (*see* **Note 3**) was added to cells.
3. The cells were incubated overnight at 37 °C and 5% CO_2.
4. On the subsequent day, the cells were transferred for live cell IF staining (*see* Subheading 3.4).

3.4 Live Cell Immunofluorescent (IF) Staining

LGCs from patients with NLF and DLF in both untreated and treated groups were stained with lipid stain BODIPY and Lyso-Tracker (lysosome marker).

1. The medium was aspirated.
2. Cells were washed with 1 mL pre-warmed DPBS with Ca^{2+} with Mag^{2+} (*see* **Note 4**).
3. 1 mL (*see* **Note 5**) fresh medium containing 2 μM BODIPY, 100 nM LysoTracker, and 1 μg/mL Hoechst 33342 (nucleus marker) was added to cells (*see* **Note 6**).
4. Right after adding, the medium plate was transferred to the incubation chamber of the confocal microscopy system for visualizing (*see* **Note 7**).

3.5 Live Cell Imaging

1. Following the transfer of the plates to the incubation chamber of the confocal microscopy, recording process was initiated within 5 min (*see* **Note 9**).
2. After putting a drop of immersion liquid on the 63X objective, plate was placed on the objective for visualization.
3. Find the proper spot to visualize (*see* **Not 8**).
4. In the final step, live imaging parameters such as image size or quality, scanning speed, optimal laser intensity, time intervals, and recording duration were configured (*see* **Note 9**) and recording was initiated.

3.5.1 Untreated Group Results

In the untreated group, live cell imaging revealed that the fusion of cholesterol-laden lipids with lysosomes is significantly reduced in the LGCs of patients with DLF (Fig. 3, Electronic Supplementary Movies 1–4) [10].

3.5.2 Treated Group Results

Intracellular lipid trafficking in the cells observed by using live cell confocal time-lapse imaging technology revealed that the association of BODYPI with lysosome was significantly delayed and reduced in the cells with DLF. Defective colocalization of BODYPI with lysosome became more obvious in the cells with DLF when they were treated (Fig. 4, Electronic Supplementary Movies 5–6) [10].

3.6 ImageJ Analysis

ImageJ software was employed for BODIPY/LysoTracker colocalization analysis. This measurement was performed using the ComDet tool of ImageJ software.

1. Launch ImageJ software and open the images of red and green channels.
2. Merge the channels.
3. Navigate to the plugin section and choose the ComDet tool.
4. Optimize the parameters such as approximate particle size and intensity threshold based on visual characteristics of selected images.
5. Quantify the colocalized particle numbers, which are visualized as yellow particles due to the merging of the red and green channels.
6. In the final step, the colocalized particle numbers divided by the nucleus numbers represent the colocalization value in selected cells.

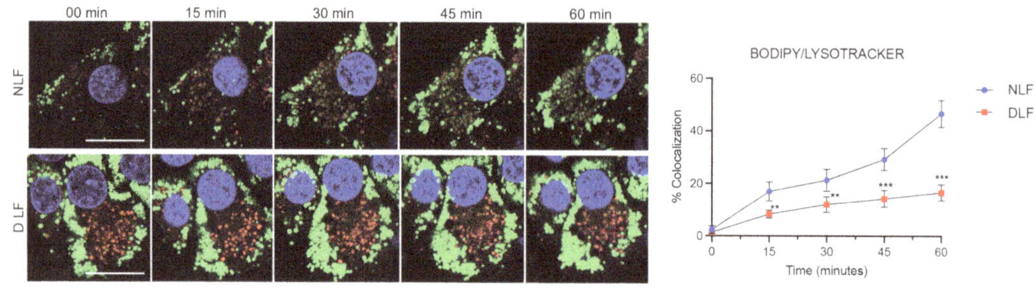

Fig. 3 Time-Laps images of the LGCs of the patients with NLF and DLF without any treatment. BODIPY (green signal), LysoTracker (red signal), and Hoechst 33342 (blue signal). Colocalization analysis is located at the right of the image. The scale bar represents 20 μm

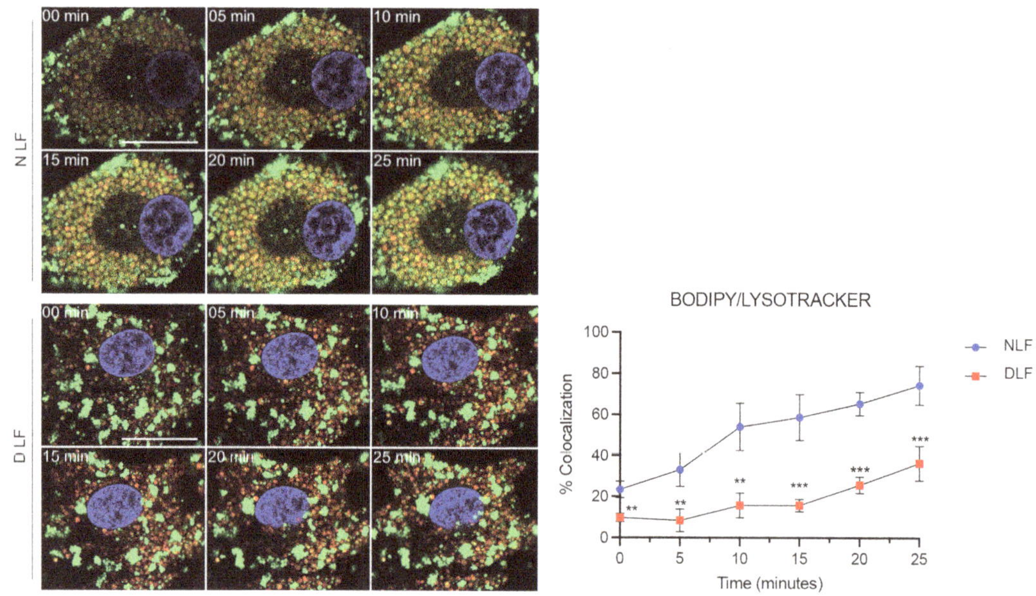

Fig. 4 Time-Laps images of the LGCs of the patients with NLF and DLF treated with the combination of CQ (60 μM) and hCG (10 IU/mL). BODIPY (green signal), LysoTracker (red signal), and Hoechst 33342 (blue signal). Colocalization analysis is located at the right of the image. The scale bar represents 20 μm

4 Notes

1. In live cell imaging (culture area of 27 mm), LGC numbers can vary from 5×10^4 to 3×10^5 depending on the viability of the cells recovered. The ideal plating density is approximately 1.5×10^5 cells in a 27 mm culture area.
2. In the ovary, granulosa cells embrace the oocyte and form spheroidal-shaped follicle. Consistent with their behavior in a natural in vivo environment, LGCs obtained from patients also tend to aggregate and form three-dimensional (3D) colonies when cultured in vitro. Due to the natural tendency of LGCs to

aggregate, their 3D structure in contrast to monolayer cells, and their relatively loose attachments, these colonies are susceptible to detachment. As a result of these challenges, during the steps of adding and, specifically, aspirating the culture medium, it is essential to handle the process gently. Otherwise, the LGC colonies may detach and be lost easily.

3. In live cell imaging, specifically, in the case of dealing with cells with weak attachment or inclined to detach, it is advisable to use the minimum effective dose of chemicals during treatments. This approach helps minimize the potential toxic effects on cells and reduces the risk of exacerbating detachment.
4. As explained in **Note 2**, LGC colonies have a loose attachment and a tendency to detach. To get rid of the colonies with extra weak attachments or unhealthy colonies that are about to detach, it is advisable to wash the cells with pre-warmed DPBS with Ca^{2+} with Mag^{2+} and shake the plate gently. The reason behind using DPBS containing Ca^{2+} and Mag^{2+} is the possible role of Ca^{2+} and Mag^{2+} in promoting cell adhesion [32–34]. Coating the plates with gelatin, collagen, poly-L-lysine, laminin, or matrigel is a common approach to enhance cell attachment. However, in the case of live cell imaging, it is important to consider the concentration of coating materials to avoid their possible blurring effect.
5. In live cell imaging, using the optimum medium volume is crucial. Low medium volume can cause nutrient depletion, reduced oxygen supply, evaporation, and drying (specifically in long-lasting imaging). Using a large amount of medium, in other words, a thicker medium can result in reduced image quality due to heightened light distortion, aberration, scattering, and absorption. In the case of observing the phenomena like endocytosis or uptake of the fluorescent-labeled vesicles, encapsulated particles, exosomes, and similar compartments, using a high medium volume or thicker medium can pose an obstacle. It may lead to divergence and delay in reaching out of the mentioned compartments to the cells attached to the plate's bottom. We conducted live cell imaging of LGCs in plates with a 27 mm bottom area and based on our experience optimal medium volume is approximately 1 mL (in case of $1–1.5 \times 10^5$ cultured cell numbers and live imaging sessions lasting no more than 1 h).
6. Using high concentrations of fluorescent dyes not only contributes to the background signal but also can lead to elevated light absorption and the potential heating of cells and damage during live cell imaging. Therefore, keeping the dye concentrations at a minimum concentration is particularly beneficial when conducting extended live imaging sessions.

7. Maintaining the cleanliness of the plate's bottom is essential for obtaining high-quality images. Consequently, it is important to avoid touching the bottom of the plate during the culture and staining steps. To ensure this, it is recommended to clean the bottom of the plate using alcohol-based solutions, such as 70% isopropanol before positioning the plate on the objective.
8. In the context of live cell imaging, it is important to note that cells are more susceptible to photobleaching because they lack the protective coating of antifade mounting solutions typically used in fixed cells. Consequently, when locating the appropriate spot, focusing, and recording, it is advisable to keep the laser intensity as low as possible to minimize the risk of photobleaching.
9. Excessive image sizes or quality, low scanning speed, high laser intensity, short time intervals, and prolonged imaging duration contribute to cellular overheating, damage, and detachment. So, they should be optimized carefully. In terms of our live cell imaging experiment, we established optimized parameters, which included an image size of 1024 × 1024 pixels, a scanning speed of 400 Hz, 1-min time intervals, and recording durations spanning 30–60 min. Regarding the laser intensity, it is advisable to maintain the laser intensity level as low as possible because of the reasons mentioned above. Moreover, in live cell imaging, it is essential to take into account that as time elapses, fluorescent dyes progressively accumulate within the targeted cellular compartments, and in parallel with this, intensity steadily increases (Electronic Supplementary Movies 1–6). Hence, following the addition of the dye-containing medium to the cells, a 5-min incubation period is sufficient to commence time-lapse recording.

References

1. Miller WL (2013) Steroid hormone synthesis in mitochondria. Mol Cell Endocrinol 379(1–2):62–73. https://doi.org/10.1016/j.mce.2013.04.014
2. Wirawan E, Vanden Berghe T, Lippens S, Agostinis P, Vandenabeele P (2012) Autophagy: for better or for worse. Cell Res 22(1):43–61. https://doi.org/10.1038/cr.2011.152
3. Singh R, Kaushik S, Wang Y, Xiang Y, Novak I, Komatsu M, Tanaka K, Cuervo AM, Czaja MJ (2009) Autophagy regulates lipid metabolism. Nature 458(7242):1131–1135. https://doi.org/10.1038/nature07976
4. Scherz-Shouval R, Elazar Z (2011) Regulation of autophagy by ROS: physiology and pathology. Trends Biochem Sci 36(1):30–38. https://doi.org/10.1016/j.tibs.2010.07.007
5. Mizushima N, Yoshimori T, Ohsumi Y (2011) The role of Atg proteins in autophagosome formation. Annu Rev Cell Dev Biol 27:107–132. https://doi.org/10.1146/annurev-cellbio-092910-154005
6. Bellot G, Garcia-Medina R, Gounon P, Chiche J, Roux D, Pouyssegur J, Mazure NM (2009) Hypoxia-induced autophagy is mediated through hypoxia-inducible factor induction of BNIP3 and BNIP3L via their BH3 domains. Mol Cell Biol 29(10):2570–2581. https://doi.org/10.1128/MCB.00166-09
7. Ogata M, Hino S, Saito A, Morikawa K, Kondo S, Kanemoto S, Murakami T,

Taniguchi M, Tanii I, Yoshinaga K, Shiosaka S, Hammarback JA, Urano F, Imaizumi K (2006) Autophagy is activated for cell survival after endoplasmic reticulum stress. Mol Cell Biol 26(24):9220–9231. https://doi.org/10.1128/MCB.01453-06

8. Singh R, Cuervo AM (2012) Lipophagy: connecting autophagy and lipid metabolism. Int J Cell Biol 2012:282041. https://doi.org/10.1155/2012/282041
9. Koga H, Kaushik S, Cuervo AM (2010) Altered lipid content inhibits autophagic vesicular fusion. FASEB J 24(8):3052–3065. https://doi.org/10.1096/fj.09-144519
10. Esmaeilian Y, Hela F, Bildik G, Iltumur E, Yusufoglu S, Yildiz CS, Yakin K, Kordan Y, Oktem O (2023) Autophagy regulates sex steroid hormone synthesis through lysosomal degradation of lipid droplets in human ovary and testis. Cell Death Dis 14(5):342. https://doi.org/10.1038/s41419-023-05864-3
11. Feng Y, He D, Yao Z, Klionsky DJ (2014) The machinery of macroautophagy. Cell Res 24(1): 24–41. https://doi.org/10.1038/cr.2013.168
12. Galluzzi L, Baehrecke EH, Ballabio A, Boya P, Bravo-San Pedro JM, Cecconi F, Choi AM, Chu CT, Codogno P, Colombo MI, Cuervo AM, Debnath J, Deretic V, Dikic I, Eskelinen EL, Fimia GM, Fulda S, Gewirtz DA, Green DR, Hansen M, Harper JW, Jaattela M, Johansen T, Juhasz G, Kimmelman AC, Kraft C, Ktistakis NT, Kumar S, Levine B, Lopez-Otin C, Madeo F, Martens S, Martinez J, Melendez A, Mizushima N, Munz C, Murphy LO, Penninger JM, Piacentini M, Reggiori F, Rubinsztein DC, Ryan KM, Santambrogio L, Scorrano L, Simon AK, Simon HU, Simonsen A, Tavernarakis N, Tooze SA, Yoshimori T, Yuan J, Yue Z, Zhong Q, Kroemer G (2017) Molecular definitions of autophagy and related processes. EMBO J 36(13):1811–1836. https://doi.org/10.15252/embj.201796697
13. Zhang Z, Zhao S, Yao Z, Wang L, Shao J, Chen A, Zhang F, Zheng S (2017) Autophagy regulates turnover of lipid droplets via ROS-dependent Rab25 activation in hepatic stellate cell. Redox Biol 11:322–334. https://doi.org/10.1016/j.redox.2016.12.021
14. Kim KM, Han CY, Kim JY, Cho SS, Kim YS, Koo JH, Lee JM, Lim SC, Kang KW, Kim JS, Hwang SJ, Ki SH, Kim SG (2018) Galpha (12) overexpression induced by miR-16 dysregulation contributes to liver fibrosis by promoting autophagy in hepatic stellate cells. J Hepatol 68(3):493–504. https://doi.org/10.1016/j.jhep.2017.10.011
15. Gomez de Cedron M, Ramirez de Molina A (2016) Microtargeting cancer metabolism: opening new therapeutic windows based on lipid metabolism. J Lipid Res 57(2):193–206. https://doi.org/10.1194/jlr.R061812
16. Lu GD, Ang YH, Zhou J, Tamilarasi J, Yan B, Lim YC, Srivastava S, Salto-Tellez M, Hui KM, Shen HM, Nguyen LN, Tan BC, Silver DL, Hooi SC (2015) CCAAT/enhancer binding protein alpha predicts poorer prognosis and prevents energy starvation-induced cell death in hepatocellular carcinoma. Hepatology 61(3):965–978. https://doi.org/10.1002/hep.27593
17. Liu Y, Takahashi Y, Desai N, Zhang J, Serfass JM, Shi YG, Lynch CJ, Wang HG (2016) Bif-1 deficiency impairs lipid homeostasis and causes obesity accompanied by insulin resistance. Sci Rep 6:20453. https://doi.org/10.1038/srep20453
18. Texada MJ, Malita A, Christensen CF, Dall KB, Faergeman NJ, Nagy S, Halberg KA, Rewitz K (2019) Autophagy-Mediated Cholesterol Trafficking Controls Steroid Production. Dev Cell 48(5):659–671 e654. https://doi.org/10.1016/j.devcel.2019.01.007
19. Gawriluk TR, Ko C, Hong X, Christenson LK, Rucker EB 3rd (2014) Beclin-1 deficiency in the murine ovary results in the reduction of progesterone production to promote preterm labor. Proc Natl Acad Sci U S A 111(40): E4194–E4203. https://doi.org/10.1073/pnas.1409323111
20. Gao F, Li G, Liu C, Gao H, Wang H, Liu W, Chen M, Shang Y, Wang L, Shi J, Xia W, Jiao J, Gao F, Li J, Chen L, Li W (2018) Autophagy regulates testosterone synthesis by facilitating cholesterol uptake in Leydig cells. J Cell Biol 217(6):2103–2119. https://doi.org/10.1083/jcb.201710078
21. Khawar MB, Liu C, Gao F, Gao H, Liu W, Han T, Wang L, Li G, Jiang H, Li W (2021) Sirt1 regulates testosterone biosynthesis in Leydig cells via modulating autophagy. Protein Cell 12(1):67–75. https://doi.org/10.1007/s13238-020-00771-1
22. Liu Q, Gao H, Yang F, Zhang H, Zeng S (2021) FSH promotes progesterone synthesis by enhancing autophagy to accelerate lipid droplet degradation in porcine granulosa cells. Front Cell Dev Biol 9:626927. https://doi.org/10.3389/fcell.2021.626927
23. Esmaeilian Y, Hela F, Bildik G, İltumur E, Yusufoglu S, Yakin K, Oktem O (2023) Discovery of autophagy as a universal mechanism for sex steroid synthesis in human ovary and testis. Autophagy Rep 2(1):2251804. https://doi.org/10.1080/27694127.2023.2251804

24. Bildik G, Akin N, Esmaeilian Y, Hela F, Yildiz CS, Iltumur E, Incir S, Karahuseyinoglu S, Yakin K, Oktem O (2020) Terminal differentiation of human granulosa cells as luteinization is reversed by activin-A through silencing of Jnk pathway. Cell Death Discov 6(1):93. https://doi.org/10.1038/s41420-020-00324-9
25. Bildik G, Akin N, Seyhan A, Esmaeilian Y, Yakin K, Keles I, Balaban B, Ata B, Urman B, Oktem O (2019) Luteal granulosa cells from natural cycles are more capable of maintaining their viability, steroidogenic activity and LH receptor expression than those of stimulated IVF cycles. Hum Reprod 34(2):345–355. https://doi.org/10.1093/humrep/dey353
26. Bildik G, Esmaeilian Y, Hela F, Akin N, Iltumur E, Yusufoglu S, Yildiz CS, Yakin K, Oktem O (2022) Cholesterol uptake or trafficking, steroid biosynthesis, and gonadotropin responsiveness are defective in young poor responders. Fertil Steril 117(5):1069–1080. https://doi.org/10.1016/j.fertnstert.2022.01.024
27. Ferraretti AP, La Marca A, Fauser BC, Tarlatzis B, Nargund G, Gianaroli L, Ewgo-POR D (2011) ESHRE consensus on the definition of 'poor response' to ovarian stimulation for in vitro fertilization: the Bologna criteria. Hum Reprod 26(7):1616–1624. https://doi.org/10.1093/humrep/der092
28. Drakopoulos P, Blockeel C, Stoop D, Camus M, de Vos M, Tournaye H, Polyzos NP (2016) Conventional ovarian stimulation and single embryo transfer for IVF/ICSI. How many oocytes do we need to maximize cumulative live birth rates after utilization of all fresh and frozen embryos? Hum Reprod 31(2):370–376. https://doi.org/10.1093/humrep/dev316
29. Yding Andersen C, Vilbour Andersen K (2014) Improving the luteal phase after ovarian stimulation: reviewing new options. Reprod Biomed Online 28(5):552–559. https://doi.org/10.1016/j.rbmo.2014.01.012
30. Kolibianakis EM, Schultze-Mosgau A, Schroer A, van Steirteghem A, Devroey P, Diedrich K, Griesinger G (2005) A lower ongoing pregnancy rate can be expected when GnRH agonist is used for triggering final oocyte maturation instead of HCG in patients undergoing IVF with GnRH antagonists. Hum Reprod 20(10):2887–2892. https://doi.org/10.1093/humrep/dei150
31. Shapiro BS, Andersen CY (2015) Major drawbacks and additional benefits of agonist trigger—not ovarian hyperstimulation syndrome related. Fertil Steril 103(4):874–878. https://doi.org/10.1016/j.fertnstert.2015.01.035
32. Riikonen T, Vihinen P, Potila M, Rettig W, Heino J (1995) Antibody against human alpha 1 beta 1 integrin inhibits HeLa cell adhesion to laminin and to type I, IV, and V collagens. Biochem Biophys Res Commun 209(1):205–212. https://doi.org/10.1006/bbrc.1995.1490
33. Shimizu Y, Mobley JL (1993) Distinct divalent cation requirements for integrin-mediated CD4+ T lymphocyte adhesion to ICAM-1, fibronectin, VCAM-1, and invasin. J Immunol 151(8):4106–4115
34. Takeichi M, Okada TS (1972) Roles of magnesium and calcium ions in cell-to-substrate adhesion. Exp Cell Res 74(1):51–60. https://doi.org/10.1016/0014-4827(72)90480-6

Methods in Molecular Biology (2025) 2879: 151–171
DOI 10.1007/7651_2024_520

Published online: 28 February 2024

Visualizing Lipophagy as a New Mechanism of the Synthesis of Sex Steroids in Human Ovary and Testis Using Immunofluorescence Staining Method

Yashar Esmaeilian, Sevgi Yusufoglu, Ece Iltumur, Gamze Bildik, and Ozgur Oktem

Abstract

Immunofluorescence, a transformative tool in cellular biology, is employed to dissect the intricate mechanisms of cholesterol trafficking in human reproductive tissues. Autophagy, a key player in cellular homeostasis, particularly lipophagy, emerges as a free cholesterol source for steroidogenesis. In this chapter, we describe a comprehensive immunofluorescence staining protocol, with details provided for the precise visualization of subcellular dynamics of mitochondria, lysosomes, and lipid droplets in ex vivo testicular tissue and primary luteal granulosa cell culture models, pivotal components in sex steroid biosynthesis. Here, we detail the culture, treatment, and immunofluorescence protocols, providing a comprehensive guide for researchers. The provided immunofluorescence toolkit serves as a valuable resource for researchers, paving way for advancements in human reproductive health to investigate the intricate interplay between autophagy, lipophagy, and cholesterol trafficking.

Key words Autophagy, Confocal microscopy, ImageJ, Immunofluorescence staining, Luteinized granulosa cells, Steroidogenesis, Testis

1 Introduction

Immunofluorescence, a powerful technique in modern cell biology, revolutionizes our ability to visualize and study cellular components. By leveraging the principles of fluorescent labeling, this method enables the precise localization and identification of specific biomolecules within cells. Its applications extend across various biological disciplines, providing researchers with invaluable insights into cellular structures and functions. Within the realm of reproductive research, immunofluorescence plays a crucial role in unravelling the intricacies of human reproductive tissues.

Steroidogenesis, the intricate biosynthetic process responsible for the production of steroid hormones, including androgens, estrogens, and progesterone, originates from the common

Authors Sevgi Yusufoglu and Ece Iltumur have equally contributed to this chapter.

precursor molecule, cholesterol [1]. This complex pathway unfolds within the mitochondria, highlighting the critical role of these organelles in the synthesis of vital hormones [2]. Despite its recognized importance, the precise mechanisms governing intracellular cholesterol trafficking for steroidogenesis remain a subject of extensive research for decades. This chapter specifically focuses on two pivotal components: human luteal granulosa cells and testicular tissue. The application of immunofluorescence in these contexts becomes paramount for gaining a deeper understanding of the complexities of autophagy-mediated cholesterol trafficking/utilization in basal and gonadotropin-stimulated sex steroid biosynthesis in human gonads using ex vivo tissue and primary cell culture models.

The primary source of steroidogenic cholesterol is obtained from circulating low-density lipoproteins (LDLs) [3, 4]. This cholesterol-rich fraction undergoes internalization through receptor-mediated endocytosis upon binding to LDL receptors [5, 6]. Subsequently, endocytic vesicles formed during this process merge with lysosomes. Within the lysosomes, the engulfed LDL proteins undergo proteolysis, liberating cholesterol esters [7]. Lysosomal acid lipase (LAL) then hydrolyzes these esters, releasing free cholesterol [2]. In steroidogenic cells, the fate of free cholesterol diverges into two main pathways. Firstly, it can be transported to mitochondria facilitated by the carrier protein StAR (steroidogenic acute regulatory protein). Alternatively, free cholesterol may be stored in lipid droplets (LDs) through re-esterification by Acyl-Coenzyme A cholesterol acyltransferase (ACAT), also known as sterol-O-acetyltransferase (SOAT1) [2]. The intricate trafficking of cholesterol within cells for steroid hormone production has been the subject of rigorous investigation for the past five decades, yet a comprehensive understanding remains elusive.

Autophagy, an evolutionarily conserved process aimed at maintaining cellular energy homeostasis, has emerged as a key player in cholesterol trafficking for steroidogenesis [8]. Notably, autophagy also regulates lipid metabolism through a specialized form known as lipophagy, wherein lipids undergo lysosomal degradation [9]. This process orchestrates the lysosomal degradation of lipid droplets (LDs), liberating cholesterol essential for steroid hormone synthesis [9–11]. These findings underscore the significance of autophagy, particularly lipophagy, as a regulatory mechanism in the dynamic interplay between cholesterol metabolism and steroidogenesis. While the autophagic process has been established as a key player in cellular homeostasis, its intricate involvement in facilitating the release of cholesterol for steroidogenesis remains an area of active exploration. The intricate crosstalk between these cellular processes provides a foundation for future research aimed at unravelling the details of this fascinating interplay and its implications for reproductive biology.

This chapter seeks to delve into the detailed mechanisms of the interplay between autophagy, specifically lipophagy, and cholesterol trafficking in steroidogenic cells within the context of human luteal granulosa cells and testicular tissue by providing highly detailed methods for immunofluorescence detection, offering a precise toolkit to visualize and dissect autophagy within the complex microenvironment of human reproductive tissues [12, 13]. Leveraging ex vivo tissue and primary cell culture models, these meticulous immunofluorescence techniques enable a granular examination of subcellular dynamics. In essence, this chapter serves as a comprehensive guide, emphasizing the pivotal role of immunofluorescence in advancing our understanding of reproductive biology and paving way for innovations in human reproductive health.

2 Materials

2.1 Culture

2.1.1 Cell Culture

1. Dulbecco's modified Eagle's medium/Nutrient Mixture F-12 Ham (DMEM/F12), Sigma-Aldrich, D8437.
2. Fetal bovine serum (FBS), heat-inactivated (HI), Gibco, 10500-064.
3. Penicillin/streptomycin/amphotericin B solution (PSA), Millipore, 516104.
4. 24-well culture plates (Sarstedt, Ref no: 83.3922).
5. 10X Dulbecco's phosphate buffered saline (DPBS) (Biowest Material: X0515-500).
6. Double distilled H_2O supplemented with 1% PSA.
7. 15 mL Falcon tubes (Greiner Ref no: 188 272-N).

2.1.2 Tissue Culture

1. Dulbecco's modified Eagle's medium/Nutrient Mixture F-12 Ham (DMEM/F12), Sigma-Aldrich, D8437.
2. Fetal bovine serum (FBS), heat-inactivated (HI), Gibco, 10500-064.
3. Penicillin/streptomycin/amphotericin B solution (PSA), Millipore, 516104.
4. Six-well culture plate (Sarstedt Ref no: 83.3920).
5. Glass petri dish.
6. Disposable scalpel (Kruuse, Cat No: 140963).

2.2 Treatment of Luteal Granulosa Cells and Testicular Tissue

2.2.1 Luteinized Granulosa Cells (LGCs)

Following ovulation, granulosa cells transform into LGCs throughout the process involving cell enlargement and increased progesterone production [14]. The LGCs were obtained from follicular aspirates of patients during the oocyte retrieval procedure, using methods previously outlined [15, 16].

2.2.2 Testicular Tissue

Testicular tissue specimens were collected from patients undergoing orchiectomy surgeries. Three primary cell types in the histology of the testes: steroidogenic Leydig cells situated between seminiferous tubules, germ cells, and Sertoli cells have been shown using hematoxylin-eosin staining [12].

2.2.3 Treatment of LGCs and Testicular Tissues

1. Chloroquine (CQ) diphosphate salt, Sigma-Aldrich, C6628.
2. OVITRELLE® recombinant human chorionic gonadotropin (hCG), 250 μg/0.5 mL, Merck.

To prepare a 100 mM stock solution, 50 mg of CQ was dissolved in 1 mL of autoclaved water in darkness. The solution was then sterilized by passing it through a 0.2 μm membrane filter, aliquoted, and stored at −20 °C.

2.3 Fixation of Luteal Granulosa Cells and Testicular Tissue

1. Glass beaker.
2. Paraformaldehyde (PFA) (Sigma Aldrich, CAS: 30525-89-4) powder.
3. Distilled water.
4. Sodium phosphate monobasic (Sigma Aldrich, CAS:7558-80-7).
5. Sodium phosphate dibasic (Sigma Aldrich, CAS: 7558-79-4).
6. Magnetic stirrer with heater (VWR, 442-1271).
7. 1 N of NaOH.
8. 0.2 M phosphate buffer (PB).
9. pH meter (VWR pHenomenal, USA).
10. Hydrochloric acid (HCl).
11. 0.2 μm of sterile filter (VWR, 10040-464).
12. 1X Dulbecco's phosphate buffered saline (DPBS) (Biowest, L0615-500).

0.2 M PB preparation: Add 4.15 grams of sodium phosphate monobasic and 22.5 grams of sodium phosphate dibasic into 1 L of double distilled water. Mix the solution completely to ensure that the components are mixed well and use a pH meter to adjust the pH for 7.4.

4% PFA preparation: In a glass beaker, add 4 g of PFA (Sigma Aldrich, CAS: 30525-89-4) powder to 50 mL of distilled water. Heat this water up to 58–60 °C to dissolve the PFA while stirring on the magnetic stirrer. Avoid boiling. Continue stirring and add 1 N of NaOH drop by drop until the solution is clear. Add 50 mL of 0.2 M PB on clearly dissolved PFA. Allow the PFA solution to cool to room temperature. Adjust the pH of the solution to around 7.4. This can be done by adding a few drops of sodium hydroxide (NaOH) or hydrochloric acid (HCl) as necessary while monitoring the pH using a pH meter. Filter the solution through a 0.2 μm sterile filter to remove any undissolved particles. Store the prepared 4% PFA solution at 4 °C.

2.3.1 Sucrose Dehydration, OCT Embedding, and Sectioning in Testicular Tissue

1. *Preparation of sucrose solutions*
   1. 0.1 M Phosphate buffer (PB).
   2. Double distilled water.
   3. pH meter (VWR pHenomenal, USA).
   4. Sucrose powder (ISOLAB CAS: 57-50-1).

0.1 M Phosphate buffer (PB): Add a 1:1 ratio of double distilled water to 0.2 M phosphate buffer to obtain 0.1 M of phosphate buffer. The preparation of 0.2 M phosphate buffer has been explained in Subheading 2.3.

Dissolve 10 g of sucrose in 90 mL of 0.1 M PB to prepare 10% of sucrose solution. For 20% of sucrose solution, dissolve 20 g of sucrose in 80 mL of 0.1 M PB. To prepare 30% of sucrose solution, dissolve 30 g of sucrose in 70 mL of 0.1 M PB.

2. *OCT Embedding*
   1. OCT (optimal cutting temperature) compound (22-118, DDK, Vigenavo, Italy).
   2. Cryomolds (Tissue Tek, CAT: 62534-25).
   3. 1X Dulbecco's phosphate buffered saline (DPBS) (Biowest, L0615-500).
3. *Cryosectioning*
   1. Adhesive microscope slide (Epredia, Ref: J2800AMNZ).
   2. Cryostat (Leica CM 1950).

2.4 Immunofluorescence Staining in Luteal Granulosa Cells and Testicular Tissue

1. 1X Dulbecco's phosphate buffered saline (DPBS) (Biowest, L0615-500).
2. Permeabilizer (0.1% Triton-X100 in DPBS).
3. Permeabilizer (0.5% Triton-X100 in DPBS).
4. Triton-X (100X) (Sigma Aldrich, T9284).
5. Magnetic stirrer.
6. DPBS-T (0.1% Tween in DPBS).
7. DPBS-T (0.3% Tween in DPBS).
8. Tween-20 (Sigma Aldrich, P1379).
9. Superblock (ScyTek laboratories, AAA125).
10. Mounting medium with DAPI (Abcam, ab104139).
11. 13 mm Coverslip (Marienfeld, 0111530).
12. Microscope Slide (Epredia).

13. Tweezer.
14. TC Dish 150, Standard (Sarstedt Ref No: 83.3903).
15. Perilipin 3 (Abcam, ab47638).
16. SQSTM1/P62 (Abcam, ab56416).
17. LC3B (Cell Signalling, 83506).
18. LC3B-Alexa Fluor 647 Conjugate (Cell Signalling, 18577).
19. MitoTracker (Invitrogen, M7514).
20. LysoTracker (Invitrogen, L7528).
21. NBD Cholesterol (Invitrogen, N1148).
22. Lamp2 (lysosome-associated membrane glycoprotein 2) (Abcam, ab25631).
23. Goat anti-Mouse Alexa Flour 594 secondary antibody (ThermoFisher, A11005).
24. Goat anti-Rabbit Alexa Flour 594 secondary antibody (ThermoFisher, A11012).
25. Goat anti-Rabbit Alexa Flour 488 secondary antibody (ThermoFisher, A11008).
26. Goat anti-Mouse Alexa Flour 488 secondary antibody (ThermoFisher, A11001).
27. Goat anti-Rabbit Alexa Flour 568 secondary antibody (ThermoFisher, A11011).
28. Nail polish.
29. PAP-PEN (hydrophobic marker to prevent the spreading of aqueous solutions).
30. Oil red O (19056 EMS).
31. 2-Propanol (Supelco UN1219).

    Permeabilizer (0.1% Triton-X100 in DPBS): The concentrated Triton-X 100 (100×) is diluted into 10% Triton X-100 solution. 1 mL of Triton-x 100 is added to 9 mL of 1X DPBS. Precise volume of 500 μl is taken from the prepared 10% Triton X-100 solution and then diluted with 50 mL of 1X DPBS. Mix the solution with magnetic stirrer.

    Permeabilizer (0.5% Triton-X100 in DPBS): Add 500 μL of Triton X-100 in 100 mL of 1X DPBS. Mix them all with a magnetic stirrer.

    DPBS-T (0.1% Tween in DPBS): Add 100 μL of Tween-20 in 100 mL of 1X DPBS. Mix them all with a magnetic stirrer.

    DPBS-T (0.3% Tween in DPBS): Add 300 μL of Tween-20 in 100 mL of 1X DPBS. Mix them all with a magnetic stirrer.

MitoTracker: In the first step, for preparing 1 mM stock solution, 50 μg MitoTracker was dissolved in 74.4 μL DMSO in darkness. In the next step, 1 mM stock solution was diluted to 1 μM by adding 1 μL of stock solution to 1 mL PBS in darkness, aliquoted, and stored at −20 °C.

LysoTracker: 1 mM stock solution was diluted to 1 μM by adding 1 μL of stock solution to 1 mL PBS in darkness, aliquoted, and stored at −20 °C.

NBD Cholesterol: For preparing 1 mg/mL stock solution, 1 mg NBD-Cholesterol was dissolved in 1 mL ethanol in darkness, aliquoted, and stored at −20 °C.

Oil Red O: A 60% isopropanol solution is prepared, wherein 0.5 g of Oil Red O is dissolved in 100 mL of the 60% isopropanol solution. This mixture is subsequently subjected to boiling at 100 °C (*see* **Note 1**).

2.5 Imaging by Confocal Microscopy

1. Confocal microscopy (Leica TCS SP8/DMI8, Germany).
2. Type F immersion liquid, Leica.

2.6 ImageJ Analysis

ImageJ software, National Institutes of Health (NIH).

3 Methods

3.1 Cell and Tissue Culture

3.1.1 Cell Seeding

1. After the isolation of luteinized granulosa cells (LGCs), they are cultured in 24-well format culture plates with 1×10^5 LGCs per well (*see* **Note 2**). Cells are seeded in a total of four different wells.
2. 2 mL DMEM/F12 supplemented with 10% FBS and 1% PSA is used as a culture medium.
3. The cells are incubated overnight at 37 °C and 5% CO_2.
4. The following day, in the case of treatment, the cells are treated with CQ and hCG (*see* Subheading 3.3).

3.1.2 Tissue Culture

Testicular tissue samples were extracted from patients undergoing orchiectomy procedures. Human testicular tissue samples were cultured to investigate the relationship of autophagy in testosterone production [12].

1. Testicular tissue samples were promptly placed into 50 mL Falcon tubes containing HEPES-buffered DMEM-F12 media supplemented with 10% FBS within a sterile culture hood.
2. Samples were sliced into 5×5 mm^2 (*see* **Note 3**) in a glass petri dish (*see* **Note 4**) with a disposable scalpel.
3. Sliced samples are taken in six-well culture plates with each well containing 3 mL of complete culture media.

4. Tissue samples are cultured both with and without drug treatment at specified concentrations for 24 h.
5. The culture media were supplemented with hCG to increase steroidogenic activity, whereas lysosomal proteolysis inhibitor CQ was used as pharmacological autophagy inhibitor at specified concentrations.
6. Keep tissue samples at 37 °C and 5% CO_2.

3.2 Treatment of Luteal Granulosa Cells and Testicular Tissue

3.2.1 Treatment of Cell Culture

1. For one control and three different groups of treatment, 2 mL of complete media is taken per Falcon tube.
2. hCG treatment was performed using 10 IU hCG in 2 mL of culture media.
3. CQ treatment was performed using 60 μM CQ in 2 mL of culture media.
4. For cotreatment with hCG and CQ, they were added at the same concentrations to 2 mL of culture media in the second and third steps.
5. The mediums in four wells of a 24-well plate are aspirated. Control and supplemented culture media are added into each well separately according to the treatment group (*see* **Note 5**).
6. Cells are treated for 24 h at 37 °C and 5% CO_2 in the incubator.

3.2.2 Treatment of Tissue Culture

1. For one control and three different groups of treatment, 3 mL of complete media is taken per Falcon tube.
2. 10 IU hCG treatment is diluted from the stock hCG solution for 3 mL of culture media.
3. 60 μM CQ treatment is diluted from the stock CQ solution for 3 mL of culture media.
4. For cotreatment with hCG and CQ, they were added at the same concentrations to 2 mL of culture media in the second and third steps.
5. Control and supplemented culture media are taken into four different wells of a six-well culture plate.
6. Sliced tissue pieces are shared equally between all groups.
7. Testicular tissue samples were treated for 24 h at 37 °C and 5% CO_2 in the incubator.

3.3 Cell Fixation

1. After 24-h of treatment, the medium is discarded.
2. Each well containing coverslip is washed using 500 μL of DPBS (*see* **Note 6**).
3. 500 μL of 4% PFA solution is added to each well (*see* **Note 6**). Following the addition of 4% PFA solution, the samples undergo a precisely timed incubation of 20 min.
4. After the 20-min incubation, each well is washed three times with 500 μL of DPBS.

3.4 Tissue Fixation and Processing

3.4.1 Fixation of Tissue with 4% Paraformaldehyde

While paraformaldehyde is a common fixative, other fixatives such as formalin, glutaraldehyde, and methanol may be preferred in certain situations based on the goals of the experiment and the type of tissue being studied.

As fixative, we prefer 4% paraformaldehyde for both testicular tissue and luteal granulosa cells. Paraformaldehyde is a powerful fixative that stabilizes the tissue's structure and prevents cross-linking proteins [17].

1. Transfer the tissue from culture media to 1X Dulbecco's DPBS in a petri dish.
2. For each 0.5 × 0.5 cm tissue piece, prepare a solution of 4% paraformaldehyde by adding 15 mL of the solution to a 15 mL Falcon tube.
3. Use tweezers to transfer the tissue from the dish to the Falcon tube.
4. Keep the tissue sample in PFA for about 24 h at 4 °C (*see* **Note 7**).

3.4.2 Sucrose Dehydration

Transferring fixed tissue to sucrose solutions before embedding in OCT compound is a common step in preparing tissue samples for frozen sectioning. The gradual replacement of water with sucrose is essential to protect the tissue from ice crystal formation, which can damage cellular structures and morphology during freezing and sectioning [18].

1. After fixation, rinse the tissue with 1X DPBS to remove excessive fixative.
2. Transfer the tissue to a 15 mL of series of sucrose solution in a 15 mL Falcon tube with increasing concentrations (10%, 20%, 30% sucrose).
3. The tissue piece is kept in 10% sucrose until it sinks to the bottom of the Falcon tube at 4 °C.
4. Take the tissue piece from 10% sucrose into 20% sucrose and keep until it sinks to the bottom of the Falcon tube at 4 °C.
5. Take the tissue piece from 20% sucrose into 30% sucrose and keep until it sinks to the bottom of the Falcon tube at 4 °C.

3.4.3 Embedding in OCT

OCT acts as a cryoprotectant that prevents tissue samples from freezing-induced damage. It provides structural support for tissue samples, ensuring their shape and orientation during the freezing process [19].

1. After sucrose dehydration, wash the tissue sample with 1X DPBS to remove the excessive sucrose solution.
2. Use cryostat's freezing section to transfer the tissue to cryomolds.

3. Fill the cryomold with OCT compound and put a tissue sample inside the OCT compound (*see* **Note 8**).
4. The OCT-embedded tissue is then rapidly frozen and then kept in a solid block of tissue until sectioning at −80 °C.

3.4.4 Frozen Sectioning

Thin tissue sections (typically 5–10 μm thick) can be cut using a cryostat, a specialized freezing microtome. In testicular tissue, 10 μm thick slides are taken in our study.

1. The cryostat temperature is set to −25 °C for sectioning (*see* **Note 9**).
2. Locate the tissue section that should be firmly attached to the cryostat's blade.
3. Use a tissue adhesive slide to fix the tissue section onto the slide.
4. Continue cutting the tissues until the desired number of sections is achieved.
5. Allow the slide to dry (takes a few minutes).
6. Until staining, keep the slides at −20 °C.

3.5 Oil Red O Staining

1. After the fixation step, 60% isopropanol is used to rinse the coverslips before Oil Red O staining is applied.
2. The coverslips are incubated with the prepared Oil Red O solution for 20 min.
3. Before the immunofluorescence staining process, coverslips are washed with 60% isopropanol for about 10–20 s (*see* **Note 10**) and running tap water, respectively.

3.6 Immunofluorescence Staining of Cells and Tissues

Costaining with multiple antibodies refers to the technique of simultaneously using two or more antibodies, each tagged with a distinct fluorochrome or label, to detect and visualize multiple specific proteins within a sample [20].

3.6.1 Double and Triple Staining of Cell

1. After the last wash, 500 μL of permeabilizer is added to each well and incubated for 15 min for permeabilization (*see* **Note 11**).
2. Following 15 min, the permeabilizer is discarded, and each well is washed with 400 μL of DPBS-T.
3. Upon removal of the DPBS-T, 400 μL of a Super Block medium is added to each well, followed by an incubation period of 20 min.
4. For double staining: Two primary antibodies are diluted in the superblock and combined in the same tube according to the supplier's recommended dilution. To guarantee specificity

when employing secondary antibodies, every primary antibody ought to originate from a distinct species (e.g., mouse, rabbit, or goat).

For triple staining: Three primary antibodies are diluted in the superblock and combined in the same tube according to the supplier-recommended dilution. LC3B antibody is conjugated with fluorochrome; therefore, there is no need for further secondary antibody for LC3B in that specific condition. Every primary antibody ought to be originated from a distinct species (e.g., mouse, rabbit, or goat).

5. A petri dish is covered with parafilm, and 20 μL of diluted antibody mix is added onto parafilm for each coverslip.
6. The coverslip inside the well is gently removed with two tweezers.
7. The coverslip is placed over the parafilm, with the side containing cells facing the side that holds the diluted antibody mixture. The petri dish should be kept overnight at 4 °C and wet (*see* **Note 12**).
8. The following day, the coverslips are taken from the petri dish with tweezers and washed by dipping them into the beaker containing DPBS-T (*see* **Note 13**).
9. Apply fluorescent dyes conjugated to species-specific secondary antibodies at a ratio of 1:100 or dilute them in 1X DPBS as advised by the supplier. They are mixed in the same tube. The coverslips should be incubated with the secondary antibody mixture for 60 min at 37 °C (*see* **Note 14**) while minimizing light exposure.
10. The coverslips are taken from the petri dish with tweezers and washed by dipping them into the beaker containing DPBS-T.
11. A droplet of mounting medium is dispensed, ranging from 5 to 10 μL (*see* **Note 15**) onto a labeled microscope slide. The coverslip should be delicately lifted from the petri dish using tweezers, ensuring that any residual drops on the slide are removed by gently wiping the edge with a napkin. Subsequently, the coverslip should be positioned with the cell side facing downward onto the drop of the mounting medium.
12. The coverslip is surrounded by four nail polish droplets, and once they are partially dried, nail polish is applied to the edges of the coverslip.
13. Visualization can be done after the drying process is completed.

3.6.2 Triple Staining of Tissues

1. Wash the samples with 1X DPBS and clean up the excessive DPBS and OCT remnants around the section by using tissue paper.

2. Surround the tissue section drawing with PAP-PEN.
3. Permeabilize the tissue with a permeabilizer for 15 min at room temperature. Drop 300–400 μL of permeabilizer per circle of sample. Aspirate permeabilizer gently (see first step in Subheading 3.6.1).
4. Wash the sample three times for 5 min with 0.3% DPBS-Tween to reduce background staining. Drop 300–400 μL of 0.3% DPBS-T per circle and wait for 5 min and repeat. Aspirate DPBS-T gently.
5. Incubate the sample with Superblock for 20 min at room temperature. 300–400 μL of superblock will be enough to cover the specimen.
6. Dilute the primary antibodies in superblock and mix them in the same tube according to the supplier's recommended dilution. Vortex the tube. LC3B antibody is conjugated with fluorochrome; therefore, there is no need for further secondary antibody for LC3B in that specific condition. Each primary antibody should be from a different species (e.g., mouse, rabbit, goat) to ensure specificity when using secondary antibodies.
7. Add diluted primary antibodies on top of the sample. 20–30 μL of diluted antibody mix will be enough to cover the sample. Incubate the sample overnight at 4 °C (see seventh step in Subheading 3.6.1).
8. Remove the primary antibody mix and wash the samples 3 times for 5 min with DPBS-T. Drop 300–400 μL of 0.3% DPBS-T per circle and wait for 5 min and repeat. Aspirate DPBS-T gently.
9. Use species-specific secondary antibodies conjugated to fluorescent dyes as 1:100 or dilute according to supplier recommended dilution in 1X DPBS. Vortex secondary antibodies and mix them well. 20–30 μL of diluted antibody mix will be enough to cover the sample.
10. Incubate the sample with secondary antibody mix for 90 min at 37 °C and reduce unnecessary exposure to light (see ninth step in Subheading 3.6.1).
11. After removing the secondary antibody solution, wash the samples three times for 5 min with DPBS-T.
12. Drop 5–10 μL of mounting medium with DAPI from the corner of the tissue section.
13. Lay the coverslip with tweezers on a microscope slide so that the mounting medium spreads thoroughly the sample.

14. Drop four drops of nail polish around the coverslip, and after these drops dry a little, surround the edges of the coverslip with nail polish.
15. Once the nail polish dries, store the sample at 4 °C.

3.7 NBD-Cholesterol Uptake Assay

LGCs are stained with NBD-cholesterol (a fluorescent analog of cholesterol) and LysoTracker (lysosome marker) [12].

1. The medium is aspirated.
2. Cells are washed with 1 mL pre-warmed DPBS.
3. 0.5 mL fresh medium containing 1 μg/mL NBD-Cholesterol and 100 nM LysoTracker (Lysosome marker) is added to each well.
4. Following washing with DPBS, cells are embedded in 4% PFA for 20 min.
5. The coverslips are taken from the wells with tweezers and washed by dipping them into the beaker containing DPBS.
6. A droplet of mounting medium is dispensed, ranging from 5 to 10 μL onto a labeled microscope slide. The coverslip is gently taken from the petri dish with tweezers and placed on the cell side facing downward onto the drop of the mounting medium.
7. The coverslip is surrounded by four nail polish droplets, and once they have partially dried, nail polish is applied to the edges of the coverslip.
8. Visualization can be done after the drying process is completed.

3.8 Imaging by Confocal Microscopy

1. Once the nail polish has dried, you can begin visualizing. If waiting for the next day or longer, store slides at +4 °C in darkness.
2. If the slides are at +4 °C, allow them to reach room temperature before imaging.
3. In the imaging process, place the slide on the 63× objective after applying a drop of immersion liquid for visualization.
4. Locate the correct spot for visualization (*see* **Note 16**).
5. Finally, configure imaging parameters—image size, quality, scanning speed, and optimal laser intensity—before capturing images.

3.8.1 Results of Triple Staining of LGCs

1. Triple staining of LGCs demonstrated a colocalization of perilipin3 (a protein associated with lipid droplet) with LC3B and lysosomes. This colocalization notably intensified following hCG treatment and showed a marked increase with the hCG and CQ combination (Fig. 1) [12].

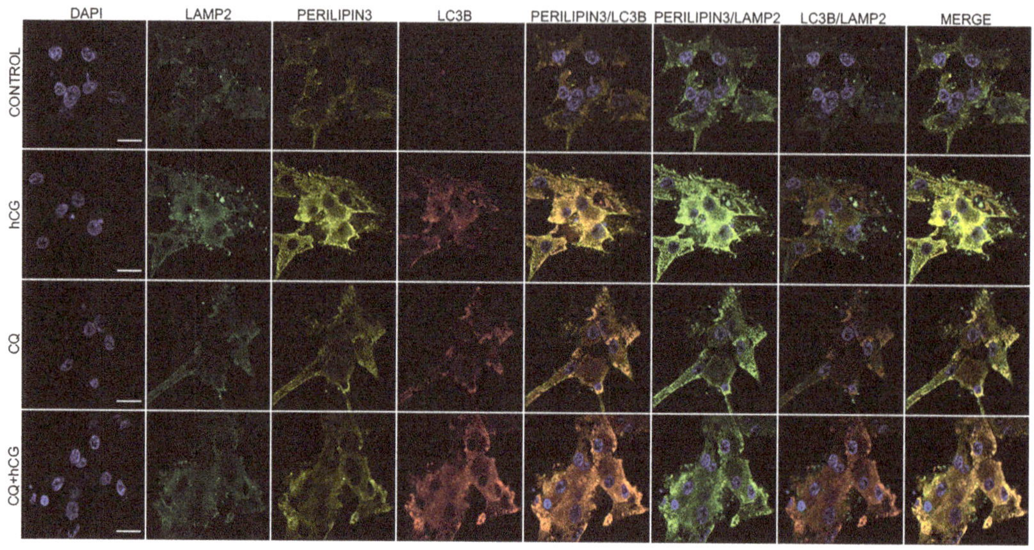

Fig. 1 Representative confocal images of luteinized granulosa cells obtained 24 h post-treatment with hCG (10 IU/mL) with or without CQ (60 μM). The cells are subjected to triple staining with LAMP2 (green signal), perilipin3 (yellow signal), and LC3B (red signal). Nuclei are stained with DAPI, and the scale bars indicate a size of 20 μm

2. Blocking autophagy pharmacologically using CQ led to NBD-cholesterol cytoplasmic accumulation and a substantial increase in its colocalization with LysoTracker. This effect was particularly amplified following treatment with hCG and CQ in combination (Fig. 2) [12].
3. We noticed a significant rise in the intensity of Oil Red O and its colocalization with MitoTracker in cells treated solely with CQ and in combination with hCG (Fig. 3) [12].
4. Dual IF staining indicated a strong colocalization of Oil Red O with the lipid droplet-associated protein perilipin3 (TIP47). This suggests that lipid accumulation predominantly takes place within LDs (Fig. 4) [12].
5. CQ treatment resulted in a significant increase in the colocalization of Oil Red O with autophagy substrates LC3 and SQSTM1/p62. This observation suggests a potential connection between the mechanisms of autophagy and lipid trafficking or degradation of LDs (Fig. 5) [12].

3.8.2 Result of Triple Staining of Testicular Tissues

Triple staining of the testicular tissue demonstrated that perilipin3 already exhibits colocalization with LC3B and LAMP2 (lysosome-associated membrane glycoprotein 2) under basal conditions. Upon hCG treatment, there was an increase in the association between perilipin3 and LAMP2. Additionally, the combination of hCG and CQ further amplified this colocalization (Fig. 6) [12].

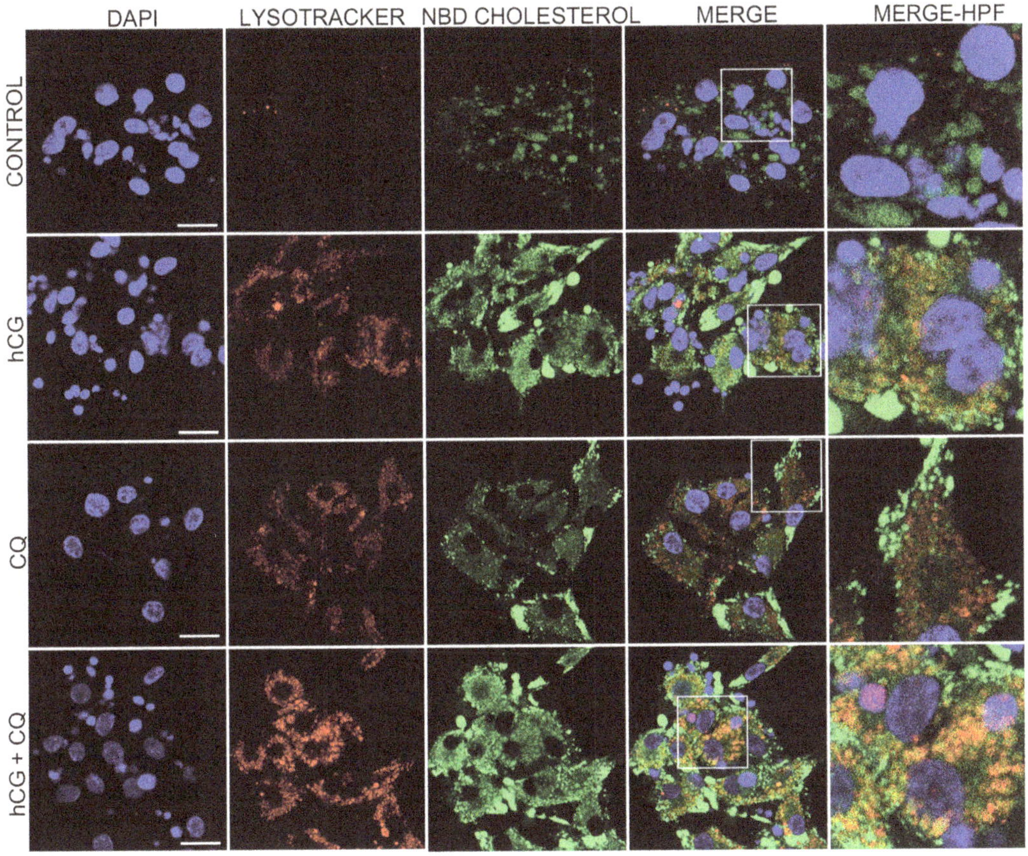

Fig. 2 Representative confocal images of luteinized granulosa cells obtained 24 h post-treatment with hCG (10 IU/mL) with or without CQ (60 μM). The cells are subjected to dual staining with LysoTracker marking the lysosomes (red signal) and NBD-cholesterol (a fluorescent analogy of cholesterol) marking cholesterol (green signal). Nuclei are stained with DAPI, and the scale bars indicate a size of 20 μm

Considering these findings, we hypothesize that autophagy might have a role in the synthesis of steroid hormones in the human LGCs and testis [12, 13].

3.9 ImageJ Analysis

ImageJ software was employed for intensity and colocalization analysis.

3.9.1 Intensity Analysis

1. Open the image in ImageJ software.
2. Measure the intensity of the selected area or florescent-stained area in the periphery of the cells.
3. Measure the vastness of the selected area.
4. The intensity of the selected area divided by the vastness of the selected area represents the mean gray value or average intensity of the selected cells.

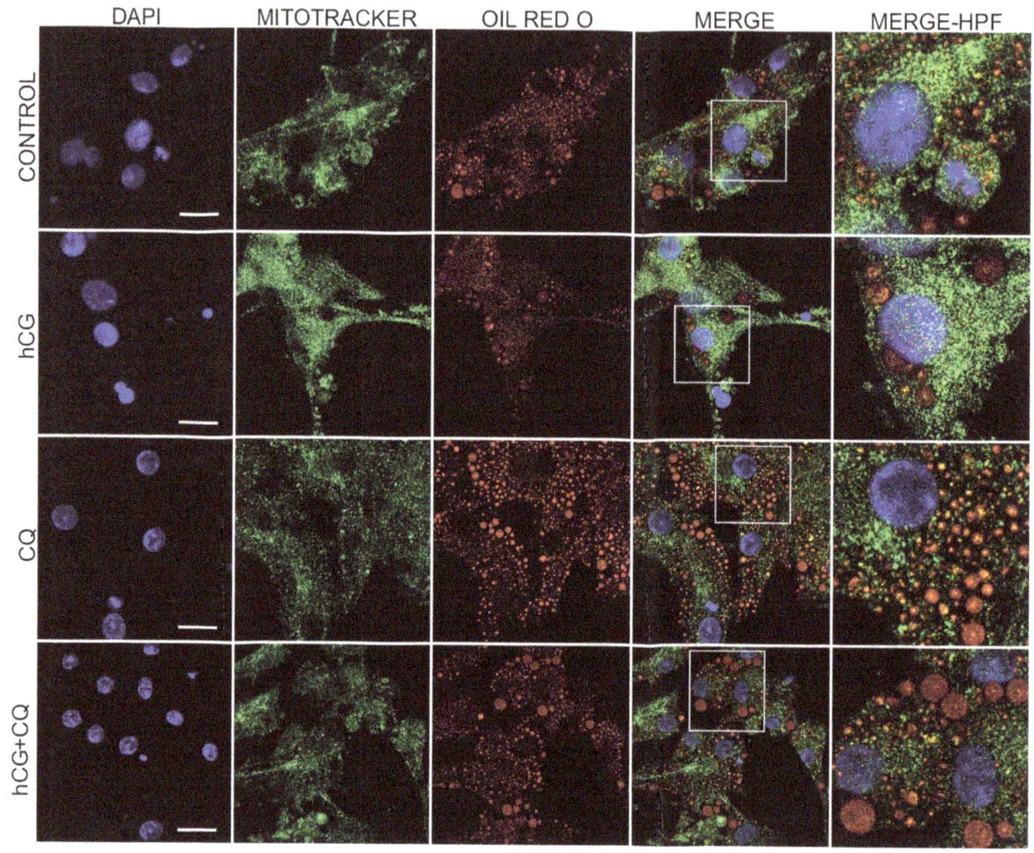

Fig. 3 Representative confocal images of luteinized granulosa cells obtained 24 h post-treatment with hCG (10 IU/mL) with or without CQ (60 μM). The cells are dual stained with MitoTracker (for mitochondria detection) (green signal) and Oil Red O (for lipid and neutral triglyceride detection) (red signal). Nuclei are stained with DAPI, and the scale bars indicate a size of 20 μm

3.9.2 Colocalization Analysis

Colocalization measurement was performed using the ComDet tool of ImageJ software.

1. Open ImageJ software and load the images from the red and green channels.
2. Merge the green and red channels.
3. Go to the plugin section and select the ComDet tool.
4. Optimize parameters of approximate particle size and intensity threshold according to the visual features of the chosen images.
5. Quantify the number of colocalized particles, which appear as yellow particles resulting from the merging of the red and green channels.
6. Finally, calculate the colocalization value in selected cells by dividing the colocalized particle count by the number of nuclei.

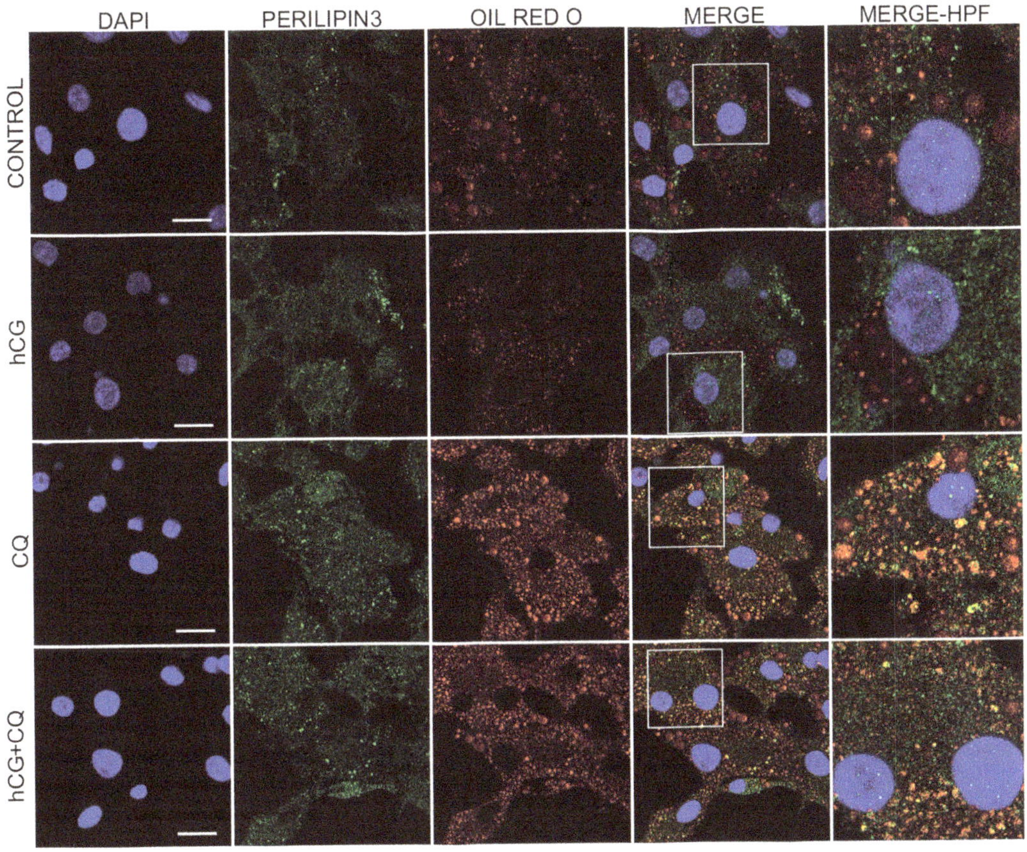

Fig. 4 Representative confocal images of luteinized granulosa cells obtained 24 h post-treatment with hCG (10 IU/mL) with or without CQ (60 μM). These cells are dual-stained with perilipin3, a protein associated with lipid droplet (green signal), and Oil Red O (red signal). Nuclei are stained with DAPI, and the scale bars indicate a size of 20 μm

4 Notes

1. To eliminate any residue in the prepared Oil Red O solution, it is recommended to use a magnetic stirrer while boiling the solution. Post-boiling, centrifugation at 500 g for 5 min is advised to isolate and utilize the supernatant remaining on top.
2. Due to the limited proliferative capacity of luteinized granulosa cells, it is necessary to initially culture these cells at a density of 100,000 cells.
3. Tissue size is an essential parameter in in vitro studies because it affects the drug diffusion rates and the equality of drug exposure.
4. The usage of a sterilized glass petri dish ensures a contamination-free surface for tissue culture which is resistant to scratches and helps smooth tissue slicing. Using plastic petri

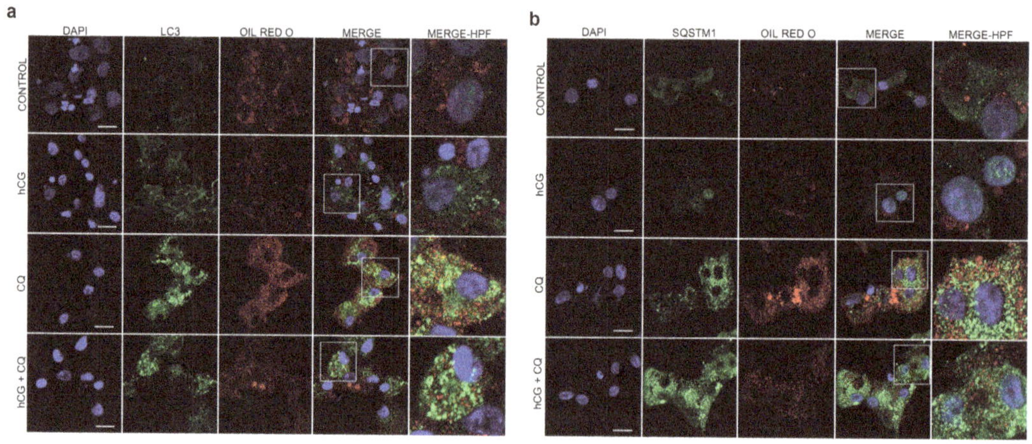

Fig. 5 Representative confocal images of luteinized granulosa cells obtained 24 h post-treatment with hCG (10 IU/mL) with or without CQ (60 μM). (**a**) The cells are dual stained with LC3 (autophagosome marker) (green signal) and Oil Red O (for lipid and neutral triglyceride detection) (red signal). (**b**) The cellular specimens undergo dual staining with SQSTM1 (autophagy substrate) (green signal), along with Oil Red O (red signal). Nuclei are stained with DAPI, and the scale bars indicate a size of 20 μm

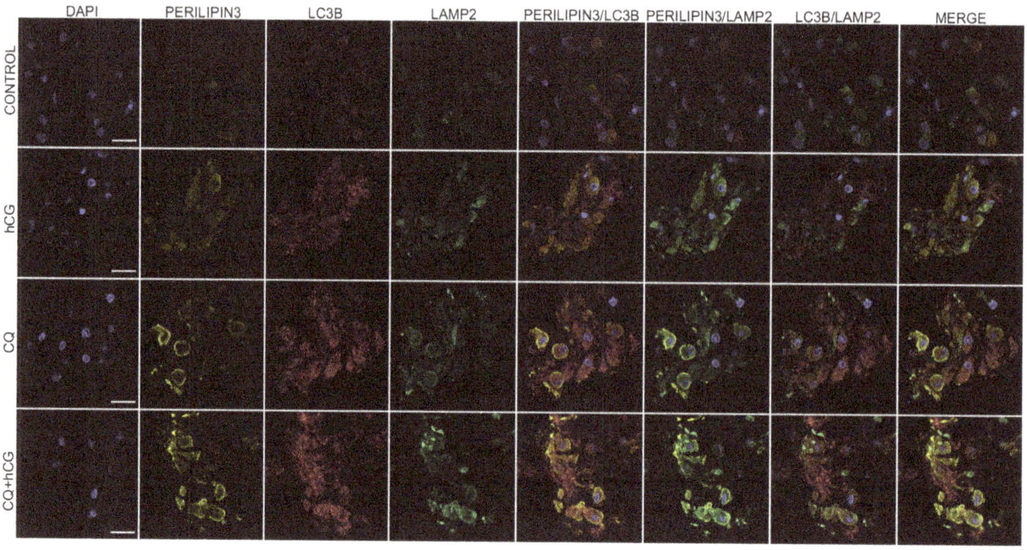

Fig. 6 Representative confocal images of cryosections of testicular tissue samples obtained 24 h post-treatment with hCG (10 IU/mL) with or without CQ (60 μM). These images display perilipin3 (yellow signal), LC3B (red signal), and LAMP2 (green signal). Cell nuclei were labeled with DAPI, and the scale bars indicate a size of 20 μm

dish in tissue slicing can lead to plastic microparticle contamination which may negatively affect cell health, growth, and the reliability of experimental outcomes.

5. LGCs, which tend to separate easily from each other, are characteristically clustered onto each other while adhering to the culture surface. Therefore, washing and media refreshing should be done gently from the corner of the culture plate.

6. DPBS and 4% PFA solutions must be reached at room temperature before starting the fixation procedure and added slowly to the well to prevent cells from detachment.
7. You should see that the tissue sinks inside the PFA which shows the tissue is fixed. Otherwise, it affects the tissue integrity, and sections of the specimen may have some debris while sectioning.
8. The tissue is oriented in a way that facilitates the desired sectioning plane. OCT compound supports tissue structure and maintains the tissue's shape in the freezing procedure which helps to study tissue structures, cellular organelles, and morphology.
9. Setting the cryostat temperature between −20 °C and −30 °C helps to keep an optimal level of tissue firmness that enables high-quality sectioning. Slight temperature adjustments may be needed depending on the tissue type to ensure optimal sectioning quality.
10. The washing step with 60% isopropanol is used to remove unbound dye and lipids and enhance the visualization of the stained lipid droplets. To prevent the washing out of Oil Red O, it is advised to keep the duration of this step short, as prolonged washing can lead to its removal.
11. Permeabilizer enables antibodies to reach and bind to intracellular targets. If the aim is to study proteins, receptors, or other molecules present on the cell membrane, permeabilization may not be necessary.
12. Petri dishes should be wrapped with parafilm to prevent the diluted antibody from drying. The inside of the petri dish should remain wet. Because of that, the wet napkin is placed into the petri dish.
13. The reason for adding non-ionic detergent Tween 20 to DPBS is to reduce the surface tension of the solution and facilitate more effective removal of unbound antibodies or other reagents. This minimizes non-specific bindings of antibodies and background fluorescence in staining.
14. The 60 min of incubation time with secondary antibodies enhances the binding affinity to primary antibodies. Incubation of less than 1 h may decrease the efficacy of secondary antibody affinity. Prolonged incubation of a secondary antibody may result in higher background fluorescence and non-specific binding of the secondary antibody to cellular components.
15. Using less than 5 μL of mounting medium may result in bubble formation, making it challenging to visualize the specimen. Conversely, an excessive volume of the mounting medium

might overflow the 12–13 mm coverslip, potentially causing coverslip displacement or sliding, which could lead to the removal or damage of cells.

16. Although in the last step of IF staining (*see* Subheading 3.6.1, **step 11**) cells are covered by the mounting medium, it is advisable to keep the laser intensity as low as possible to minimize the risk of photobleaching during the finding of the appropriate spot and focusing.

References

1. Bloch K (1965) The biological synthesis of cholesterol. Science 150(3692):19–28. https://doi.org/10.1126/science.150.3692.19
2. Miller WL (2013) Steroid hormone synthesis in mitochondria. Mol Cell Endocrinol 379(1–2):62–73. https://doi.org/10.1016/j.mce.2013.04.014
3. Azhar S, Leers-Sucheta S, Reaven E (2003) Cholesterol uptake in adrenal and gonadal tissues: the SR-BI and 'selective' pathway connection. Front Biosci 8:S998–S1029. https://doi.org/10.2741/1165
4. Tabas I (2002) Consequences of cellular cholesterol accumulation: basic concepts and physiological implications. J Clin Invest 110(7): 905–911. https://doi.org/10.1172/JCI16452
5. Luo J, Jiang L, Yang H, Song BL (2017) Routes and mechanisms of post-endosomal cholesterol trafficking: a story that never ends. Traffic 18(4):209–217. https://doi.org/10.1111/tra.12471
6. Kwon HJ, Abi-Mosleh L, Wang ML, Deisenhofer J, Goldstein JL, Brown MS, Infante RE (2009) Structure of N-terminal domain of NPC1 reveals distinct subdomains for binding and transfer of cholesterol. Cell 137(7):1213–1224. https://doi.org/10.1016/j.cell.2009.03.049
7. Subramanian K, Balch WE (2008) NPC1/NPC2 function as a tag team duo to mobilize cholesterol. Proc Natl Acad Sci USA 105(40): 15223–15224. https://doi.org/10.1073/pnas.0808256105
8. Wirawan E, Vanden Berghe T, Lippens S, Agostinis P, Vandenabeele P (2012) Autophagy: for better or for worse. Cell Res 22(1): 43–61. https://doi.org/10.1038/cr.2011.152
9. Dong H, Czaja MJ (2011) Regulation of lipid droplets by autophagy. Trends Endocrinol Metab 22(6):234–240. https://doi.org/10.1016/j.tem.2011.02.003
10. Shen WJ, Azhar S, Kraemer FB (2016) Lipid droplets and steroidogenic cells. Exp Cell Res 340(2):209–214. https://doi.org/10.1016/j.yexcr.2015.11.024
11. Weidberg H, Shvets E, Elazar Z (2009) Lipophagy: selective catabolism designed for lipids. Dev Cell 16(5):628–630. https://doi.org/10.1016/j.devcel.2009.05.001
12. Esmaeilian Y, Hela F, Bildik G, Iltumur E, Yusufoglu S, Yildiz CS, Yakin K, Kordan Y, Oktem O (2023) Autophagy regulates sex steroid hormone synthesis through lysosomal degradation of lipid droplets in human ovary and testis. Cell Death Dis 14(5):342. https://doi.org/10.1038/s41419-023-05864-3
13. Esmaeilian Y, Hela F, Bildik G, İltumur E, Yusufoglu S, Yakin K, Oktem O (2023) Discovery of autophagy as a universal mechanism for sex steroid synthesis in human ovary and testis. Autophagy Rep 2(1):2251804. https://doi.org/10.1080/27694127.2023.2251804
14. Bildik G, Akin N, Esmaeilian Y, Hela F, Yildiz CS, Iltumur E, Incir S, Karahuseyinoglu S, Yakin K, Oktem O (2020) Terminal differentiation of human granulosa cells as luteinization is reversed by activin-A through silencing of Jnk pathway. Cell Death Discov 6(1):93. https://doi.org/10.1038/s41420-020-00324-9
15. Bildik G, Akin N, Seyhan A, Esmaeilian Y, Yakin K, Keles I, Balaban B, Ata B, Urman B, Oktem O (2019) Luteal granulosa cells from natural cycles are more capable of maintaining their viability, steroidogenic activity and LH receptor expression than those of stimulated IVF cycles. Hum Reprod 34(2):345–355. https://doi.org/10.1093/humrep/dey353
16. Bildik G, Esmaeilian Y, Hela F, Akin N, Iltumur E, Yusufoglu S, Yildiz CS, Yakin K, Oktem O (2022) Cholesterol uptake or trafficking, steroid biosynthesis, and gonadotropin responsiveness are defective in young poor responders. Fertil Steril 117(5):1069–1080.

https://doi.org/10.1016/j.fertnstert.2022.01.024

17. Qin Y, Jiang W, Li A, Gao M, Liu H, Gao Y, Tian X, Gong G (2021) The combination of paraformaldehyde and glutaraldehyde is a potential fixative for mitochondria. Biomol Ther 11(5). https://doi.org/10.3390/biom11050711
18. Ruan JL, Tulloch NL, Muskheli V, Genova EE, Mariner PD, Anseth KS, Murry CE (2013) An improved cryosection method for polyethylene glycol hydrogels used in tissue engineering. Tissue Eng Part C Methods 19(10):794–801. https://doi.org/10.1089/ten.TEC.2012.0460
19. Zhang W, Sakashita S, Taylor P, Tsao MS, Moran MF (2015) Comprehensive proteome analysis of fresh frozen and optimal cutting temperature (OCT) embedded primary non-small cell lung carcinoma by LC-MS/MS. Methods 81:50–55. https://doi.org/10.1016/j.ymeth.2015.02.008
20. Suzuki T, Matsuzaki T, Hagiwara H, Aoki T, Takata K (2007) Recent advances in fluorescent labeling techniques for fluorescence microscopy. Acta Histochem Cytochem 40(5):131–137. https://doi.org/10.1267/ahc.07023

Methods in Molecular Biology (2025) 2879: 173–181
DOI 10.1007/7651_2024_531

Published online: 06 April 2024

Characterization of Bitter Taste Receptor-Dependent Autophagy in Oral Epithelial Cells

Nisha Singh, Saeid Ghavami, and Prashen Chelikani

Abstract

Microbial dysbiosis is an important trigger in the development of oral diseases. Oral keratinocytes or gingival epithelial cells (GECs) offer protection against various microbial insults. Recent studies suggest that GECs expressed higher level of bitter taste receptor 14 (T2R14) compared to other taste receptors and toll-like receptors and act as innate immune sentinels. Macroautophagy or autophagy is a cellular conserved process involved in the regulation of host innate immune responses against microbial infection. Here, we describe a robust method for evaluation of T2R14-dependent autophagy flux in GECs. Autophagy flux was detected using Western blot analysis in GECs and further was confirmed using Acridine Orange-dependent flow cytometry analysis.

Key words Bitter taste receptor, Autophagy flux inhibition, Autophagosome, Oral epithelial cells, Western blot, Flow cytometry, Acridine Orange, Serum starvation

1 Introduction

Autophagy is a cellular process involved in breaking down macromolecules and organelles via the lysosomal degradation pathway [1, 2]. Autophagy participates in the formation of autophagosomes, which transport cytoplasmic proteins and organelles to the lysosomal degradation machinery [3, 4]. The formation of autophagosomes is regulated by key protein complexes such as mammalian target of rapamycin complex 1 (mTORC1) and Beclin 1. Conversely, AMP-dependent protein kinase (AMPK), a primary sensor of cellular energy levels, serves as a key activator of autophagy [5].

In recent years, autophagy has emerged as a crucial regulator of innate immune functions which includes cytokine secretion, immune cell differentiation, and clearance of pathogens [6, 7]. Previous investigations have clearly addressed the role of nutrient-sensing G protein-coupled receptor (GPCR) in the regulation of the autophagy pathway [8, 9]. However, the involvement of majority of taste-sensing GPCRs including the 25 T2Rs in the autophagy process is not well understood. Studies over the past decade suggest an important physiological and tissue-specific role for T2Rs including host-microbe interactions and innate immune responses [10]. For example, activation of T2Rs resulted in increased nitric oxide (NO) production and enhanced phagocytosis of *Escherichia coli* and *Staphylococcus aureus* [11]. Our transcriptomic analysis of gingival epithelial cells (GECs) showed 21,448 genes, and among them, T2R14 or TAS2R14 was found to be expressed at high levels in different GECs [12]. The expression of TAS2R14 was notably elevated compared to the other 24 TAS2R genes (***$p<0.001$), as well as 8 TLR genes (***$p<0.001$).

Only few studies investigated the connection between T2Rs and autophagy [13, 14]. A previous study on airway smooth muscles cells (ASMCs) suggested that bitter agonists quinine and chloroquine resulted in microtubule-associated light chain protein (LC3β-II) accumulation, and treatment of autophagy flux inhibitor bafilomycin A1 (Baf A1) and 3-methyl adenine (3-MA) leads to significant reduction of LC3β-II accumulation [13]. Oral keratinocyte or GECs produce many innate immune markers including nitric oxide and antimicrobial peptides (AMPs) against various oral pathogenic bacteria. Our recent investigation suggested that some cariogenic bacteria and their products show robust calcium responses in GECs, and this effect is dependent on both T2R14 and autophagy protein 7 [14]. These results suggest that T2R14 is involved in determining the direction of autophagy flux in GECs via lysosomal degradation of autophagosome.

In a previous study, we reported that T2R14 modulates Gram-positive bacterial internalization (e.g., *Streptococcus mutans* and *Staphylococcus aureus*) in GECs [15]. It is known that *S. aureus* utilizes the autophagosomes as a replicative niche [16, 17]. We showed that the internalization of *S. aureus* is T2R14 dependent in GECs; therefore, it is important to analyze how cariogenic bacteria use the autophagy machinery for internalization and survival after evading the host immune system.

Here, we described two complementary methods for analyzing autophagy flux in GEC (*see* Fig. 1). The autophagic flux was characterized using Western blot analysis of the autophagic markers SQSTM1 (encoding sequestosome 1 and is also known as p62) and LC3β-II [18–20]. We also used Acridine Orange staining of acidic vesicular organelles (AVO) by flow cytometry to measure autophagy flux in GECs upon serum starvation, which is a physiological induction of autophagy [14, 21].

2 Materials

2.1 Cell Line

Oral gingival epithelial cell line GEC and the T2R14-knockout GEC cells were described in our previous studies [12, 14, 15].

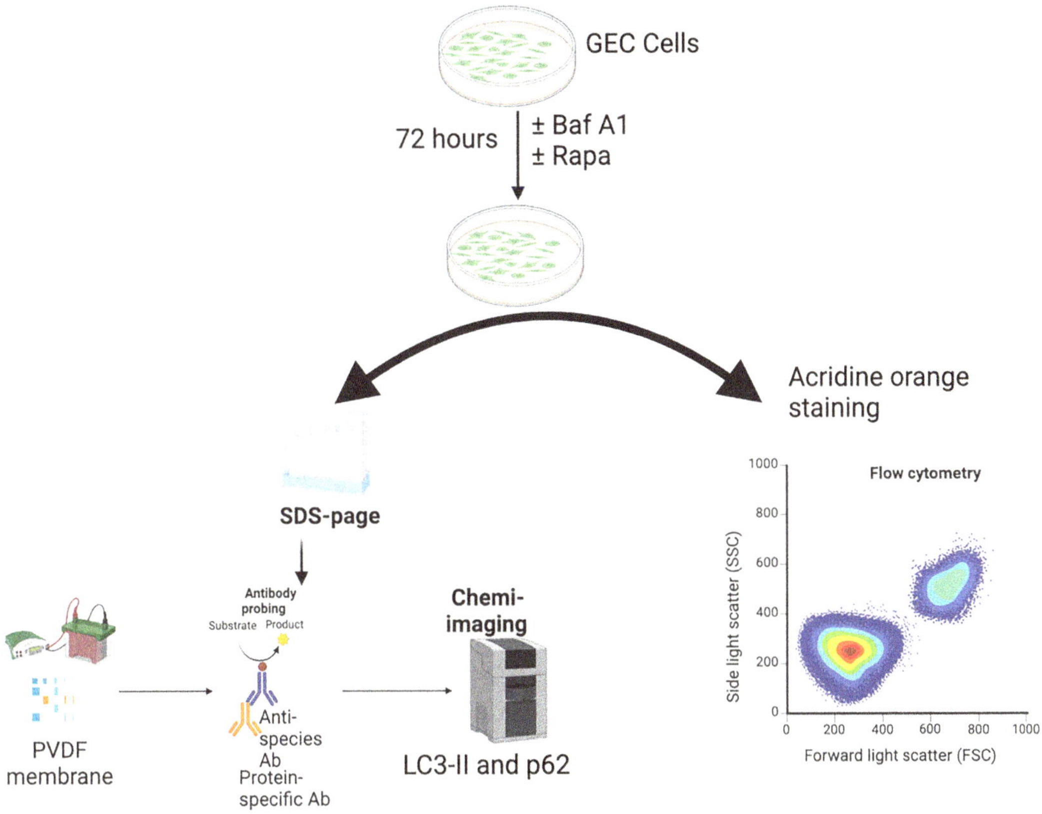

Fig. 1 Graphical abstract: Schematic showing the methodologies (Western blot and flow cytometry) used for the assessment of autophagy flux in GEC (created with Biorender). GEC: Gingival epithelial cells, BafA1: Bafilomycin, Rapa: Rapamycin, LC3-II: microtubule-associated light chain protein, p62: sequestosome 1

2.2 Cell Culture and Treatments

1. Keratinocyte growth medium-2 (KGM-2) for GEC cell culture was purchased from promo cell (Heidelberg, Germany Cat # C20011).
2. Acridine Orange (CAS Number: 65-61-2), bafilomycin A1 (Cat#B1793-10UG), and rapamycin (53210-100UG) were purchased from Sigma-Aldrich Co., Canada.
3. Antibodies used in the study: mouse SQSTM1/p62 antibody (5114) was purchased from Cell Signaling Technology (Ontario, Canada). LC3β antibody (catalog number L7543) and monoclonal anti-β-actin (catalog number #A5441) were obtained from Sigma Aldrich, Oakville, Ontario, Canada. The goat anti-rabbit IgG-HRP conjugate (catalog number #17-6515) and goat anti-mouse IgG-HRP conjugate (catalog number #A-10668) were purchased from Bio-Rad, Mississauga, Ontario, Canada.
4. Flow cytometry: Beckman Coulter CytoFlex LX Digital Flow Cytometry Analyzer Four Laser System at the University of Manitoba, Flow cytometry Core facility.

3 Methods

3.1 Experiments for Western Blot Analysis

1. OKF6 cells, both WT and T2R14 KO (40% confluent), were cultured in 10 cm tissue culture dishes (*see* **Note 2**).
2. Treat the cells with media alone or media containing bafilomycin (10 nM) and rapamycin (1000 nM) for 72 h (*see* **Note 2**).
3. After 72 h, cell culture supernatants were collected in 15 mL tubes and cells were washed two times with 1X PBS. Finally, cell scraper was used to collect the GEC cells.
4. Cells containing media were centrifuged (350xg) and washed twice with phosphate buffer saline (PBS).
5. Cells were lysed in lysis buffer containing (1% Triton X 100 and Tris buffer) using sonicator (3–5 pulses) and centrifugation at 10,000 g for 10 min at 4°C to get rid of the cell debris. Supernatants were collected, and protein concentrations were determined using the DC protein assay (Bio-Rad, Canada).
6. Western blot analysis using cell lysates (10 μg total protein per well) for LC3β-II (1:2500) and p62 (1:1000), markers for autophagy flux were performed. 15% SDS-PAGE gels were run for separating the protein lysates.
7. Separated proteins were transferred to PVDF membrane for 90 min at 4 °C with a constant voltage of 100V in transfer buffer (25 mM Tris, 192 mM glycine, pH 8.3, and 20% methanol).
8. Block the nonspecific protein by blocking the PVDF membrane in 5% skim milk in TBST (20 mM Tris, 150 mM NaCl, 0.1% Tween 20, pH 7.6) at room temperature with constant agitation for 1 h.
9. Incubate the membrane with LC3β-II (1:2500) and p62 (1: 1000) diluted in TBST buffer at 4 °C with constant agitation overnight.
10. The following day, rinse the membrane three times for 10 min each with TBST buffer. Then, incubate the membrane with an horseradish peroxidase (HRP)-conjugated antibody (anti-mouse IgG, anti-rabbit IgG, or anti-goat IgG) diluted in TBST for 1 h at room temperature.
11. Capture the antibody reaction by chemiluminescence using ECL Plus and normal ECL (Bio-Rad, Canada) (*see* **Note 4**).
12. Incubate the blots sequentially at room temp for 2 h with monoclonal anti-β-actin antibody (Sigma) diluted at 1: 10,000 and process as described above.
13. Standardize the densitometry readings of positive bands by comparing them to the corresponding value of β-actin. Chemiluminescence was assessed using the Bio-Rad ChemiDoc MP

imaging system, and the blot densities were analyzed using Image Lab version 5.2.1 software from Bio-Rad.

14. Statistical significance was calculated using two-way ANOVA with Tukey's post-hoc test (*see* **Note 1**).

3.2 Experiments with Acridine Orange (AO) Flow Cytometry Analysis

1. OKF6 cells, both WT and T2R14 KO (20,000), were seeded in 12-well plates under serum-starved conditions (*see* **Note 1**).
2. Treat the cells with media alone or media containing Baf A1 (10nM) and rapamycin (1000 nM) for 72 h (*see* **Note 3**).
3. After 72 h, cells were trypsinized with 0.25% Trypsin-EDTA (Invitrogen, Canada) for 10 min at 37 °C. Trypsin treated cells were neutralized with DMEM F12 media containing 10% serum and centrifuged for 4 min at 300 g to get the cell pellets (*see* **Note 5**).
4. OKF6 cells were stained with Acridine Orange (final concentration of 1 μg/mL) and incubated at 37 °C for 10 min in dark (*see* **Note 6**).
5. Stained cells were washed once with PBS, and the flow cytometry analysis using cytoflex flow cytometer was performed for evaluating the autophagy flux (Red and Green fluorescence) [21, 22] (*see* **Note 7**).
6. Acquisition was done by using channels in the red (PECy5) and green (FITC) lasers and recording 10,000 events.
7. Acridine Orange flow plots were analyzed on FlowJo Version 10 software (TreeStar, Ashland, OR)

3.3 Western Blot Analysis of Autophagy Flux in GEC

Our previous studies suggest that T2R14 is primarily involved in the mediation of innate immune responses to bacterial infection in GECs [12, 15]. To determine whether and how oral bacteria utilize the autophagic machinery of GEC, it is necessary to characterize the role of T2R14 in modulating autophagic flux [14].

To assess the impact of T2R14 on autophagy flux in GECs, oral keratinocyte cells (OKF6 Wt for wild type and OKF6 T2R14 KO for knockout) were utilized as previously described [15]. Autophagic flux was assessed via Western blot analysis of the autophagic markers p62 and LC3β-II. Both cell lines underwent treatment with an autophagy flux inhibitor, bafilomycin A1 (Baf A1, 10 nM), and an autophagy flux inducer, rapamycin (1000 nM), for 72 h. This treatment of GECs did not result in any noteworthy alteration in cell viability. After 72 h of treatment, cells were lysed and LC3 lipidation and p62 degradation were analyzed by Western blot. These results suggested that in the basal condition (i.e., without Baf A1 or Rapa treatment), more LC3β-II ($*p<0.05$) and p62 accumulated in T2R14 KO cells relative to WT, suggesting that fewer autophagosomes were being degraded in the T2R14 KO GECs (Fig. 2a–c). To further confirm this, both cells were treated

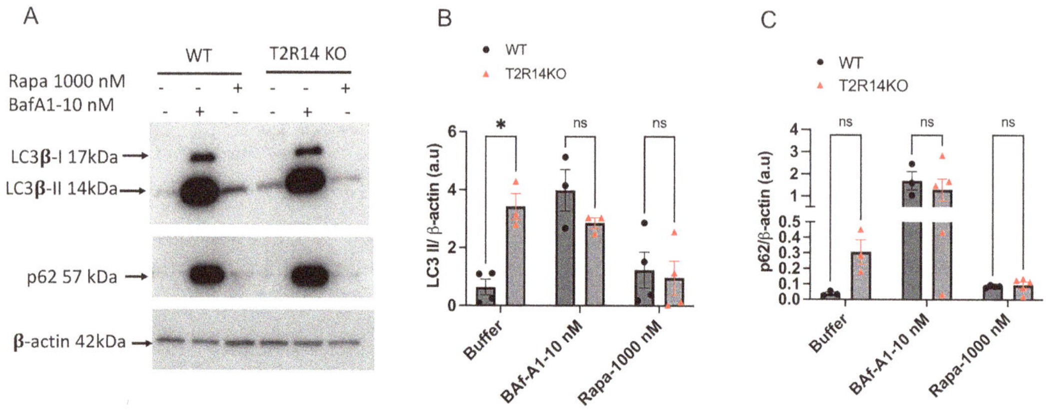

Fig. 2 T2R14-mediated autophagy flux in GEC. A. Western blot analysis of autophagy flux in GEC. (**a**) Western blot detection of LC3β-II and p62 autophagy markers in WT and T2R14KO cells upon incubation with autophagy blocker bafilomycin (Baf A1-10 nM) and activator rapamycin (Rapa 1000 nM) for 72 h. (**b**, **c**) Quantification of relative levels of LC3β-II/actin and p62/actin by densitometric analysis. The bar graphs were generated using graph pad prism 9.0. Statistical significance was calculated using two-way ANOVA with Tukey's post-hoc test; *$p<0.05$, **$p<0.01$. The data for B and C represent SEM of three to four independent experiments

with autophagy flux inhibitor Baf A1 for 72 h. Baf A1 treatment leads to further increase in LC3β-II and p62 accumulation in both WT and T2R14KO GECs (Fig. 2a–c). Treatment of GECs with Baf A1 blocked lysosomal degradation of autophagosome efficiently and thereby led to an accumulation of autophagic vesicles. Next, the cells were treated with rapamycin, an autophagic flux inducer, and changes in LC3β-II and p62 were assessed. LC3β-II and p62 accumulation was reduced in T2R14 KO cells relative to WT under conditions of induced autophagic flux (Fig. 2a–c). Overall, on the basis of LC3β-II and p62, the presence of T2R14 in GECs down-regulates autophagy flux under basal conditions.

3.4 Acridine Orange Staining for Detecting Acidic Vesicular Organelles (AVO) by Flow Cytometry Analysis in GEC

Acridine Orange staining is a quick and robust assay for monitoring and detecting AVO [22, 23]. WT and T2R14KO GECs cell lines were serum starved for 72 h to induce autophagy [20, 24]. Cells were additionally subjected to treatment with Baf A1 (10 nM) and Rapa (1000 nM) for 72 h to assess autophagy flux (measured through red and green fluorescence) using a CytoFLEX flow cytometer [22]. Our findings indicate that under conditions of nutrient deprivation, which promote autophagy, T2R14KO cells exhibit a significant increase in red fluorescence compared to WT cells, while no change in green fluorescence is noted (Fig. 3). Moreover, treatment of WT cells with the autophagy flux inhibitor Baf A1 results in a complete loss of red fluorescence (Fig. 3). Notably, BafA1 treatment inhibits vacuolar H(+)-ATPase (V-ATPase), preventing lysosomal acidification and consequently eliminating red fluorescence. Intriguingly, the autophagy flux inducer Rapa and

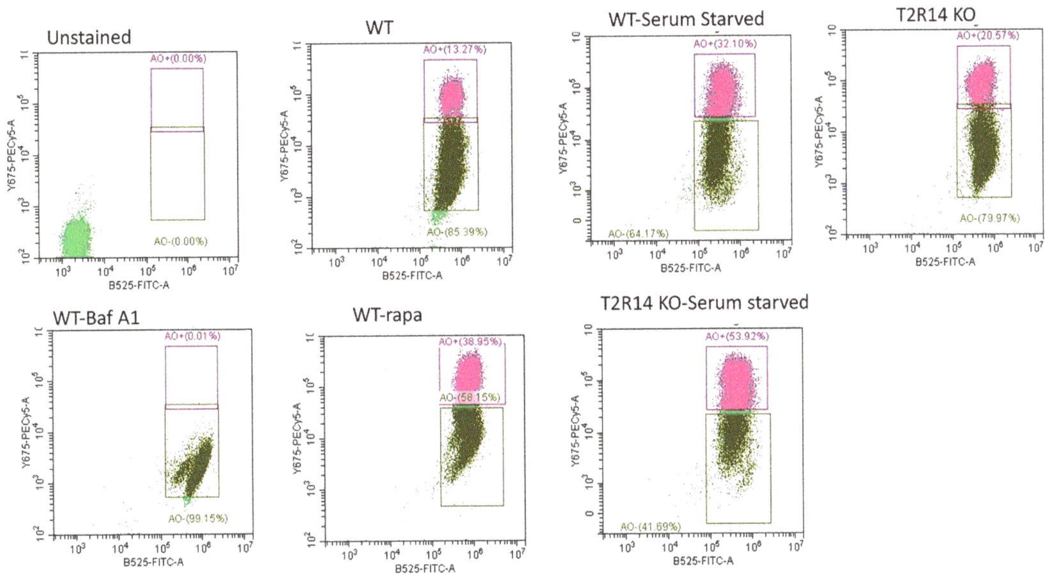

Fig. 3 Acridine Orange staining for detecting acidic vesicular organelles (AVO) by flow cytometry analysis in GEC. Flow cytometry plots depicting red, green, and red/green fluorescence ratio were utilized to assess autophagy flux in both WT and in T2R14KO GECs under serum-starved conditions and normal growth conditions. Representative flow plots illustrate the acquisition of data on red and green fluorescence in Acridine Orange-stained GECs (WT and T2R14 KO) subjected to serum starvation or treated with Baf A1 (10 nM) and Rapa (1000 nM) for 72 h to detect autophagy flux using red and green lasers. Acridine Orange flow plots were analyzed using FlowJo Version 10 software (TreeStar, Ashland, OR)

serum starvation lead to an increase in red fluorescence, indicating that stimuli inducing autophagy boost the number of acidic vesicular organelles (AVO) (Fig. 3). Hence, based on AVO, the presence of T2R14 in GECs influences the abundance of actively acidic lysosomes.

4 Notes

1. It is recommended to have at least three replicates for each condition, although having a minimum of three replicates is necessary for statistical analysis. Ideally, it would be beneficial to conduct four or more replicates for each condition. Additionally, ensure that cells are serum-starved and grown in 1% growth factor as specified in the Acridine Orange staining protocol.
2. OKF6 cell culture media preparation and cultures should be performed in a Biosafety cabinet.
3. After exposure to Baf A1 and rapa treatments, monitor the OKF6 cell morphology under the light microscope for 72 h.
4. Avoid long exposure of blots for LC3β-II for overexposure of bands especially in Baf A1 and rapa treatments.

5. OKF6 cells are very adherent cells, so use 0.25% trypsin EDTA for cells trypsinization and immediately neutralize the cells with DMEM F12 with 10% serum containing media and spin down the cells at 300× *g* for 5 min.
6. The tubes containing the Acridine Orange dye were placed in a rack and covered with foil paper because Acridine Orange is light sensitive and can easily degrade.
7. Acridine Orange-based autophagy flux flow cytometry should be performed within 30 min of staining.

Acknowledgments

This work was supported by Grant No. PJT-159731 from the Canadian Institute of Health Research (CIHR) to PC.

Author Contributions N.S conducted the experiments and analyzed the data. N.S and P.C wrote the manuscript. P.C, N.S, and S. G contributed to the conceptualization, design, and data interpretation. All authors critically revised the manuscript.

References

1. Germic N, Frangez Z, Yousefi S, Simon HU (2019) Regulation of the innate immune system by autophagy: neutrophils, eosinophils, mast cells, NK cells. Cell Death Differ 26(4): 703–714
2. Siapoush S, Rezaei R, Alavifard H, Hatami B, Zali MR, Vosough M et al (2023) Therapeutic implications of targeting autophagy and TGF-beta crosstalk for the treatment of liver fibrosis. Life Sci 329:121894
3. Xie Z, Klionsky DJ (2007) Autophagosome formation: core machinery and adaptations. Nat Cell Biol 9(10):1102–1109
4. Alizadeh J, da Silva Rosa SC, Weng X, Jacobs J, Lorzadeh S, Ravandi A et al (2023) Ceramides and ceramide synthases in cancer: Focus on apoptosis and autophagy. Eur J Cell Biol 102(3):151337
5. Kim J, Kundu M, Viollet B, Guan KL (2011) AMPK and mTOR regulate autophagy through direct phosphorylation of Ulk1. Nat Cell Biol 13(2):132–141
6. Deretic V, Levine B (2018) Autophagy balances inflammation in innate immunity. Autophagy 14(2):243–251
7. Deretic V (2021) Autophagy in inflammation, infection, and immunometabolism. Immunity 54(3):437–453
8. Wauson EM, Zaganjor E, Lee AY, Guerra ML, Ghosh AB, Bookout AL et al (2012) The G protein-coupled taste receptor T1R1/T1R3 regulates mTORC1 and autophagy. Mol Cell 47(6):851–862
9. O'Rourke EJ, Kuballa P, Xavier R, Ruvkun G (2013) omega-6 Polyunsaturated fatty acids extend life span through the activation of autophagy. Genes Dev 27(4):429–440
10. Medapati MR, Bhagirath AY, Singh N, Chelikani P (2022) Pharmacology of T2R mediated host-microbe interactions. Handb Exp Pharmacol 275:177–202
11. Gopallawa I, Freund JR, Lee RJ (2021) Bitter taste receptors stimulate phagocytosis in human macrophages through calcium, nitric oxide, and cyclic-GMP signaling. Cell Mol Life Sci 78(1):271–286
12. Medapati MR, Singh N, Bhagirath AY, Duan K, Triggs-Raine B, Batista EL Jr et al (2021) Bitter taste receptor T2R14 detects quorum sensing molecules from cariogenic Streptococcus mutans and mediates innate immune responses in gingival epithelial cells. FASEB J 35(3):e21375

13. Pan S, Sharma P, Shah SD, Deshpande DA (2017) Bitter taste receptor agonists alter mitochondrial function and induce autophagy in airway smooth muscle cells. Am J Physiol Lung Cell Mol Physiol 313(1):L154–LL65
14. Singh N, Ulmer B, Medapati M, Schroth R, Ghavami S, Chelikani P (2024) Bitter taste receptor T2R14 and autophagy flux in gingival epithelial cells. bioRxiv. https://doi.org/10.1101/2024.01.17.576143
15. Medapati MR, Bhagirath AY, Singh N, Schroth RJ, Bhullar RP, Duan K et al (2021) Bitter taste receptor T2R14 modulates gram-positive bacterial internalization and survival in gingival epithelial cells. Int J Mol Sci 22(18):9920
16. O'Keeffe KM, Wilk MM, Leech JM, Murphy AG, Laabei M, Monk IR et al (2015) Manipulation of autophagy in phagocytes facilitates staphylococcus aureus bloodstream infection. Infect Immun 83(9):3445–3457
17. Mulcahy ME, O'Brien EC, O'Keeffe KM, Vozza EG, Leddy N, McLoughlin RM (2020) Manipulation of autophagy and apoptosis facilitates intracellular survival of staphylococcus aureus in human neutrophils. Front Immunol 11:565545
18. Alizadeh J, Kochan MM, Stewart VD, Drewnik DA, Hannila SS, Ghavami S (2021) Inhibition of autophagy flux promotes secretion of chondroitin sulfate proteoglycans in primary rat astrocytes. Mol Neurobiol 58(12):6077–6091
19. Dastghaib S, Shojaei S, Mostafavi-Pour Z, Sharma P, Patterson JB, Samali A et al (2020) Simvastatin induces unfolded protein response and enhances temozolomide-induced cell death in glioblastoma cells. Cells 9(11):2339
20. Shojaei S, Koleini N, Samiei E, Aghaei M, Cole LK, Alizadeh J et al (2020) Simvastatin increases temozolomide-induced cell death by targeting the fusion of autophagosomes and lysosomes. FEBS J 287(5):1005–1034
21. Hajiahmadi S, Lorzadeh S, Iranpour R, Karima S, Rajabibazl M, Shahsavari Z et al (2023) Temozolomide, simvastatin and acetylshikonin combination induces mitochondrial-dependent apoptosis in GBM cells, which is regulated by autophagy. Biology (Basel) 12(2):302
22. Thome MP, Filippi-Chiela EC, Villodre ES, Migliavaca CB, Onzi GR, Felipe KB et al (2016) Ratiometric analysis of acridine orange staining in the study of acidic organelles and autophagy. J Cell Sci 129(24):4622–4632
23. Klionsky DJ, Abdel-Aziz AK, Abdelfatah S, Abdellatif M, Abdoli A, Abel S et al (2021) Guidelines for the use and interpretation of assays for monitoring autophagy (4th edition) (1). Autophagy 17(1):1–382
24. Ghavami S, Yeganeh B, Zeki AA, Shojaei S, Kenyon NJ, Ott S et al (2018) Autophagy and the unfolded protein response promote profibrotic effects of TGF-beta(1) in human lung fibroblasts. Am J Physiol Lung Cell Mol Physiol 314(3):L493–L504

Methods in Molecular Biology (2025) 2879: 183–206
DOI 10.1007/7651_2024_516

Published online: 06 March 2024

Involvement of Autophagic Machinery in Neuropathogenesis: Targeting and Relevant Methods of Detection

Nourhan Sayed, Alaa Emam Ali, Doaa Mokhtar Elsherbiny, and Samar S. Azab

Abstract

The exquisite balance between cellular prosurvival and death pathways is extremely necessary for homeostasis. Different forms of programmed cell death have been widely studied and reported such as apoptosis, necroptosis, pyroptosis, and autophagy. Autophagy is a catabolic process important for normal cellular functioning. The main aim of this machinery is to degrade the misfolded or damaged proteins, unuseful organelles, and pathogens, which invade the cells, thereby maintaining cellular homeostasis and assuring the regular renewal of cell components. This prosurvival function of autophagy highlights its importance in many human diseases, as the disturbance of this tightly organized process ultimately causes detrimental effects. Interestingly, neurons are particularly susceptible to damage upon the presence of any alteration in the basal level of the autophagic activity; this could be due to their high metabolic demand, post-mitotic nature, and the contribution of autophagy in the different fundamental functions of neurons. Herein, we have reported the role of autophagy in different CNS disorders such as Parkinson's disease, Alzheimer's disease, Huntington's disease, and epilepsy, besides the pharmacological agents targeting autophagy. Due to the significant contribution of autophagy in the pathogenesis of many diseases, it is crucial to develop effective methods to detect this dynamic process. In this chapter, we have summarized the most frequently employed techniques in studying and detecting autophagy including electron microscopy, fluorescence microscopy, Western blotting, intracellular protein degradation, and sequestration assay.

Key words Autophagy, CNS Homeostasis, Microscopy, Neuropathogenesis, Sequestration assay, Western blotting

1 Introduction

Delicate intracellular balance between prosurvival signals and programmed cell death pathways is extremely important for cellular homeostasis [1]. Various types of programmed cell death (PCD) have been extensively studied and reported such as apoptosis, necroptosis, pyroptosis, and autophagy [2–4]. For example, apoptosis, the well-known form of PCD, is evident to be triggered through two main distinct molecular pathways, the extrinsic and the intrinsic pathways. Briefly, the extrinsic pathway is initiated by binding extracellular ligands such as tumor necrosis factor (TNF-α) or Fas ligand (Fas-L) to death receptors on plasma membranes. These receptors are capable of activating caspase 8 (an initiator caspase) via the interaction with various adaptor proteins. Then through an autocatalytic activation cascade, active caspase 8 promotes the activation of the effector caspase, caspase 3, leading to fragmentation of DNA, damaging the cytoskeleton and finally phagocytosis takes place [5, 6]. The second arm of apoptosis, the intrinsic pathway, is initiated when the cell is subjected to a stressful condition which subsequently leads to the activation of proapoptotic proteins, Bcl-2-associated X protein (Bax), and BCL-2 antagonist/killer1 (Bak), resulting in the formation of pores in the mitochondrial membrane and the subsequent release of cytochrome c, then cytochrome c binds to apoptotic peptidase activating factor 1 (APAF-1) forming the apoptosome, in which procaspase 9 is activated. After that, caspase 9 activates caspase 3 leading to cell death [5, 7]. On the other hand, the caspase-independent programmed necrosis, necroptosis, is regarded to have a role in the pathogenesis of various human diseases [8, 9]. Necroptosis relies on the activation of receptor-interacting protein kinase 1 (RIPK-1) that induces receptor-interacting protein kinase 3 (RIPk-3) phosphorylation. Then RIPK1/RIPK-3 complex phosphorylates a key molecule in the necroptosis pathway which is the pseudo-kinase mixed-lineage kinase domain-like (MLKL) that acts on the destabilization of plasma membrane leading to disrupted ion influx followed by cellular swelling and membrane lysis [7, 8]. Regarding pyroptosis, an inflammation-based cell death, it is operated upon inflammasome NLR family pyrin domain-containing 3 (NLRP3) activation due to different upstream signals. In turn, NLRP3 assembly causes the activation of caspase-1, an inflammatory caspase. The functional active caspase-1 then cleaves pro-interleukin 1β (pro-IL-1β) and pro-IL-18 to generate their mature forms. Moreover, caspase-1 works on Gasdermin D (GSDMD) producing the N-terminal GSDMD fragment. This fragment is capable of disrupting the plasma membrane stability by forming pores in the membrane leading to cell lysis and the release of the cellular contents including the active cytokines IL-1β and IL-18 [2, 10]. Autophagy is generally considered a prosurvival process inside the cell due to its effective role in eliminating the misfolded proteins, damaged organelles, and pathogens. The main feature of the autophagic pathway is the formation of the double-membraned vesicle, autophagosomes, and then its fusion to lysosomes to digest the autophagosome contents. However, its excessive activation or its dysregulation can eventually cause cell death [11, 12]. Figure 1 illustrates the key molecules involved in the above-mentioned types of programmed cell death: apoptosis, pyroptosis, necroptosis, and autophagy. Herein, we will focus on the autophagic machinery as well as the role of autophagy in different CNS disorders and highlight the

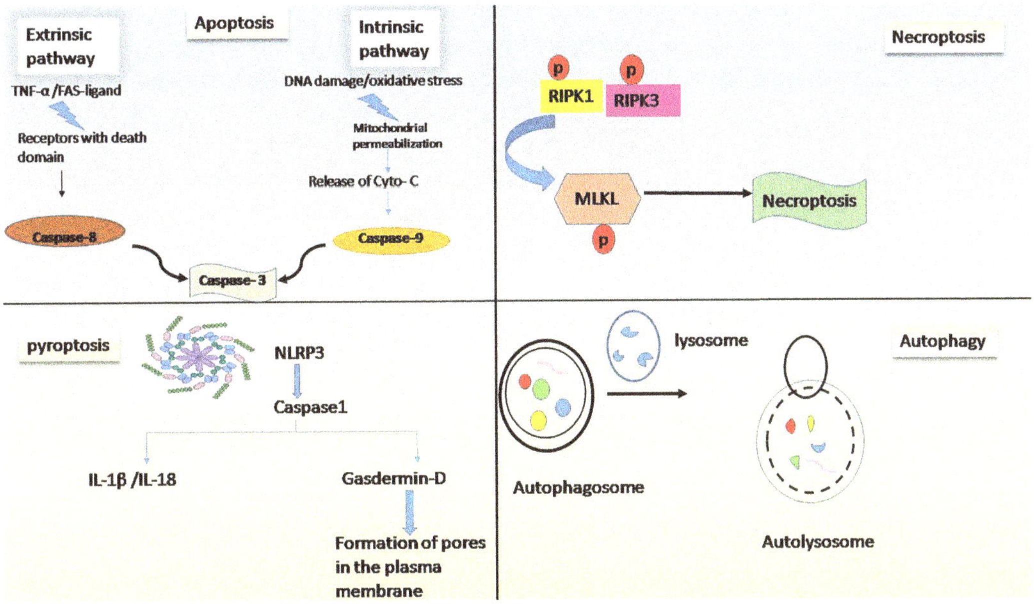

Fig. 1 The key molecules in four types of programmed cell death pathways: apoptosis, pyroptosis, necroptosis, and autophagy. Abbreviations: *TNF-α* tumor necrosis factor α, *Cyto- C* cytochrome C, *IL-1β* interleukin-1β, *IL-18* interleukin-18, *RIPK1* receptor-interacting protein kinase 1, *RIPK3* receptor-interacting protein kinase 3, *MLKL* pseudo-kinase mixed-lineage kinase domain-like

most common practical methods used to study and detect autophagy.

2 Autophagic Machinery

Autophagy is a catabolic process that is essential for normal functioning of the cells. The main mission of this machinery is to degrade the misfolded or damaged proteins and unuseful organelles in order to maintain the cellular homeostasis and aid in the regular renewal of cell components [11]. Another beneficial consequence of autophagy is that cells can reuse the nutrients such as amino acids after the turnover of the proteins by this machinery [13]. Moreover, it can be regarded as a cellular adaptation to different forms of stressors, since the cells urge this self-eating habit to overcome possible damaging signals. This prosurvival function of autophagy highlights its importance in many human diseases, as the delicate disturbance of this tightly organized process ultimately causes detrimental effects [14, 15].

Autophagy can be broadly classified into three main categories: microautophagy, chaperone-mediated autophagy, and macroautophagy. In brief, microautophagy means that lysosomes directly engulf a small portion of cytoplasm for degradation by lysosomal enzymes. Chaperone-mediated autophagy (CMA) is a more selective process in which cytoplasmic proteins containing KFERQ motif are recognized by heat shock cognate 71 Kda (Hsc70) protein and then delivered to lysosomal membrane by binding to lysosomal-associated membrane protein 2A (LAMP2A) [13]. Macroautophagy is considered to be the most important mode of autophagy, since its role in the turnover of cell components is quantitatively significant compared to the other two modes [16]. Macroautophagy generally proceeds in three distinct steps. First, autophagosomes form in which a double-membrane vesicle is formed and encloses the targeted protein or organelle. Second, autophagosomes–lysosome fusion occurs. Third, the degradation of the auto lysosomal contents by lysosomal enzymes and the discharge of the remaining fatty acids and amino acids back to the cytoplasm take place [17]. Mechanistically, the macroautophagy is initiated through the activation of Unc-51-like kinase 1 (ULK1) complex followed by activation of PI3K- III complex so that the double-membraned vesicle starts to develop. Then, elongation of the membrane proceeds till the formation of autophagosomes that engulf the cargo for digestion. The autophagosome membrane is distinguished from other membranes inside the cell by the presence of microtubule-associated protein light chain 3 (LC3-II) molecule. LC3-II is generated in two steps: first, ATG-4 converts pro-LC3 to soluble LC3-I, and then LC3-I is conjugated to phosphatidylethanolamine (PE) through the action of ATG3 and ATG7 producing membrane-bound LC3-II [15]. LC3-II molecule allows the interaction of cargo receptors such as SQSTM1/P62 [18, 19]. Figure 2 describes in brief the major steps in the pathway of macroautophagy.

3 Role of Autophagy in Different CNS Disorders

Since autophagy has been proved to engage in preserving the cell homeostasis and in the protection of cells against different forms of stressors, its impairment greatly influences the health and the survival of cells [20]. The neurons, in particular, are highly sensitive to any deviation that might occur in the normal autophagic activity level. This could be, on the one hand, due to the contribution of autophagy in many neuronal activities such as synaptic plasticity, neurotransmission, and dendrite spine formation [21]. On the other hand, the unique features that characterize the neurons, such as the high metabolic demands [22] and the inability to perform mitosis, make the neurons vulnerable to damage and death in case of the presence of any defect in the autophagic machinery [23]. Therefore, this section will shed light on the role of autophagy in different CNS diseases, which is summarized in Table 1.

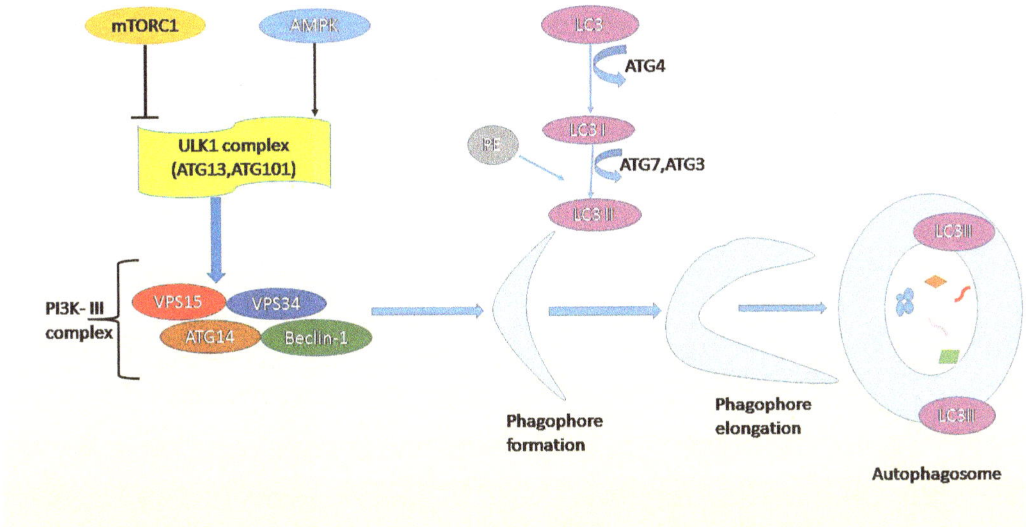

Fig. 2 The major steps in the pathway of macroautophagy. Abbreviations: *mTORC1* mammalian target of rapamycin complex 1, *AMPK* AMP-activated protein kinase, *ULK1* Unc-51-like kinase 1, *PI3K- III* phosphoinositide 3-kinase, *VPS15* vacuolar protein sorting 15, *VPS34* vacuolar protein sorting 34, *LC3* Microtubule-associated protein light chain 3, *PE* phosphatidylethanolamine

3.1 Autophagy in Parkinson's Disease

Parkinson's disease (PD), one of the most prevalent neurodegenerative diseases, has been the focus for the investigators to examine the underlying molecular and cellular changes taking place in the brain of PD patients. It is evident that oxidative stress, mitochondrial dysfunction, and presence of protein aggregates are closely related to the initiation of neurodegeneration and damaging of the dopaminergic neurons, which are the main basic abnormalities observed in the brain of PD patients [24]. Autophagy, being the normal cellular machinery for clearance of damaged organelles and aggregated proteins, is definitely involved in the pathophysiology of this disease [25]. Consequently, it is not surprising that many studies have shed light on the possible role of autophagy in PD and have tried to uncover the gene mutations and abnormalities occurring in the autophagic pathway.

Among the abnormalities that take place in PD, misfolded or aggregated α-synuclein participates in the appearance of dysfunctional dopaminergic neurons and is responsible for the clinical symptoms. α-Synuclein protein is a basic constituent of the cytoplasmic inclusions known as "Lewy bodies," a key pathological feature in the PD brain [26]. In general, α-synuclein protein is routinely cleared from the cell by macroautophagy and CMA, with the observation of the upper hand for CMA machinery [25, 27]. However, the rising levels of α-synuclein disrupt the macroautophagy by suppressing the last step in the autophagic

Table 1
Role of autophagy in the pathogenesis of different CNS disorders

Disease	Pathogenic hallmark	Role of autophagy
Parkinson's disease	Misfolded or aggregated α-synuclein is a basic constituent of the cytoplasmic inclusions known as "Lewy bodies," a key pathological feature in the dysfunctional dopaminergic neurons [26]	α-Syn overexpression decreased the abundance of the v-SNARE protein SNAP29, a member of the SNARE complex mediating autophagolysosome fusion [28] A53T and A30P synuclein mutants bound to the receptor LAMP-2A on the lysosomal membrane and they act as uptake blockers, thus inhibiting their own degradation by CMA and also the degradation of the other substrates [29] Mutation of *PARK2* and *PARK6 genes* results in impaired mitophagy [33]
Alzheimer's disease	Two pathological hallmarks are evident, extracellular plaques of amyloid β (Aβ) and intracellular neurofibrillary tangles of phosphorylated Tau protein [41]	Defective autophagy causes the increase in Aβ production from APP through the upregulated action of BACE1 and γ secretase [46–48]
Huntington's disease	Intracellular accumulation of mutated huntingtin protein is a characteristic feature observed in the cortex and striatum regions [55]	Mutant huntingtin protein binds p62 and also the polyubiquitinated proteins leading to the failure of their recognition by autophagosomes [59]
Epilepsy	Imbalance between excitatory and inhibitory signals in the brain [60]	Reduction in the activity of autophagy leads to decreased GABA receptor presentation on the plasma membrane due to sequestration of GABARAP by increased p62-containing aggregates [63]

Abbreviations: *α-Syn* α-synuclein, *SNARE* soluble N-ethylmaleimide sensitive factor attachment protein receptors, *CMA* chaperone-mediated autophagy, *LAMP-2A* lysosomal-associated membrane protein 2A, *Aβ* amyloid β, *APP* amyloid precursor protein, *BACE1* β secretase 1, *AD* Alzheimer's disease, *GABA* γ-aminobutyric acid, *GABARAP* γ-aminobutyric acid receptor-associated proteins

route in which autophagosomes merge with lysosomes to generate autolysosomes where the cargoes are digested. The obstruction of fusion is discovered to be due to the diminished level of SNAP29, one of the crucial players in the autophagosome–lysosome incorporation. Due to the failure of lysosomal fusion, the number of autophagosomes increases by time and they tend to fuse to plasma membranes and release their contents in the extracellular space, instigating severe toxic effects [28]. Moreover, mutant forms of α-synuclein protein, A53T and A30P, hamper the CMA process. These mutants bind to LAMP-2A receptors found on the lysosomal membrane but fail to enter inside the lysosome for degradation [25, 29, 30]. This inhibitory action on the receptor prevents the normal functioning of CMA machinery leading to the accumulation of not only α-synuclein but also other proteins routinely degraded by CMA, and thus neurotoxicity occurs [30]. As a compensatory mechanism, cells tend to activate the macroautophagy to get rid of the aggravated levels of toxic proteins and damaged organelles accumulated after CMA inhibition. But unfortunately, the excessive activation of macroautophagy might not favor cell survival but rather might cause cell death [25, 31]. Typically, detection of autophagosomes in the normal brain is very minimal due to the rapid turnover of these vesicles. However, an aberrant increase in the number of autophagosomes observed in the brain of PD patients with dysfunctional clearance of these autophagosomes leads to neuronal cell death [32]. Collectively, maintaining the macroautophagy activity at a certain optimal level is vitally important for typical functioning of cells.

Moreover, mutation in the genes affecting mitophagy (mitochondrial specific autophagy), such as *PARK2* encoding Parkin and *PARK6* encoding PINK1, has been reported to contribute to the pathogenesis of PD [33]. Mutation in *PARK2* gene has been discovered in 50% of autosomal recessive and 10–15% of sporadic early onset PD [25]. At steady state, Parkin, a cytosolic ubiquitin E3 ligase, has the ability to discriminate between normal and impaired mitochondria since it is attracted to the mitochondria with reduced or no membrane potential [34]. PINK1, a mitochondrial membrane–anchored kinase, is normally cleaved in mitochondria and then released to cytosol. On the other hand, depolarization of mitochondria interferes with this cleavage leading to the accumulation of PINK1 [25, 35]. Then, accumulated PINK1 recruits Parkin to damaged mitochondria executing ubiquitination of certain substrates on mitochondria [25, 35, 36]. These ubiquitinated substrates are recognized by autophagic adapters such as p62/SQSTM1, which is a ubiquitin- binding protein that mediates the interaction with LC3 to finally recruit the cargo to the autophagosomes [19]. The role of p62 in PINK-/Parkin-mediated mitophagy is controversial, with some studies supporting that it is essential in completing the autophagic process and others reporting that other autophagic adapters may be involved [37–39]. Taken together, flawed mitophagy caused by *PARK2* and *PARK6* mutation is associated with the development of PD [40]. Emphasizing that preserving perfectly working mitochondria and getting rid of the defective ones are unexceptionally essential for preventing toxic ROS accumulation and for maintaining cellular homeostasis.

3.2 Autophagy in Alzheimer's Disease

Alzheimer's disease (AD) is a neurodegenerative disease that is accompanied by a progressive impairment in learning and memory functions. Investigations on the pathological features in the brains of AD patients revealed that two pathological hallmarks are widely evident—extracellular plaques of amyloid β (Aβ) and intracellular neurofibrillary tangles of phosphorylated Tau protein

[41]. Accumulation of Aβ is extremely toxic to neuronal cells due to its devastating actions on different cellular components, including increasing oxidative stress, mitochondrial dysfunction, inflammation, and synaptic dysfunction [42, 43]. Autophagy, which plays a crucial indispensable role in cleaning the cells from toxicants, is also involved in eliminating accumulated Aβ. Moreover, autophagy contributes to the production of Aβ [44]. Aβ is formed due to the action of β and γ secretase on amyloid precursor protein (APP) [45]. Defective autophagy enhanced the production of Aβ through multiple ways. First, β secretase is usually degraded within the lysosomes. Moreover, autophagosomes aid in transporting BACE1 from the distal axons toward the soma for lysosomal clearance. However, defective autophagy blocked this way of transport, thus, the accumulated non-degraded BACE1-boosted-APP processing, and therefore the increased production of Aβ [46, 47]. In addition, impaired autophagy upregulates the expression of the component proteins of γ secretase complex, thereby promoting its activity and further increasing Aβ production [48].

Altered autophagy in AD pathogenesis is evident in the different steps of this pathway. Decreased expression of beclin-1 (a protein required for autophagy) was detected in the brains of AD patients [49]. In addition, p62 expression is also reduced, thus impeding the interaction with LC3 and further impairing the completion of the autophagic process [50]. Also, the movement of autophagosomes from the terminals of axons toward the soma, where the lysosomes normally perform their function, is reduced in the AD brain leading to a significant accumulation of autophagosomes in dystrophic neurons [51]. Finally, a dysfunctional proteolysis within the lysosomes has been also reported [51, 52].

3.3 Autophagy in Huntington's Disease

Huntington's disease (HD) is an inherited neurodegenerative disease with severe symptoms such as involuntary motor movement, cognitive impairment, difficulty in swallowing, and psychiatric symptoms [53, 54]. This brain disorder is caused by a mutation in the gene encoding for huntingtin protein where CAG trinucleotide is repeated more than 36 times encoding for an abnormal protein with long polyglutamine tail [54]. Intracellular accumulation of this mutated misfolded protein is a characteristic feature observed in the cortex and striatum regions of HD brain [55]. The onset of the disease symptoms usually starts at middle age as it is known that the efficiency of the cellular degradation mechanisms declines gradually with aging [56, 57]. Losing the control over the aggravated levels of damaged organelles, toxic substances, and aggregated misfolded proteins disturbs cellular homeostasis and produces neurotoxicity that worsens and progresses when survival mechanisms such as autophagy are impaired [58].

The autophagic pathway impairment in HD was investigated by Martinez-Vicente et al, who combined several methodological procedures to test the efficiency of the various steps of the autophagy process. They found that autophagosomes form properly and fuse into the lysosomes in the usual manner and also proteolytic activity of lysosomes was not affected by the disease. However, examination of the cells using electron microscope revealed that autophagosomes are devoid of cytoplasmic content such as lipid droplets and mitochondria. The researchers suggested that these empty vacuoles failed in the entrapment of the cargo inside the autophagosomes. This could be explained by an aberrant interaction between mutant huntingtin protein and p62; in addition, the mutant protein binds to the polyubiquitinated proteins leading to the failure of their recognition by autophagosomes [59].

3.4 Autophagy in Epilepsy

Epilepsy is a chronic neurological disease characterized by recurrent seizures provoked due to imbalance between excitatory and inhibitory signals in the brain [60]. Multiple mechanisms are considered to be involved in the pathogenesis of epilepsy. Of these mechanisms is autophagy [60, 61]. Autophagy failure contributes to the imbalance between the excitatory and inhibitory signals in the brain, leading to the provoking of seizures [61]. Since Atg7, an E1-like ligase that participates in autophagosomes biogenesis, knock-out mice experienced spontaneous recurrent seizures as observed by McMahon et al, they stated that blocking of autophagy alone is sufficient to promote epileptogenesis [62]. Moreover, reduction in the activity of autophagy decreases the transfer of $GABA_A$ receptors to the surface of plasma membrane, hindering the inhibitory input to the neurons. This is due to the sequestration of γ-aminobutyric acid receptor-associated proteins (GABARAP) (responsible for GABA receptor trafficking) by increased p62-containing aggregates [63]. In addition, impaired mitophagy and the subsequent increase in the level of damaged mitochondria may trigger epileptogenesis [64].

Mammalian target of rapamycin (mTOR) is a key regulator for several cellular processes including growth, proliferation, and autophagy, and its activation inhibits the induction of autophagy [65]. The inhibition of the autophagic machinery observed in different animal models of epilepsy can be explained by the detected hyperactivation of mTOR signaling [60]. Excessive activation of mTOR has been linked to a high incidence of epilepsy in humans and animal models [62]. However, mTOR signaling has multiple effects on the neuronal cells other than the autophagic pathway such as its role in neurogenesis, synaptic plasticity, and other metabolic issues, which may all contribute to the development of seizures independently from autophagy inhibition [66]. Therefore, it is better to directly assess the markers for autophagy along with the markers of the mTOR signaling pathway to detect the status of autophagy in a given experimental model of epilepsy and determine the effect of a certain treatment intervention on the development of seizures.

4 Examples of Pharmacological Modulatory Agents Targeting Autophagy

4.1 Rapamycin

Rapamycin, a well-known immunosuppressant, is a powerful inducer of autophagy as it inhibits the activity of mTOR [67]. Rapamycin exerted promising neuroprotective effects on different PD pharmacological models by rescuing the lysosomal biogenesis and thus protecting the dopaminergic neurons from damage [68]. Moreover, it exhibited beneficial effects and attenuated the neurodegeneration induced by mutant huntingtin protein in fly and mouse model of Huntington's disease [69]. Furthermore, rapamycin modulated a variety of seizure models and epilepsies and reduced the frequency and duration of seizures [70, 71]. However, the inhibition of mTOR has a diverse range of effects inside the cells rather than the regulation of autophagy, so it is possible that the neuroprotection of rapamycin is exerted by multiple interconnected pathways [65].

4.2 Lithium

Lithium is a commonly used mood stabilizer that is particularly used to manage the symptoms of bipolar disorder [72]. Its pharmacological action depends on the inhibition of inositol monophosphatase, thus reducing the amount of available cellular inositol. This decline in the level of inositol has been shown to be responsible for the autophagy enhancing effect for lithium drugs, representing a pathway for autophagy process regulation apart from the mTOR signaling pathway [72, 73]. In preclinical studies including HD cell models and HD *Drosophila* model, the inductive effect of lithium on autophagy was proved to be beneficial in clearing the cells from the mutant huntingtin protein [74]. Moreover, lithium was capable of reducing aggregation of tau protein in SH-SY5Y cells by boosting the autophagy machinery [75]. In addition, lithium showed a neuroprotective effect in an in vitro model of Parkinson's disease, in which lithium treatment protected the cells from apoptosis and also reduced ROS level. These protective effects were diminished by pretreatment with autophagy inhibitors, indicating that induction of autophagy was responsible for the overall neuroprotection offered by lithium [76].

4.3 Rilmenidine

Rilmenidine is a drug used for the treatment of hypertension by acting centrally on imidazoline-1 (I1R) receptors and exerts a sympatholytic activity [77]. The activation of imidazoline receptors causes the reduction of cAMP and thus induces autophagy [78]. The neuroprotective effect of rilmenidine was examined in a mouse model of HD and was proved to decrease the mutant huntingtin protein and attenuate the severity of symptoms

[79]. Moreover, rilmenidine is intended to be used chronically and was proved for its safety, which makes it a good candidate for the management of HD in humans [77].

4.4 Calcium Channel Blockers

Calcium channel blockers are commonly used as antihypertensive drugs. However, they were proved to exert multiple beneficial effects in a diverse range of diseases including brain disorders [80]. It was evidenced that calcium channel blockers are able to induce autophagy in spite of lacking any inhibitory effect on mTOR through the regulation of cAMP [81]. An example of calcium channel blockers that was reported to enhance autophagy is verapamil. Verapamil boosted the clearance of mutant huntingtin in PC12 cells, and this effect is abolished in cells deficient of *Atg5* gene that is essential for autophagy [81]. Moreover, felodipine, one of the calcium channel blockers, provided a neuroprotective effect in a mouse model of PD and reduced A53T α-synuclein mutant aggregation [82].

5 Detection of Autophagy

Traditionally, researchers tend to quantify the number of autophagosomes to determine the autophagic state of the cells. However, the number of autophagosomes alone is not sufficient and probably leads to misinterpretation of the results, since the increase in the number of autophagosomes may represent either the induction of autophagy or blockage of autophagosome–lysosome fusion. So, the term "autophagic flux" then appears to be more suitable for correct estimation of the cell autophagic activity. Autophagic flux can be simply defined as the rate of autophagic degradation, which covers the three steps of autophagy from the autophagosome formation to its fusion with lysosomes ending with the digestion of the constituents. Therefore, actual detection of autophagic flux may probably require a combination of different methods for proper evaluation of this dynamic process [83].

In the following section, we will discuss some of the common methods used to detect autophagy. Electron microscope, which is mainly used for visualization of autophagosomes, has been the first method to be used for this purpose. Also, the fluorescence microscopy has been manipulated by different techniques to measure autophagic flux. Western blotting is used to identify and quantify the key molecules in the autophagic pathway. While the intracellular protein degradation assay has been exploited to distinguish the percentage of proteins degraded via the different forms of autophagy, the sequestration assay ensures the proper functionality of the autophagic process by examining whether the autophagosomes truly engulf cargo for subsequent degradation. Table 2 summarizes the advantages and disadvantages of these methods.

Table 2
Advantages and disadvantages of different methods for autophagy detection

Method	Advantages	Disadvantages
Electron microscopy	Provides detailed morphological information about the engulfed material by the autophagosomes [85]	Not suitable for quantitative measurement of autophagic flux [88] Sample preparation for TEM analysis is quite complex [89] Needs expert investigator [90]
Fluorescence microscopy	LC3-tagged tandem can be used to measure autophagic flux (such as using more than one fluorescent protein or keima) [89, 94] Can be used to detect autophagy for specific organelles (such as detection of mitophagy) [89]	Transfection is required and transfection itself may induce autophagy [88, 90] Self-aggregation of GFP puncta [88] Counting puncta is required, which needs time and effort and also it is a subjective task [90]
Western blotting	It is technically easily achievable due to the availability of reliable commercial antibodies [90]	To measure autophagic flux, comparison between LC3-II/LC3-I ratio should be done in the presence and in the absence of lysosomal degradation inhibitors [90] Caution should be taken when using lysosomal degradation inhibitors since they will be nonspecific (also inhibit lysosomes) in case of the use of very high concentration or for prolonged time [112]
Intracellular protein degradation	Reliable method for detection of autophagic flux [90, 113] Provides a method for CMA detection [102] Applicable to be used in different cell lines [113] Enables processing several samples at the same time [113]	Inability to distinguish between degradation of proteins via CMA and microautophagy [103] The need to use radiolabeled materials [113]
Sequestration assay	Suitable for quantitative analysis [108] Exploit an endogenous substance in the assay [107] Validate the functionality of autophagosomes (their ability to enclose cargo for subsequent degradation not just the formation of the double membraned vesicle "autophagosome") [99, 108]	Caution should be taken during electrodisruption of cells to separate the cytosolic material from the sedimented autophagosomes [106] Can be used to directly quantify the bulk non-selective form of autophagy [107] The need to use a degradation inhibitor (except in case of use of raffinose as an exogenous probe) [106]

Abbreviations: *TEM* transmission electron microscope, *CMA* chaperone-mediated autophagy, *LC3* microtubule-associated protein light chain 3, *GFP* green-fluorescent protein

5.1 Electron Microscopy

Electron microscopy has been traditionally used to detect several cellular activities including autophagy. In the late 1950s, mammalian autophagy was firstly discovered by electron microscopists while they were studying the lysosomes. Since then, great technical advances have occurred in the field of electron microscopy, and also our knowledge of molecular biology profoundly expanded, which made possible the clear visualization of the double membrane vesicle “autophagosome”. However, autolysosome, which is the result of fusion of autophagosomes with lysosomes, is more difficult to analyze and visualize compared to the distinct autophagosome [84]. Although autophagy is a dynamic process and preferentially needs live cell imaging analysis, transmission electron microscope (TEM) can provide snapshots of the process and also detailed insights of the engulfed organelle. To further confirm the identity of the engulfed object, certain approaches can be utilized such as immunogold–electron microscopy (TEM with specific immunogold labeling that enables labeling of LC3 along with visualization by electron microscope) and in addition, correlative light and electron microscopy (which is an integrated system of fluorescent microscope and electron microscope) can also be used as an identifying alternative [85–87]. In spite of these valuable morphological and structural information obtained by TEM, it is not an optimal method to quantify the autophagic flux [88]. In addition, sample preparation for TEM analysis is quite complex and needs an expert investigator [89, 90].

5.2 Fluorescence Microscopy

Another advanced strategy to visualize autophagosomes is to exploit a biomarker known for autophagy and tag it to a fluorescent–protein biosensor. This technology allows for the real-time monitoring of molecular changes in live cells [89]. The most commonly used autophagic biomarker in fluorescence microscopy is LC3. This protein exists in two forms: (i) cytosolic LC3-I which undergoes a conjugation reaction with PE to produce (ii) LC3-II, the lipidated form. LC3-II is then incorporated into the autophagosomal membrane. Consequently, lipidation of LC3-I is a hallmark of the autophagy induction [83, 91], and detection of this biomarker can be done by using a fluorescent protein. The most commonly used fluorescent protein in this case is green fluorescent protein (GFP). The appearance of GFP-LC3-II as green fluorescent punctae represents the number of autophagosomes, so quantification of autophagosomes is possible. However, since GFP is an acid labile protein and therefore degraded inside the lysosomes, it is not applicable to detect the autolysosomes by using the GFP as a fluorescent biosensor [92]. In contrast, the mCherry protein is another fluorescent biosensor that can survive inside the lysosomes as it is stable in the acidic media; therefore, mCherry can be exploited to visualize the autolysosomes. Moreover, combining both proteins GFP and mCherry is a useful way to detect both the autophagosomes and autolysosomes. The mCherry-GFP-LC3 shows a difference in the fluorescence between the autophagosomes and autolysosomes. The appearance of spots with both mCherry (red) and GFP (green) were counted as autophagosomes, while spots with only mCherry (red) were counted as autolysosomes. However, in some cases, spots showing both green and red fluorescence are not actually autophagosomes, but they are defective autolysosomes with neutral pH [92]. This malfunctioning autolysosomes is due to defective vacuolar-type H^+-ATPase (V-ATPase), an enzyme responsible for acidification of the intracellular compartments [92, 93]. To overcome this possible cause of misinterpretation, co-localization assay of mCherry-GFP-LC3-II protein with lysosomal marker is experimentally required as an additional step [92]. Another similar approach involves the use of a fluorescent protein (Keima) that fluoresces at two different wavelengths depending upon the pH of the surrounding medium. So fluorescence emission at 543 nm which is preferred at acidic pH and the fluorescence emission at 458 nm which is preferred at neutral pH are used as a ratio to indicate autolysosomes formation [94]. The use of a single fluorescent protein can have certain advantages over the mCherry-GFP-LC3 fusions including reduction of intramolecular fluorescence transfer and self-aggregation of LC3 [88].

The GFP-LC3 labeling method has been applied to in vivo studies by generating GFP-LC3 transgenic mice. These mice successfully expressed GFP-LC3, with the observation of GFP punctae in almost all tissues after24 hours of fasting period. However, in the brain, GFP-LC3 is not observed even after 48 h of starvation. This could be explained as an absence of autophagy control by nutritional conditions inside the brain, or another explanation is also evident, which is the very rapid turnover of autophagosomes in the brain [84, 95]. Although the application of this method has been widely and successfully used to assess the number of autophagosomes, some potential experimental errors can occur. For example, GFP-LC3 (also the endogenous LC3) can be easily aggregated in the case of overexpression or co-expression with other aggregate prone proteins, which makes it difficult to distinguish between true autophagosomes and GFP-LC3 aggregates [88, 96]. Therefore, some precautions should be followed to reduce this expected aggregation; one of these precautions is the use of stable GFP-LC3 transformants to select clones with appropriate level of GFP-LC3 expression [84]. Furthermore, the usage of a c-terminal glycine mutant of GFP-LC3 (LC3ΔG) as a negative control has been reported to aid in canceling the effect of the false aggregates [97]. The LC3ΔG mutant cannot be conjugated to the PE (therefore cannot form LC3- II and is not able to incorporate into the autophagosomal membrane), so if autophagosome number is truly increased, an increase in wild-type GFP-LC3 punctae will be observed with no increase in mutant LC3ΔG punctae [84]. Another issue to be taken into consideration while investigating the cells using this technique is whether the increase in the number of autophagosomes results from the induction of autophagy or because of the blockage of fusion of autophagosomes to lysosomes (blocking the autophagic flux). To distinguish between these two cases (the case of induction of autophagy and the case of blocking the fusion of autophagosomes to lysosomes), lysosomal inhibitor such as chloroquine or bafilomycin A1 needs to be further applied, and then the numbers of autophagosomes in the presence and in the absence of the lysosomal inhibitor are compared. If the number of autophagosomes tends to increase furthermore after applying the lysosomal inhibitor, then induction of autophagy takes place. While if no change occurs to the number of autophagosomes after the addition of lysosomal inhibitor, then probably blocking the fusion of autophagosomes to lysosomes takes place [89]. Moreover, another issue that may mislead the interpretation of the experimental results is the autophagy-inducing effect of both the transfection protocol and the overexpression of GFP-LC3 itself [96]. Therefore, some researchers recommend using this technique only for qualitative measurements due to the multiple sources of experimental errors [88].

Alternatively, staining of endogenous LC3 may bypass a lot of the problems faced during the usage of exogenous GFP-LC3. Normally in the basal state, LC3 is extremely diffused inside the cell, while by induction of autophagy, LC3 appears in the cytoplasm as punctate. Quantification of LC3-positive puncta has been used to count the number of autophagosomes. Additional information can be gained from this experiment by co-staining LC3 with other specific antibodies to determine the autophagic substrate. For instance, using BNIP3 as a mitochondrial marker with LC3-positive autophagosomes indicates mitophagy. Also co-staining can be performed using cell-permeable organelle-specific dyes such as MitoTracker and ER-Tracker instead of using specific antibodies [88].

5.3 Western Blotting

Western blotting is a common and widely used method to identify and quantify proteins in a complex mixture obtained from cellular or tissue lysate [98]. Most commonly, the LC3 protein is chosen in this method as a key marker and a representative for the autophagic pathway, since it is present on the phagophore membrane and remains attached to the membrane till the formation of mature autophagosomes [99]. The conversion of LC3-I (unconjugated form) to LC3-II (conjugated form) indicates autophagosome formation and incorporation of LC3-II to autophagosomal membrane, so LC3-II/ LC3-I ratio can be a useful information that can be obtained from Western blotting. However, it is not enough to get this ratio to estimate the autophagic flux status. Since, the increase in LC3-II/LC3-I ratio does not always imply autophagy induction but rather may occur due to failure of autophagosomes fusion to the lysosomes (blockage in the autophagic pathway after LC3 conjugation step). To untangle this issue, the LC3-II/LC3-I ratio should be determined in the presence and in the absence of lysosomal degradation inhibitors. So as mentioned before, if this ratio increased in the presence of the lysosomal inhibitor, then autophagy is induced; however, if the value of the ratio is almost the same in the presence and in the absence of lysosomal inhibitor, then failure of autophagosome fusion to the lysosomes might be the reason for accumulated LC3-II [90].

Other molecules in the autophagic pathway can also be detected using Western blotting technique. For example, SQSTM1/P62 is an adaptor protein which interacts with poly-ubiquitinated proteins to target their degradation by autolysosomes and it was observed that the level of p62 increases in response to impaired autophagy [90, 100]. However, since p62 has multiple roles in the cell and is involved in different processes such as cellular death, inflammation, and oxidative stress [37, 101], it is not recommended to solely rely on measuring the level of p62 to determine the autophagic flux.

5.4 Intracellular Protein Degradation

Since CMA plays a pivotal role in a diverse range of diseases, the assessment of its activity in cultured cells is experimentally applicable by evaluating the intracellular protein degradation. Pulse and chase experiments provide a useful tool to reach this target and evaluate the rate of protein degradation through the lysosomes and distinguish between degradation via either CMA or macroautophagy inside the cell [102]. This experiment is performed in two main phases: the pulse phase and the chase phase. In the pulse phase, cultured cells are supplied with radiolabeled amino acids to be used as building blocks for the newly synthesized proteins [103]. Regarding the duration of incubation of radiolabeled amino acids with cultured cells, it is advisable to be not less than 48 h in order to permit labeling of the long-lived proteins since CMA and the other autophagic pathways are usually incorporated in the breaking down of proteins having long half-life [102]. Then in the next phase, chase phase, cells are extensively washed with Hank's solution and supplemented with an excess of unlabeled amino acids; so cells preferentially utilize the unlabeled amino acids in the formation of new proteins during the chase period, allowing only the measurement of the free radiolabeled amino acids resulting from the degradation of proteins previously synthesized in the pulse phase (preventing the reincorporation of labeled amino acids in new proteins) [102]. After that, samples are taken at different time points and the free radiolabeled amino acids are measured to determine the rate of protein degradation [103]. In order to evaluate the percent of proteins degraded via CMA or other types of autophagy or even by other proteolytic pathways such as proteasomes, it is recommended to perform the same experiment in the presence of a specific inhibitor for macroautophagy such as 3-methyladenine (3-phosphatidylinositol-3-kinase (PI3K) inhibitor) [103] and repeat the same steps in the presence of a universal inhibitor for lysosomal proteolysis such as ammonium chloride (a weak base that rises the lysosomal pH interfering with lysosomal protease activity) [104]. Then, by calculating the percent of proteins degraded at a certain time point in the presence of ammonium chloride and in the absence of ammonium chloride, lysosomal proteolysis (proteolysis done by autophagic machinery) can be distinguished from proteolysis done by other pathways inside the cells such as protein degradation by proteasomes [103]. Moreover, the use of 3-methyladenine (3-phosphatidylinositol-3-kinase (PI3K) inhibitor) that mostly blocks macroautophagy activity while CMA and microautophagy activity remains unchanged provides a good choice to identify the percent of proteins degraded specifically by CMA and not macroautophagy [103].

Some technical points should be taken into consideration while applying the method of intracellular protein degradation for successful and efficient evaluation of overall CMA activity. First of all, the choice of the amino acid to be radiolabeled is critical, for example, leucine (the most commonly used amino acid for radiolabeling) is proved to have intrinsic inhibitory effect on macroautophagy in certain types of cells, which would affect the correct interpretation of the experimental results. In such case, valine could be a better alternative. However, very low concentration of leucine can be used as the concentration required for radiolabeling in this experiment is much lower than that described to have inhibitory effect on macroautophagy [102]. Moreover, it is better to use ammonium chloride in combination with leupeptin to extend the duration of its inhibitory effect on autophagy since its action is limited to 12 h only [102]. In addition, one of the pitfalls of using 3-methyladenine in blocking macroautophagy is that prolonged inhibition of macroautophagy leads to activation of CMA, thereby misleading the interpretation of the results [105]. Also, it is not applicable to distinguish between proteins degraded via CMA and those degraded by microautophagy [103].

5.5 Sequestration Assay

Quantification of sequestered material inside the autophagosomes for subsequent lysosomal clearance has been crucial in validating the efficiency of the macroautophagy process [99]. Measurement of the amount of certain substances actually engulfed and enclosed in the autophagosomes provides valuable evidence for the successful functionality of the autophagy process [99]. Failure of this critical sequestration step has been reported in the pathogenesis of certain diseases such as Huntington's disease, where empty vacuoles (deficient of any enclosed material) are discovered [59]. Therefore, scientists tried to develop methods to analyze the activity of this key step. The principle of this assay is to select an endogenous protein, or introduce an exogenous substance to the cells, and measure separately the amount of this substance in the soluble form inside the cytosol and the amount of the same substance in the sedimented form inside the autophagosomes per unit time [106]. The separation of the cytosolic material from the sedimented autophagosomes is easily achievable by electro disruption of the cells followed by centrifugation [106, 107]. An important step in this assay is to add a protease inhibitor to block the digestion of the targeted substance inside the lysosome allowing for the measurement of the intact sequestered material [106]. The selected indicator for the cargo sequestration assay should be necessarily processed only via the autophagosome-lysosome system without the interference of any other metabolic pathways or sequestration inside any cellular organelle rather than autophagosomes to allow the accurate measurement of the amount engulfed by autophagosomes [106]. Moreover, the non-selectivity characteristic of macroautophagy permits the use of an indicator that also represents the sequestration of other autophagy substrates [107, 108]. Since the autophagosomes enclose the indicator substance along with other cargo by the same rate, the presence of the indicator substance inside the autophagosomes represents, also, the sequestration rate of other autophagy substrates [99, 107]. This was proved by Kopitez et al. when they found that seven cytosolic enzymes are sequestered and delivered to the lysosomes with the same rate regardless of their varying half-lives [109]. Endogenous proteins such as lactate dehydrogenase (LDH) or exogenous substances such as di- and trisaccharides are usually used in this type of assay [106].

LDH is a cytosolic enzyme that cells can terminate its activity through entrapment inside the autophagosomes and subsequent digestion. Thus according to assay principle, measuring the LDH activity in the whole cell (including the LDH in the cytosol and the autophagosomes) and also measuring the LDH activity in the sedimented portion (autophagosomes) gives us the possibility to calculate the percent of sequestered LDH per unit time, reflecting the rate of autophagy and the efficiency of the engulfment process. A key step in this assay is to add a protease inhibitor such as leupeptin to preserve the LDH activity after the entrapment inside the autophagosomes permitting the accurate assessment of the engulfment process [106, 110, 111].

Exogenous probes, such as radiolabeled di- and trisaccharides are also commonly used. [3H] Raffinose is usually preferred in this type of assay than sucrose and lactose. This is because sucrose and lactose are not only sequestered inside the autophagosomes but also tend to accumulate inside the mitochondria, thus using digitonin which selectively extracts the sequestered [14C] sucrose from autophagosomes and lysosomes but not from mitochondria is required [106]. Accordingly, the absence of [3H] raffinose in mitochondria enables us to bypass this extraction step [106]. Moreover, raffinose is not degraded inside the autophagosomes; therefore building up of the accumulated sequestered raffinose in the autophagosomes can be measured, per unit time, without the need to use a protease inhibitor [99].

6 Summary

To sum up, the indispensable role of autophagy in the various physiological functions in the cells forces the researchers to explore the possible involvement of dysfunctional autophagy process in the pathogenesis of a wide range of human diseases including the CNS disorders. Its fundamental role in clearing the cells from the misfolded aggregated proteins and damaged organelles has directed the researchers to exploit this cellular machinery in the development of novel therapeutic agents. Nonetheless, maintaining the level of autophagy activity to an optimal normal level is essential for developing a successful therapeutic agent as the excessive activation of autophagy leads paradoxically to undesirable cellular death. Accordingly, the accurate assessment of the proper functionality of autophagy has been the focus of many studies. Combining different techniques to monitor this dynamic process and accurately quantify the autophagic flux are very important in the process of developing new therapies.

References

1. Flusberg DA, Sorger PK (2015) Surviving apoptosis: life-death signaling in single cells. Trends Cell Biol 25(8):446–458. https://doi.org/10.1016/j.tcb.2015.03.003
2. Bertheloot D, Latz E, Franklin BS (2021) Necroptosis, pyroptosis and apoptosis: an intricate game of cell death. Cell Mol Immunol 18(5):1106–1121. https://doi.org/10.1038/s41423-020-00630-3
3. Schwartz LM (2021) Autophagic cell death during development – ancient and mysterious. Front Cell Dev Biol 9:656370. https://doi.org/10.3389/fcell.2021.656370
4. D'Arcy MS (2019) Cell death: a review of the major forms of apoptosis, necrosis and autophagy. Cell Biol Int 43(6):582–592. https://doi.org/10.1002/cbin.11137
5. Goldar S, Khaniani MS, Derakhshan SM, Baradaran B (2015) Molecular mechanisms of apoptosis and roles in cancer development and treatment. Asian Pac J Cancer Prev 16(6):2129–2144. https://doi.org/10.7314/apjcp.2015.16.6.2129
6. Elmore S (2007) Apoptosis: a review of programmed cell death. Toxicol Pathol 35(4):495–516. https://doi.org/10.1080/01926230701320337
7. Bedoui S, Herold MJ, Strasser A (2020) Emerging connectivity of programmed cell death pathways and its physiological implications. Nat Rev Mol Cell Biol 21(11):678–695. https://doi.org/10.1038/s41580-020-0270-8
8. Walsh CM (2014) Grand challenges in cell death and survival: apoptosis vs. necroptosis. Front Cell Dev Biol 2:3. https://doi.org/10.3389/fcell.2014.00003
9. Khoury MK, Gupta K, Franco SR, Liu B (2020) Necroptosis in the pathophysiology of disease. Am J Pathol 190(2):272–285.

https://doi.org/10.1016/j.ajpath.2019.10.012
10. Que X et al (2024) Fantastic voyage: the journey of NLRP3 inflammasome activation. Genes Dis 11(2):819–829. https://doi.org/10.1016/j.gendis.2023.01.009
11. Glick D, Barth S, Macleod KF (2010) Autophagy: cellular and molecular mechanisms. J Pathol 221(1):3–12. https://doi.org/10.1002/path.2697
12. Bialik S, Dasari SK, Kimchi A (2018) Autophagy-dependent cell death – where, how and why a cell eats itself to death. J Cell Sci 131(18). https://doi.org/10.1242/jcs.215152
13. Mizushima N, Komatsu M (2011) Autophagy: renovation of cells and tissues. Cell 147(4):728–741. https://doi.org/10.1016/j.cell.2011.10.026
14. Kroemer G, Mariño G, Levine B (2010) Autophagy and the integrated stress response. Mol Cell 40(2):280–293. https://doi.org/10.1016/j.molcel.2010.09.023
15. Parzych KR, Klionsky DJ (2014) An overview of autophagy: morphology, mechanism, and regulation. Antioxid Redox Signal 20(3):460–473. https://doi.org/10.1089/ars.2013.5371
16. Ueno T, Komatsu M (2020) Monitoring autophagy flux and activity: principles and applications. BioEssays 42(11):e2000122. https://doi.org/10.1002/bies.202000122
17. Kocaturk NM et al (2019) Autophagy as a molecular target for cancer treatment. Eur J Pharm Sci 134:116–137. https://doi.org/10.1016/j.ejps.2019.04.011
18. Onorati AV, Dyczynski M, Ojha R, Amaravadi RK (2018) Targeting autophagy in cancer. Cancer 124(16):3307–3318. https://doi.org/10.1002/cncr.31335
19. Pankiv S et al (2007) p62/SQSTM1 binds directly to Atg8/LC3 to facilitate degradation of ubiquitinated protein aggregates by autophagy. J Biol Chem 282(33):24131–24145. https://doi.org/10.1074/jbc.M702824200
20. Stavoe AKH, Holzbaur ELF (2019) Autophagy in neurons. Annu Rev Cell Dev Biol 35:477–500. https://doi.org/10.1146/annurev-cellbio-100818-125242
21. Valencia M, Kim SR, Jang Y, Lee SH (2021) Neuronal autophagy: characteristic features and roles in neuronal pathophysiology. Biomol Ther (Seoul) 29(6):605–614. https://doi.org/10.4062/biomolther.2021.012
22. Magistretti PJ, Allaman I (2015) A cellular perspective on brain energy metabolism and functional imaging. Neuron 86(4):883–901. https://doi.org/10.1016/j.neuron.2015.03.035
23. Aranda-Anzaldo A (2012) The post-mitotic state in neurons correlates with a stable nuclear higher-order structure. Commun Integr Biol 5(2):134–139. https://doi.org/10.4161/cib.18761
24. Poewe W et al (2017) Parkinson disease. Nat Rev Dis Primers 3:17013. https://doi.org/10.1038/nrdp.2017.13
25. Lynch-Day MA et al (2012) The role of autophagy in Parkinson's disease. Cold Spring Harb Perspect Med 2(4):a009357. https://doi.org/10.1101/cshperspect.a009357
26. Meade RM, Fairlie DP, Mason JM (2019) Alpha-synuclein structure and Parkinson's disease – lessons and emerging principles. Mol Neurodegener 14(1):29. https://doi.org/10.1186/s13024-019-0329-1
27. Vogiatzi T, Xilouri M, Vekrellis K, Stefanis L (2008) Wild type alpha-synuclein is degraded by chaperone-mediated autophagy and macroautophagy in neuronal cells. J Biol Chem 283(35):23542–23556. https://doi.org/10.1074/jbc.M801992200
28. Tang Q et al (2021) Alpha-Synuclein defects autophagy by impairing SNAP29-mediated autophagosome-lysosome fusion. Cell Death Dis 12(10):854. https://doi.org/10.1038/s41419-021-04138-0
29. Cuervo AM et al (2004) Impaired degradation of mutant alpha-synuclein by chaperone-mediated autophagy. Science 305(5688):1292–1295. https://doi.org/10.1126/science.1101738
30. Martinez-Vicente M et al (2008) Dopamine-modified alpha-synuclein blocks chaperone-mediated autophagy. J Clin Invest 118(2):777–788. https://doi.org/10.1172/jci32806
31. Stefanis L et al (2001) Expression of A53T mutant but not wild-type alpha-synuclein in PC12 cells induces alterations of the ubiquitin-dependent degradation system, loss of dopamine release, and autophagic cell death. J Neurosci 21(24):9549–9560. https://doi.org/10.1523/jneurosci.21-24-09549.2001
32. Cheung ZH, Ip NY (2009) The emerging role of autophagy in Parkinson's disease. Mol Brain 2:29. https://doi.org/10.1186/1756-6606-2-29
33. Ichimiya T et al (2020) Autophagy and autophagy-related diseases: a review. Int J Mol Sci 21(23). https://doi.org/10.3390/ijms21238974

34. Geisler S et al (2010) PINK1/Parkin-mediated mitophagy is dependent on VDAC1 and p62/SQSTM1. Nat Cell Biol 12(2):119–131. https://doi.org/10.1038/ncb2012
35. Narendra D, Tanaka A, Suen DF, Youle RJ (2008) Parkin is recruited selectively to impaired mitochondria and promotes their autophagy. J Cell Biol 183(5):795–803. https://doi.org/10.1083/jcb.200809125
36. Gegg ME et al (2010) Mitofusin 1 and mitofusin 2 are ubiquitinated in a PINK1/parkin-dependent manner upon induction of mitophagy. Hum Mol Genet 19(24):4861–4870. https://doi.org/10.1093/hmg/ddq419
37. Katsuragi Y, Ichimura Y, Komatsu M (2015) p62/SQSTM1 functions as a signaling hub and an autophagy adaptor. FEBS J 282(24): 4672–4678. https://doi.org/10.1111/febs.13540
38. Lee JY et al (2010) HDAC6 controls autophagosome maturation essential for ubiquitin-selective quality-control autophagy. EMBO J 29(5):969–980. https://doi.org/10.1038/emboj.2009.405
39. Narendra D et al (2010) p62/SQSTM1 is required for Parkin-induced mitochondrial clustering but not mitophagy; VDAC1 is dispensable for both. Autophagy 6(8): 1090–1106. https://doi.org/10.4161/auto.6.8.13426
40. Moon HE, Paek SH (2015) Mitochondrial dysfunction in Parkinson's disease. Exp Neurobiol 24(2):103–116. https://doi.org/10.5607/en.2015.24.2.103
41. Uddin MS et al (2018) Autophagy and Alzheimer's disease: from molecular mechanisms to therapeutic implications. Front Aging Neurosci 10:04. https://doi.org/10.3389/fnagi.2018.00004
42. Jellinger KA (2010) Basic mechanisms of neurodegeneration: a critical update. J Cell Mol Med 14(3):457–487. https://doi.org/10.1111/j.1582-4934.2010.01010.x
43. Pelucchi S, Gardoni F, Di Luca M, Marcello E (2022) Synaptic dysfunction in early phases of Alzheimer's disease. Handb Clin Neurol 184: 417–438. https://doi.org/10.1016/b978-0-12-819410-2.00022-9
44. Nilsson P, Saido TC (2014) Dual roles for autophagy: degradation and secretion of Alzheimer's disease Aβ peptide. BioEssays 36(6): 570–578. https://doi.org/10.1002/bies.201400002
45. Iliyasu MO, Musa SA, Oladele SB, Iliya AI (2023) Amyloid-beta aggregation implicates multiple pathways in Alzheimer's disease: understanding the mechanisms. Front Neurosci 17:1081938. https://doi.org/10.3389/fnins.2023.1081938
46. Yuan M et al (2022) Impaired autophagy in amyloid-beta pathology: a traditional review of recent Alzheimer's research. J Biomed Res 37(1):30–46. https://doi.org/10.7555/jbr.36.20220145
47. Feng T et al (2017) Autophagy-mediated regulation of BACE1 protein trafficking and degradation. J Biol Chem 292(5):1679–1690. https://doi.org/10.1074/jbc.M116.766584
48. Ohta K et al (2010) Autophagy impairment stimulates PS1 expression and gamma-secretase activity. Autophagy 6(3):345–352. https://doi.org/10.4161/auto.6.3.11228
49. Pickford F et al (2008) The autophagy-related protein beclin 1 shows reduced expression in early Alzheimer disease and regulates amyloid beta accumulation in mice. J Clin Invest 118(6):2190–2199. https://doi.org/10.1172/jci33585
50. Du Y, Wooten MC, Gearing M, Wooten MW (2009) Age-associated oxidative damage to the p62 promoter: implications for Alzheimer disease. Free Radic Biol Med 46(4):492–501. https://doi.org/10.1016/j.freeradbiomed.2008.11.003
51. Boland B et al (2008) Autophagy induction and autophagosome clearance in neurons: relationship to autophagic pathology in Alzheimer's disease. J Neurosci 28(27): 6926–6937. https://doi.org/10.1523/jneurosci.0800-08.2008
52. Colacurcio DJ, Pensalfini A, Jiang Y, Nixon RA (2018) Dysfunction of autophagy and endosomal-lysosomal pathways: roles in pathogenesis of down syndrome and Alzheimer's disease. Free Radic Biol Med 114:40–51. https://doi.org/10.1016/j.freeradbiomed.2017.10.001
53. Walker FO (2007) Huntington's disease. Lancet 369(9557):218–228. https://doi.org/10.1016/s0140-6736(07)60111-1
54. Han T et al (2024) Microglial response to aging and neuroinflammation in the development of neurodegenerative diseases. Neural Regen Res 19(6):1241–1248. https://doi.org/10.4103/1673-5374.385845
55. Cortes CJ, La Spada AR (2014) The many faces of autophagy dysfunction in Huntington's disease: from mechanism to therapy. Drug Discov Today 19(7):963–971. https://doi.org/10.1016/j.drudis.2014.02.014
56. Cuervo AM (2003) Autophagy and aging–when "all you can eat" is yourself. Sci Aging

Knowl Environ 2003(36):pe25. https://doi.org/10.1126/sageke.2003.36.pe25
57. Vernace VA, Schmidt-Glenewinkel T, Figueiredo-Pereira ME (2007) Aging and regulated protein degradation: who has the UPPer hand? Aging Cell 6(5):599–606. https://doi.org/10.1111/j.1474-9726.2007.00329.x
58. Wong SQ, Kumar AV, Mills J, Lapierre LR (2020) Autophagy in aging and longevity. Hum Genet 139(3):277–290. https://doi.org/10.1007/s00439-019-02031-7
59. Martinez-Vicente M et al (2010) Cargo recognition failure is responsible for inefficient autophagy in Huntington's disease. Nat Neurosci 13(5):567–576. https://doi.org/10.1038/nn.2528
60. Limanaqi F et al (2020) mTOR-related cell-clearing Systems in Epileptic Seizures, an update. Int J Mol Sci 21(5). https://doi.org/10.3390/ijms21051642
61. Ali NH et al (2023) Autophagy and autophagy signaling in epilepsy: possible role of autophagy activator. Mol Med 29(1):142. https://doi.org/10.1186/s10020-023-00742-2
62. McMahon J et al (2012) Impaired autophagy in neurons after disinhibition of mammalian target of rapamycin and its contribution to epileptogenesis. J Neurosci 32(45):15704–15714. https://doi.org/10.1523/jneurosci.2392-12.2012
63. Hui KK et al (2019) GABARAPs dysfunction by autophagy deficiency in adolescent brain impairs GABA(A) receptor trafficking and social behavior. Sci Adv 5(4):eaau8237. https://doi.org/10.1126/sciadv.aau8237
64. Goto J et al (2011) Regulable neural progenitor-specific Tsc1 loss yields giant cells with organellar dysfunction in a model of tuberous sclerosis complex. Proc Natl Acad Sci USA 108(45):E1070–E1079. https://doi.org/10.1073/pnas.1106454108
65. Khalil MI, Ali MM, Holail J, Houssein M (2023) Growth or death? Control of cell destiny by mTOR and autophagy pathways. Prog Biophys Mol Biol 185:39–55. https://doi.org/10.1016/j.pbiomolbio.2023.10.002
66. Zhu H, Wang W, Li Y (2022) Molecular mechanism and regulation of autophagy and its potential role in epilepsy. Cells 11(17). https://doi.org/10.3390/cells11172621
67. Li J, Kim SG, Blenis J (2014) Rapamycin: one drug, many effects. Cell Metab 19(3):373–379. https://doi.org/10.1016/j.cmet.2014.01.001
68. Dehay B et al (2010) Pathogenic lysosomal depletion in Parkinson's disease. J Neurosci 30(37):12535–12544. https://doi.org/10.1523/jneurosci.1920-10.2010
69. Ravikumar B et al (2004) Inhibition of mTOR induces autophagy and reduces toxicity of polyglutamine expansions in fly and mouse models of Huntington disease. Nat Genet 36(6):585–595. https://doi.org/10.1038/ng1362
70. Ali SO, Shahin NN, Safar MM, Rizk SM (2019) Therapeutic potential of endothelial progenitor cells in a rat model of epilepsy: role of autophagy. J Adv Res 18:101–112. https://doi.org/10.1016/j.jare.2019.01.013
71. Giorgi FS et al (2015) The role of autophagy in epileptogenesis and in epilepsy-induced neuronal alterations. J Neural Transm (Vienna) 122(6):849–862. https://doi.org/10.1007/s00702-014-1312-1
72. Motoi Y, Shimada K, Ishiguro K, Hattori N (2014) Lithium and autophagy. ACS Chem Neurosci 5(6):434–442. https://doi.org/10.1021/cn500056q
73. Vakifahmetoglu-Norberg H, Xia HG, Yuan J (2015) Pharmacologic agents targeting autophagy. J Clin Invest 125(1):5–13. https://doi.org/10.1172/jci73937
74. Sarkar S et al (2005) Lithium induces autophagy by inhibiting inositol monophosphatase. J Cell Biol 170(7):1101–1111. https://doi.org/10.1083/jcb.200504035
75. Uddin MN et al (2021) Strain-specific clearance of seed-dependent tau aggregation by lithium-induced autophagy. Biochem Biophys Res Commun 543:65–71. https://doi.org/10.1016/j.bbrc.2020.12.113
76. Hou L et al (2015) Lithium protects dopaminergic cells from rotenone toxicity via autophagy enhancement. BMC Neurosci 16:82. https://doi.org/10.1186/s12868-015-0222-y
77. Kawecka-Jaszcz K et al (2006) Rilmenidine–its antihypertensive efficacy, safety and impact on quality of life in perimenopausal women with mild to moderate essential hypertension. Blood Press 15(1):51–58. https://doi.org/10.1080/08037050600565969
78. Sarkar S, Ravikumar B, Floto RA, Rubinsztein DC (2009) Rapamycin and mTOR-independent autophagy inducers ameliorate toxicity of polyglutamine-expanded huntingtin and related proteinopathies. Cell Death Differ 16(1):46–56. https://doi.org/10.1038/cdd.2008.110

79. Rose C et al (2010) Rilmenidine attenuates toxicity of polyglutamine expansions in a mouse model of Huntington's disease. Hum Mol Genet 19(11):2144–2153. https://doi.org/10.1093/hmg/ddq093
80. Popović N et al (2020) Verapamil and Alzheimer's disease: past, present, and future. Front Pharmacol 11:562. https://doi.org/10.3389/fphar.2020.00562
81. Williams A et al (2008) Novel targets for Huntington's disease in an mTOR-independent autophagy pathway. Nat Chem Biol 4(5):295–305. https://doi.org/10.1038/nchembio.79
82. Siddiqi FH et al (2019) Felodipine induces autophagy in mouse brains with pharmacokinetics amenable to repurposing. Nat Commun 10(1):1817. https://doi.org/10.1038/s41467-019-09494-2
83. Singh B, Bhaskar S (2019) Methods for detection of autophagy in mammalian cells. Methods Mol Biol 2045:245–258. https://doi.org/10.1007/7651_2018_190
84. Mizushima N, Yoshimori T, Levine B (2010) Methods in mammalian autophagy research. Cell 140(3):313–326. https://doi.org/10.1016/j.cell.2010.01.028
85. Chakraborty J, Caicci F, Roy M, Ziviani E (2020) Investigating mitochondrial autophagy by routine transmission electron microscopy: seeing is believing? Pharmacol Res 160:105097. https://doi.org/10.1016/j.phrs.2020.105097
86. Mattoscio D, Raimondi A, Tacchetti C, Chiocca S (2017) Immunogold electron microscopy of the autophagosome marker LC3. Bio Protoc 7(24):e2648. https://doi.org/10.21769/BioProtoc.2648
87. Gudmundsson S et al (2019) Correlative light and electron microscopy of autophagosomes. Methods Mol Biol 1880:199–209. https://doi.org/10.1007/978-1-4939-8873-0_12
88. Sharifi MN, Mowers EE, Drake LE, Macleod KF (2015) Measuring autophagy in stressed cells. Methods Mol Biol 1292:129–150. https://doi.org/10.1007/978-1-4939-2522-3_10
89. Kim H, Seong J (2021) Fluorescent protein-based autophagy biosensors. Materials (Basel) 14(11). https://doi.org/10.3390/ma14113019
90. Barth S, Glick D, Macleod KF (2010) Autophagy: assays and artifacts. J Pathol 221(2):117–124. https://doi.org/10.1002/path.2694
91. Tanida I, Ueno T, Kominami E (2008) LC3 and autophagy. Methods Mol Biol 445:77–88. https://doi.org/10.1007/978-1-59745-157-4_4
92. Kim S, Choi S, Kang D (2020) Quantitative and qualitative analysis of autophagy flux using imaging. BMB Rep 53(5):241–247. https://doi.org/10.5483/BMBRep.2020.53.5.046
93. Forgac M (2007) Vacuolar ATPases: rotary proton pumps in physiology and pathophysiology. Nat Rev Mol Cell Biol 8(11):917–929. https://doi.org/10.1038/nrm2272
94. Katayama H et al (2011) A sensitive and quantitative technique for detecting autophagic events based on lysosomal delivery. Chem Biol 18(8):1042–1052. https://doi.org/10.1016/j.chembiol.2011.05.013
95. Mizushima N et al (2004) In vivo analysis of autophagy in response to nutrient starvation using transgenic mice expressing a fluorescent autophagosome marker. Mol Biol Cell 15(3):1101–1111. https://doi.org/10.1091/mbc.e03-09-0704
96. Kuma A, Matsui M, Mizushima N (2007) LC3, an autophagosome marker, can be incorporated into protein aggregates independent of autophagy: caution in the interpretation of LC3 localization. Autophagy 3(4):323–328. https://doi.org/10.4161/auto.4012
97. Tanida I et al (2008) Consideration about negative controls for LC3 and expression vectors for four colored fluorescent protein-LC3 negative controls. Autophagy 4(1):131–134. https://doi.org/10.4161/auto.5233
98. Meftahi GH et al (2021) Applications of western blot technique: from bench to bedside. Biochem Mol Biol Educ 49(4):509–517. https://doi.org/10.1002/bmb.21516
99. Seglen PO et al (2015) Macroautophagic cargo sequestration assays. Methods 75:25–36. https://doi.org/10.1016/j.ymeth.2014.12.021
100. Bjørkøy G et al (2005) p62/SQSTM1 forms protein aggregates degraded by autophagy and has a protective effect on huntingtin-induced cell death. J Cell Biol 171(4):603–614. https://doi.org/10.1083/jcb.200507002
101. Wu Y et al (2020) p62/SQSTM1 accumulation due to degradation inhibition and transcriptional activation plays a critical role in silica nanoparticle-induced airway inflammation via NF-κB activation. J Nanobiotechnol 18(1):77. https://doi.org/10.1186/s12951-020-00634-1
102. Kaushik S, Cuervo AM (2009) Methods to monitor chaperone-mediated autophagy.

Methods Enzymol 452:297–324. https://doi.org/10.1016/s0076-6879(08)03619-7
103. Patel B, Cuervo AM (2015) Methods to study chaperone-mediated autophagy. Methods 75:133–140. https://doi.org/10.1016/j.ymeth.2015.01.003
104. Klionsky DJ, Cuervo AM, Seglen PO (2007) Methods for monitoring autophagy from yeast to human. Autophagy 3(3):181–206. https://doi.org/10.4161/auto.3678
105. Kaushik S, Massey AC, Mizushima N, Cuervo AM (2008) Constitutive activation of chaperone-mediated autophagy in cells with impaired macroautophagy. Mol Biol Cell 19(5):2179–2192. https://doi.org/10.1091/mbc.e07-11-1155
106. Zhang XW et al (2021) Autophagic flux detection: significance and methods involved. Adv Exp Med Biol 1208:131–173. https://doi.org/10.1007/978-981-16-2830-6_9
107. Engedal N, Luhr M, Szalai P, Seglen PO (2019) Measurement of bulk autophagy by a cargo sequestration assay. Methods Mol Biol 1880:307–313. https://doi.org/10.1007/978-1-4939-8873-0_20
108. Luhr M et al (2017) A simple cargo sequestration assay for quantitative measurement of nonselective autophagy in cultured cells. Methods Enzymol 587:351–364. https://doi.org/10.1016/bs.mie.2016.09.064
109. Kopitz J et al (1990) Nonselective autophagy of cytosolic enzymes by isolated rat hepatocytes. J Cell Biol 111(3):941–953. https://doi.org/10.1083/jcb.111.3.941
110. Luhr M, Szalai P, Engedal N (2018) The lactate dehydrogenase sequestration assay – a simple and reliable method to determine bulk Autophagic sequestration activity in mammalian cells. J Vis Exp 137. https://doi.org/10.3791/57971
111. Orhon I, Reggiori F (2017) Assays to monitor autophagy progression in cell cultures. Cell 6(3). https://doi.org/10.3390/cells6030020
112. Klionsky DJ, Elazar Z, Seglen PO, Rubinsztein DC (2008) Does bafilomycin A1 block the fusion of autophagosomes with lysosomes? Autophagy 4(7):849–850. https://doi.org/10.4161/auto.6845
113. Luhr M, Sætre F, Engedal N (2018) The long-lived protein degradation assay: an efficient method for quantitative determination of the Autophagic flux of endogenous proteins in adherent cell lines. Bio Protoc 8(9): e2836. https://doi.org/10.21769/BioProtoc.2836

Methods in Molecular Biology (2025) 2879: 207–217
DOI 10.1007/7651_2024_517

Published online: 06 March 2024

Assessment of Autophagy in *Leishmania* Parasites

Somtochukwu S. Onwah, Jude E. Uzonna, and Saeid Ghavami

Abstract

Leishmaniasis is a neglected tropical disease caused by numerous species of *Leishmania* parasites, including *Leishmania major*. The parasite is transmitted by several species of sandfly vectors and infects myeloid cells leading to a myriad of inflammatory responses, immune dysregulations, and disease manifestations. Every cell undergoes autophagy, a self-regulated degradative process that permits the cells to recycle damaged or worn-out organelles in order to maintain cellular health and homeostasis. Studies have shown that *Leishmania* modulates their host cell autophagic machinery and there are indications that the parasite-specific autophagic processes may be valuable for parasite virulence and survival. However, the role of autophagy in *Leishmania* is inconclusive because of the limited tools available to study the *Leishmania*-specific autophagic machinery. Here, we describe methods to study and definitively confirm autophagy in *Leishmania major*. Transmission electron microscopy (TEM) allowed us to visualize *Leishmania* autophagosomes, especially those containing damaged mitochondrial content, as well as dividing mitochondria with ongoing fusion/fission processes. Flow cytometry enabled us to identify the amount of acridine orange dye accumulating in the acidic vacuolar compartments in *Leishmania major* by detecting fluorescence in the red laser when autophagic inhibitors or enhancers were included. These methods will advance studies that aim to understand autophagic regulation in *Leishmania* parasites that could provide insights into developing improved therapeutic targets against leishmaniasis.

Key words Acridine Orange, Autophagy, Cutaneous leishmaniasis, *Leishmania major*, TEM

1 Introduction

Autophagy is a natural "in-house" cleaning process that allows a stressed cell to breakdown cellular components and recycle/reuse some parts for efficient cellular processes [1, 2]. Many cells require autophagy to recycle damaged or stressed intracellular components for the purpose of maintaining homeostasis and cell health through both selective and non-selective degradative mechanisms [3, 4]. The autophagic process is activated when the cell senses a need for nutrients, growth, or energy [5–7] or to eliminate invasive pathogens [8–10]. All cells are known to engage in basal levels of autophagy to perform basic hemostatic processes that include protein or organelle turnover [8, 11]. In mammals, the induction of autophagy is a complicated process that involves a myriad of autophagy-like proteins, ATGs, which induce the formation of the autophagic organelle, autophagosome. Once the autophagosomes mature, they engulf many intracellular materials, transporting them to the lysosome for further degradation [12]. Autophagy evolved to help eliminate invading intracellular pathogens by releasing reactive oxygen species and degradative proteolytic enzymes. It directs the imbalance in the nutrient availabilities and cations, which causes the destruction of organisms [13]. In line with this, the continuous persistence of intracellular pathogens is associated with manipulating the host autophagic pathway [14, 15]. Indeed, the establishment of many infections and human disease pathologies has been linked to an aberrant autophagic machinery [16–18]. Thus, a good understanding of the autophagic processes and targeting the host autophagic machinery could lead to the development of novel therapeutic strategies for controlling disease progression.

Leishmaniasis is a neglected disease that has become a global health burden, with an estimated one million new cases occurring annually and over one billion individuals living in the endemic continents (such as Africa, Asia, the Americas, and the Middle East) at risk [19–21]. The disease, which is caused by several species of the protozoan parasite belonging to the genus *Leishmania*, displays diverse clinical manifestations. Transmission of the parasite to their mammalian hosts is through the bite of an infected sandfly vector [19, 21]. Cutaneous leishmaniasis is the more common and mild form of the disease and presents as a self-resolving lesion, often resulting in complicated and disfiguring outcomes. The more fatal and severe form of the disease, visceral leishmaniasis, is caused by *L. donovani* and systemically affects multiple organs, which if left untreated can lead to death [19, 21]. Besides the clinical manifestations, *Leishmania* infections result in chronic inflammation, immune cell dysregulation, and cytokine imbalances [22]. Despite the myriad of research aimed at understanding interactions between the host cells and *Leishmania* parasites, there is still no approved vaccine against the disease because this complex interaction is poorly understood. Therefore, drug treatment remains the only viable option to control the disease, but the current treatment modalities are plagued with various challenges [23, 24]. A clear understanding of the parasite's cellular biologic processes and molecules that influence and/or drive protective responses against the host could advance the development of novel therapeutic strategies against leishmaniasis [25, 26].

In addition to evading host immune responses, *Leishmania* parasites have been shown to hijack the host cell's autophagic machinery [8]. It has been suggested that the continued survival of *Leishmania* parasites in their host is due to their ability to modulate host mTOR-regulated autophagy processes [27], which helps them to obtain essential nutrients in their harsh intracellular host environment [28]. Interestingly, as their mammalian host cells, *Leishmania* parasites have been reported to undergo autophagic processes, and nutrient deprivation was observed to enhance autophagy in *Leishmania* [29, 30]. Furthermore, the transition of promastigotes (insect) to amastigotes (mammalian) life forms of the parasite depends on autophagy [31], suggesting that just like mammalian cells, autophagic processes are indeed critical for parasite homeostasis. Interestingly, the expression of autophagy-like proteins observed in mammalian cells has also been identified in *Leishmania* parasites [32]. ATG4 and ATG8 are some of the well-characterized autophagic proteins that have been identified to play a role in the survival and virulence of *Leishmania* parasites [33–36]. The presence of these key autophagic proteins in *Leishmania* suggests ample avenues to target either the host or parasite autophagic machinery for anti-leishmanial interventions. However, not many studies have explored *Leishmania*'s autophagic responses extensively as the exact mechanisms of autophagy in *Leishmania* remain elusive. Therefore, to help understand the mechanisms of autophagy in *Leishmania*, tools geared at measuring autophagy in the parasites are needed.

Here, we describe two complementary methods for detecting autophagy in *Leishmania*. Transmission electron microscopy (TEM) has long been the tool that accurately detects autophagic accumulation and organelles in mammalian cells because the exact ultrastructural composition is revealed in nanoscales [37]. This chapter reports the visual identification of specific autophagic organelles and structures in *Leishmania major* by TEM. During autophagy, there is an increase in the formation of acidic vacuolar organelles (AVO), such as lysosomes and auto phagolysosomes [38]. Acridine orange (AO) is known to accumulate in these acidic compartments and fluoresces bright red [38]. However, since the use of multiple methods to detect autophagy is encouraged, [39] we used flow cytometry to detect and validate the accumulation of AO in the AVO of *Leishmania major*. Although TEM is the gold standard for accessing autophagy because of the high-resolution images and morphometric output [37, 39, 40], in assessing autophagic flux, a time-lapse observation is critical to deduce autophagic mechanisms, which is not achievable with TEM [40]. Live cell imaging through fluorescence microscopy is an alternative tool that measures autophagic flux [41]. Also, identification of autophagic proteins such as the accumulation of MAP 1LC3/LC3-II in Western blots is considered an indication of autophagosomal flux especially in the presence of inhibitors of the autophagic process such as bafilomycin A1 [42]. Assessing autophagic flux in *Leishmania* has been challenging because there are no anti-autophagic protein antibodies such as anti-LC3 or anti-p62. Thus, accessing autophagic flux by measuring the accumulation of acridine orange in the AVO of *Leishmania* over time is a technique that could be critical to unravel the mechanisms of autophagy in *Leishmania*.

2 Materials

Aseptic techniques were applied in the preparation of *Leishmania* parasites in axenic culture, and all work with the parasites was carried out inside a biosafety hood in a biosafety level 2 laboratory.

2.1 Leishmania Parasites Growth and Media

1. *Leishmania major* MHOM/80/Friedlin strain.
2. Twenty percent inactivated-fetal bovine serum (FBS).
3. Medium 199 (M199) incomplete: mix 9.5 g M199 powder (Gibco, Cat# 31100035), 2.2 g $NaHCO_3$ (Sigma Cat# S6014-25G), 25 mL of 1 M HEPES (Gibco, Cat# 15630080), 10 mM adenine in 50 mM HEPES, 0.25% hemin in 50% triethanolamine, 0.1% biotin in 95% ethanol, 0.25 mg/mL biopterin in PBS, and distilled water up to a total volume of 1 L.
4. T25 Biolite™ culture flasks (ThermoFischer, Cat# 130189).
5. Sterile pipette tips and pipettors.
6. Incubator set at 27 °C.
7. Penicillin-streptomycin (Gibo™ Cat # 15140122).
8. Water bath set at 37 °C.
9. Centrifuge, max speed 4000 rpm (ThermoFischer Cat# 75007200).
10. Optical microscope.
11. Glass vacuum filtration apparatus.
12. 15 mL Falcon tubes.
13. 1 L and 200 mL sterile bottles.

2.2 Fixing Parasites for Electron Microscopy

1. 6% Glutaraldehyde stock stored at 4 °C (*see* **Note 1**).
2. 0.2 M Sorensen's phosphate buffer stored at 4 °C.
3. 0.5 M Sorensen's phosphate buffer store at room temperature (RT) and should be a clear solution.
4. Sucrose powder.
5. Fixation solution: 0.1 M Sorensen's phosphate buffer with 3% glutaraldehyde (prepare a 1:1 mixture of 0.2 M Sorensen's buffer and 6% glutaraldehyde (*see* **Notes 2** and **3**).
6. Sucrose solution: prepare 5% sucrose in 0.1 M Sorensen's phosphate buffer for instance a final volume of 25 mL, mix 1.25 g of sucrose with 5 mL of 0.5 M Sorensen's phosphate buffer up to 25 mL of ddH_2O (*see* **Note 2**).
7. Centrifuge (Fisher Scientific Cat# 13864456).
8. Fifteen 1.5 mL Eppendorf tubes.

9. Philips-CMC-300 KV Transmission Electronic Microscope (The Netherlands), Histology Lab, University of Manitoba, Canada.
10. 1X Phosphate buffered saline (PBS).

2.3 Experiments with Acridine Orange

1. Acridine orange hydrochloride solution (Sigma, Cat# A8097).
2. ddH_2O.
3. 5 mL conical glass tubes.
4. Tube racks.
5. Foil paper.
6. 100 nM Bafilomycin.
7. 2% FBS.
8. 1X Phosphate buffered saline (PBS).
9. 24-well plate.

2.4 Flow Cytometry

1. Beckman Coulter CytoFlex LX Digital Flow Cytometry Analyzer Four Laser System at the University of Manitoba, Flow Cytometry Core Facility.
2. FlowJo Version 10 software (TreeStar, Ashland, OR).

3 Methods

3.1 Preparing Parasite Complete Media

1. M199 incomplete media is prepared and filtered using the vacuum filtration system inside a biosafety hood. After filtration, the media is poured into a 1 L sterile bottle.
2. For 100 mL complete M199 media, add 80 mL of the incomplete M199 media in a 200 mL sterile bottle alongside 20 mL of FBS (heat-inactivated) and 1 mL of 1000 U/mL penicillin-streptomycin to prevent bacteria contamination.
3. Shake gently by swirling the bottle in a circular motion to mix.

3.2 Preparing Leishmania *Axenic Cultures*

1. *L. major* parasites MHOM/80/Friedlin strain stored in cryo vials in liquid nitrogen are swirled in 37 °C water bath until thawed (*see* **Note 4**).
2. 10 mL of complete M199 media is poured into a 15 mL Falcon tube and the thawed contents of the cryo vial from **step 1** above are added.
3. Centrifuge at 3000 rpm for 15 min.
4. Label T25 tubes and add fresh 10 mL complete M199 medium.

5. Discard supernatant from **step 3**, pipette out the pellets, and add them into the T25 flash containing complete M199 medium.
6. Visually check to see parasites in the flask using an optical microscope (*see* **Note 5**).
7. Incubate at 27 °C for 3 days.
8. Under an optical microscope, check that the parasites are healthy and motile and have their characteristic slender body with a flagellum.
9. Passage the parasites by taking 100 μL into a new T25 flask containing 10 mL of complete M199 medium and further incubate at 27 °C. At the specified days post-passage, the parasites are ready for downstream experiments.

3.3 Fixation of Parasites for Transmission Electron Microscopy Imaging

1. At day 3 and 7 post-passage, the parasites were transferred into a 15 mL Falcon tube inside a biosafety cabinet.
2. Centrifuge at 3000 rpm for 7 min (*see* **Note 6**).
3. Discard supernatant and transfer pellets containing the parasites into 1.5 mL Eppendorf tubes.
4. Resuspend the parasites in PBS and gently stir by pipetting up and down. Do not vortex.
5. Spin at 3000 rpm for 7 min.
6. Discard the supernatant and add 1 mL of the fixation solution. Let the parasites be fully resuspended for proper fixation by gently pipetting up and down.
7. Fix the parasites for 3 h inside a fume hood at RT.
8. Pellet the parasites by spinning at 3000 rpm for 7 min, remove as much of the fixative as possible, and replace them with 1 mL of 5% sucrose in 0.1 Sorensen's buffer. Keep parasites at 4 °C.
9. Fixed parasites are embedded onto the stage of an electron microscope following the instructions of the transmission electron microscopy manual. A trained histologist should identify parasites containing double membrane vacuoles, mitochondria in the parasite, ER in the parasite, and the cytosolic membrane after capturing images at 34000× magnification.

3.4 Acridine Orange Stain and Flow Cytometry Acquisition

1. At day 3, passaged parasites were counted using a hemocytometer or a cell counter.
2. In some wells, plate 1×10^6 parasites in 1 mL complete M199 medium and in other wells plate 1×10^6 parasites containing incomplete M199 medium supplemented with 100 U/mL penicillin, 100 μg/mL streptomycin, and 1% FBS in a 24-well plate (*see* **Note 7**).

3. After 48 h post-day 3 and day 7 cultures, 100 nM bafilomycin was added to some of the wells that contained complete M199 medium.
4. At the time point to measure autophagy (3 and 7 days), parasites in individual wells of the 24-well plate were transferred to a 5 mL glass conical tube.
5. Wash parasites twice in PBS by spinning at 3000 rpm for 10 min.
6. Discard supernatants and add 1 μL of 1 mg/mL acridine orange solution to the pellets in the dark at RT for 10 min (*see* **Note 8**).
7. Wash parasites twice in PBS by spinning at 3000 rpm for 10 min and resuspended in 200 μL of PBS for acquisition in the cytoflex machine.
8. Acquisition was done by opening channels in the red (PECy5) and green (FITC) lasers and recording 20,000 events.
9. Further analysis of acridine orange accumulating in the lysosomal vesicles was assessed by the ratio of the mean florescence intensity (MFI) in the PECy5 channel to the FITC channel by the FlowJo software.

3.5 Results

3.5.1 Visualizing Autophagic Organelles by TEM

The TEM images of fixed *L. major* captured double membranous autophagosomes including those containing damaged mitochondrial contents, as well as mitochondria with an ongoing fusion/fission process (Fig. 1). These images suggest that *Leishmania* undergoes basal autophagic processes (quality control processes) that enable the removal of damaged mitochondria.

3.5.2 Accumulation of Acridine Orange in the Lysosomal Vacuoles by Flow Cytometry

The strategy employed in the acquisition of the AO-stained parasites by flow cytometry was to gate on live parasite from the forward and live scatter axis plots. Following gating on single parasites (singlets), the red and green florescence in the PECy5 and FITC channels were gated, respectively (Fig. 2a). The frequency and ratios of the mean fluorescent intensity of the AO stained parasites in the red and green florescent channels without treatment, with 100 nM bafilomycin (autophagy inhibition) and in nutrient starvation condition (1% FBS, autophagy enhancer), were assessed (Fig. 2b, c). In the parasites, basal autophagic processes were observed in the no-treatment group because when autophagy was inhibited by bafilomycin, there was a decrease in the mean florescent intensity (MFI) of parasites gated in the red channel. However, in nutrient-deprivation condition (1% FBS) which enhances autophagy, there was a significant increase in the MFI of the red florescence compared to the untreated parasite groups (Fig. 2b, c).

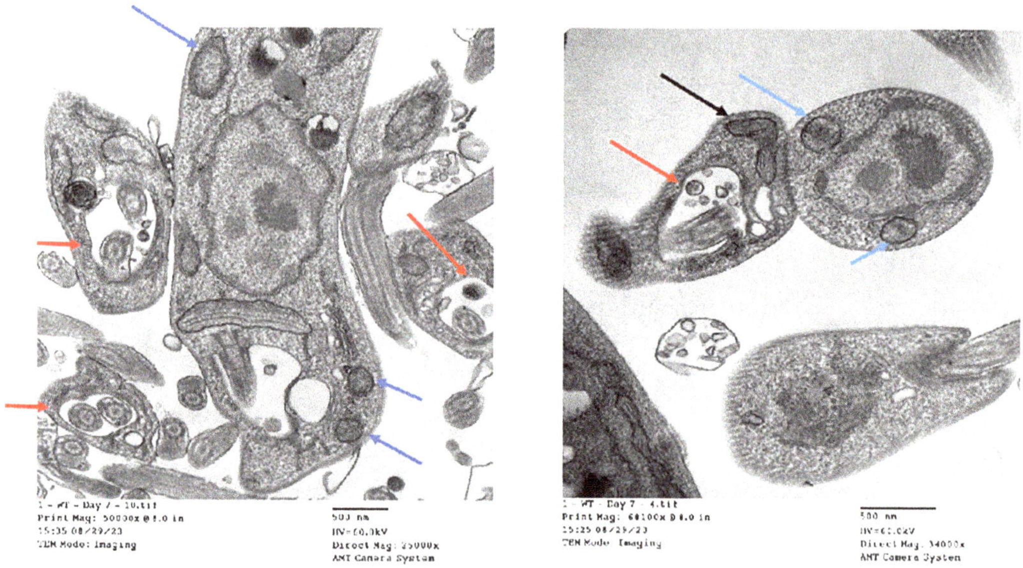

Fig. 1 Images of autophagic organelles in *Leishmania major* by TEM. Fixed *Leishmania major* at day 7 of axenic culture growth showing double membrane vacuoles representing autophagosome (**a** & **b**, blue arrows); autophagosome containing damaged mitochondrial (**a** & **b**, red arrows) and dividing mitochondria with an ongoing fusion/fission process (**b**, black arrows). Images were taken at a magnification of 34,000X

4 Notes

1. Use 6% glutaraldehyde stock if it completely clear but once it starts turning yellow, discard and make a fresh batch.
2. Prepare fresh on the day of fixation.
3. Work in the fume hood.
4. All parasite media preparation and cultures should be carried out in a biosafety hood.
5. Morphologically, the parasites may have an oval shape with less pronounced flagella when brought straight out from liquid nitrogen. Occasional movements may be observed with these parasites at this time.
6. Too high spins and longer periods can damage the parasite's ultrastructure.
7. The parasites were plated at the same time in two different plates to indicate day 3 and day 7 growth cultures. This was done to prevent the risk of contamination since the experiment was performed on different days.
8. The tubes were placed in a rack and covered with foil paper but alternatively can be kept in a dark cupboard because acridine orange is light sensitive and can easily degrade.

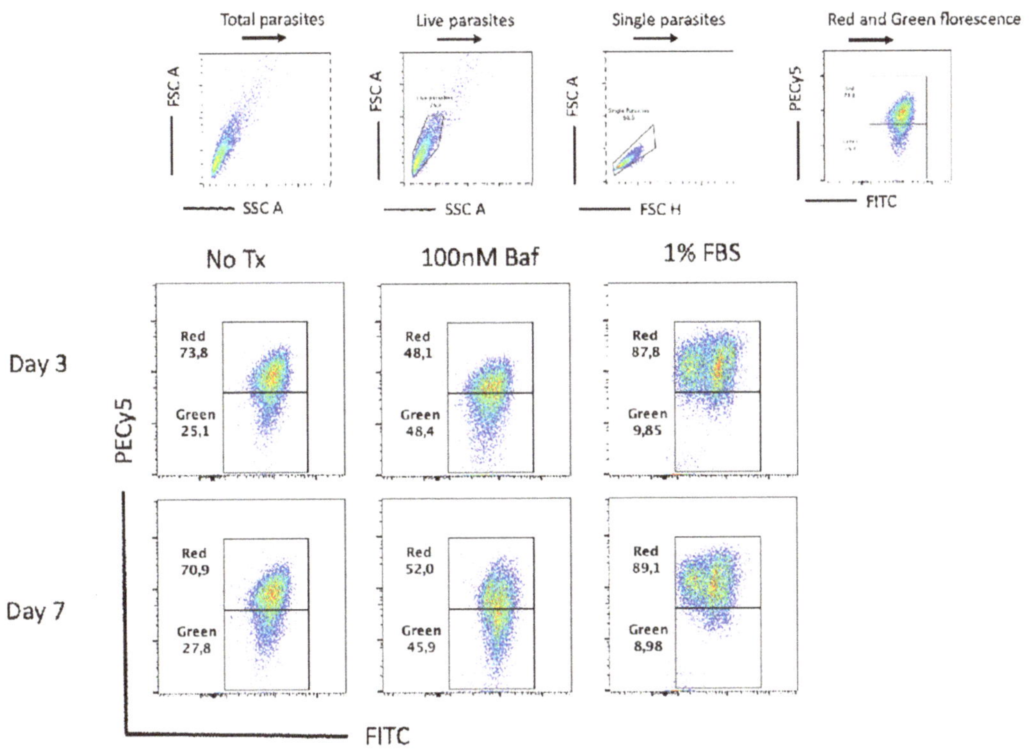

Fig. 2 Detection of acridine orange accumulation in the lysosomal vacuole in *Leishmania major* by flow cytometry. Axenic cultures of *Leishmania major* at day 3 and 7 were stained with acridine orange and the amount of red florescence detected by flow cytometry indicates autophagic activity occurring in the lysosomes. (**a**) Gating strategy (**b**, **c**) flow plots and graphs demonstrating the ratio of the MFI of the red florescence to green florescence when the parasites were not treated, given autophagic inhibitors (Bafilomycin, Baf), or exposed to a nutrient-deprivation model (1% FBS)

Acknowledgments

SG conceptualized and designed the study. SSO performed the experiments. SG and SSO analyzed the data. SSO wrote the manuscript. SSO, SG, JU reviewed and edited the manuscript. JU funded and SG supervised the experiments.

Funding

This study was funded in part by Canadian Institutes for Health Research (CIHR).

References

1. Alizadeh J, da Silva Rosa SC, Weng X, Jacobs J, Lorzadeh S, Ravandi A, Vitorino R, Pecic S, Zivkovic A, Stark H, Shojaei S, Ghavami S (2023) Ceramides and ceramide synthases in cancer: focus on apoptosis and autophagy. Eur J Cell Biol 102:151337

2. Behrooz AB, Cordani M, Donadelli M, Ghavami S (2023) Metastatic outgrowth via the two-way interplay of autophagy and metabolism. Biochim Biophys Acta Mol basis Dis:166824. https://doi.org/10.1016/j.bbadis.2023.166824
3. Hajiahmadi S, Lorzadeh S, Iranpour R, Karima S, Rajabibazl M, Shahsavari Z, Ghavami S (2023) Temozolomide, simvastatin and Acetylshikonin combination induces mitochondrial-dependent apoptosis in GBM cells, which is regulated by autophagy. Biology (Basel) 12:302
4. Alizadeh J, Kavoosi M, Singh N, Lorzadeh S, Ravandi A, Kidane B, Ahmed N, Mraiche F, Mowat MR, Ghavami S (2023) Regulation of autophagy via carbohydrate and lipid metabolism in cancer. Cancers (Basel) 15:2195
5. Klionsky DJ (2007) Autophagy: from phenomenology to molecular understanding in less than a decade. Nat Rev Mol Cell Biol 8: 931–937
6. Maiuri MC, Zalckvar E, Kimchi A, Kroemer G (2007) Self-eating and self-killing: crosstalk between autophagy and apoptosis. Nat Rev Mol Cell Biol 8:741–752
7. Martelli A, Omrani M, Zarghooni M, Citi V, Brogi S, Calderone V, Sureda A, Lorzadeh S, da Silva Rosa SC, Grabarek BO, Staszkiewicz R, Los MJ, Nabavi SF, Nabavi SM, Mehrbod P, Klionsky DJ, Ghavami S (2022) New visions on natural products and cancer therapy: autophagy and related regulatory pathways. Cancers (Basel) 14:5839
8. Guhe V, Ingale P, Tambekar A, Singh S (2023) Systems biology of autophagy in leishmanial infection and its diverse role in precision medicine. Front Mol Biosci 10:1113249
9. Yeganeh B, Rezaei Moghadam A, Alizadeh J, Wiechec E, Alavian SM, Hashemi M, Geramizadeh B, Samali A, Bagheri Lankarani K, Post M, Peymani P, Coombs KM, Ghavami S (2015) Hepatitis B and C virus-induced hepatitis: apoptosis, autophagy, and unfolded protein response. World J Gastroenterol 21:13225–13239
10. Yeganeh B, Ghavami S, Kroeker AL, Mahood TH, Stelmack GL, Klonisch T, Coombs KM, Halayko AJ (2015) Suppression of influenza A virus replication in human lung epithelial cells by noncytotoxic concentrations bafilomycin A1. Am J Physiol Lung Cell Mol Physiol 308: L270–L286
11. Dalvand A, da Silva Rosa SC, Ghavami S, Marzban H (2022) Potential role of TGFB and autophagy in early crebellum development. Biochem Biophys Rep 32:101358
12. Nakatogawa H (2020) Mechanisms governing autophagosome biogenesis. Nat Rev Mol Cell Biol 21:439–458
13. Galán JE, Cossart P (2005) Host-pathogen interactions: a diversity of themes, a variety of molecular machines. Curr Opin Microbiol 8:1–3
14. Campoy E, Colombo MI (2009) Autophagy in intracellular bacterial infection. Biochim Biophys Acta 1793:1465–1477
15. Yeganeh B, Ghavami S, Rahim MN, Klonisch T, Halayko AJ, Coombs KM (2018) Autophagy activation is required for influenza A virus-induced apoptosis and replication. Biochim Biophys Acta, Mol Cell Res 1865:364–378
16. Chen T, Tu S, Ding L, Jin M, Chen H, Zhou H (2023) The role of autophagy in viral infections. J Biomed Sci 30:5
17. Golovkine GR, Roberts AW, Morrison HM, Rivera-Lugo R, McCall RM, Nilsson H, Garelis NE, Repasy T, Cronce M, Budzik J, Van Dis E, Popov LM, Mitchell G, Zalpuri R, Jorgens D, Cox JS (2023) Autophagy restricts Mycobacterium tuberculosis during acute infection in mice. Nat Microbiol 8:819–832
18. Jiang P, Mizushima N (2014) Autophagy and human diseases. Cell Res 24:69–79
19. Mathison BA, Bradley BT (2023) Review of the clinical presentation, pathology, diagnosis, and treatment of Leishmaniasis. Lab Med 54: 363–371
20. World Health Organization (2022) Fifteenth meeting of the strategic and technical advisory group for neglected tropical diseases. WHO, Geneva
21. Alvar J, Vélez ID, Bern C, Herrero M, Desjeux P, Cano J, Jannin J, de Boer M (2012) Leishmaniasis worldwide and global estimates of its incidence. PLoS One 7:e35671
22. Scott P, Novais FO (2016) Cutaneous leishmaniasis: immune responses in protection and pathogenesis. Nat Rev Immunol. https://doi.org/10.1038/nri.2016.72
23. Roatt BM, de Oliveira Cardoso JM, De Brito RCF, Coura-Vital W, de Oliveira Aguiar-Soares RD, Reis AB (2020) Recent advances and new strategies on leishmaniasis treatment. Appl Microbiol Biotechnol 104:8965–8977
24. Dinc R (2022) Leishmania vaccines: the current situation with its promising aspect for the future. Korean J Parasitol 60:379–391
25. Mou Z, Li J, Boussoffara T, Kishi H, Hamana H, Ezzati P, Hu C, Yi W, Liu D, Khadem F, Okwor I, Jia P, Shitaoka K, Wang S, Ndao M, Petersen C, Chen J, Rafati S, Louzir H, Muraguchi A, Wilkins JA,

Uzonna JE (2015) Identification of broadly conserved cross-species protective Leishmania antigen and its responding CD4+ T cells. Sci Transl Med 7:1–12

26. Mou Z, Barazandeh AF, Hamana H, Kishi H, Zhang X, Jia P, Ikeogu N, Onyilagha C, Gupta G, Uzonna JE (2020) Identification of a protective Leishmania antigen Dihydrolipoyl dehydrogenase and its responding CD4+ T cells at clonal level. J Immunol 205:1355–1364
27. Thomas SA, Nandan D, Kass J, Reiner NE (2018) Countervailing, time-dependent effects on host autophagy promotes intracellular survival of Leishmania. J Biol Chem 293:2617–2630
28. Esch KJ, Schaut RG, Lamb IM, Clay G, Lima ÁLM, do Nascimento PRP, Whitley EM, Jeronimo SMB, Sutterwala FS, Haynes JS, Petersen CA (2015) Activation of autophagy and nucleotide-binding domain leucine-rich repeat-containing-like receptor family, pyrin domain-containing 3 inflammasome during Leishmania infantum-associated glomerulonephritis. Am J Pathol 185:2105–2117
29. Cull B, Prado Godinho JL, Fernandes Rodrigues JC, Frank B, Schurigt U, Williams RA, Coombs GH, Mottram JC (2014) Glycosome turnover in Leishmania major is mediated by autophagy. Autophagy 10:2143–2157
30. Besteiro S, Williams RAM, Coombs GH, Mottram JC (2007) Protein turnover and differentiation in Leishmania. Int J Parasitol 37:1063–1075
31. Vanrell MC, Losinno AD, Cueto JA, Balcazar D, Fraccaroli LV, Carrillo C, Romano PS (2017) The regulation of autophagy differentially affects Trypanosoma cruzi metacyclogenesis. PLoS Negl Trop Dis 11:e0006049
32. Williams RAM, Woods KL, Juliano L, Mottram JC, Coombs GH (2009) Characterization of unusual families of ATG8-like proteins and ATG12 in the protozoan parasite Leishmania major. Autophagy 5:159–172
33. Williams RAM, Smith TK, Cull B, Mottram JC, Coombs GH (2012) ATG5 is essential for ATG8-dependent autophagy and mitochondrial homeostasis in Leishmania major. PLoS Pathog 8:e1002695
34. Williams RAM, Mottram JC, Coombs GH (2013) Distinct roles in autophagy and importance in infectivity of the two ATG4 cysteine peptidases of Leishmania major. J Biol Chem 288:3678–3690
35. Giri S, Shaha C (2019) Leishmania donovani parasite requires Atg8 protein for infectivity and survival under stress. Cell Death Dis 10:808
36. Besteiro S, Williams RAM, Morrison LS, Coombs GH, Mottram JC (2006) Endosome sorting and autophagy are essential for differentiation and virulence of Leishmania major. J Biol Chem 281:11384–11396
37. Ylä-Anttila P, Vihinen H, Jokitalo E, Eskelinen E-L (2009) Monitoring autophagy by electron microscopy in Mammalian cells. Methods Enzymol 452:143–164
38. Thomé MP, Filippi-Chiela EC, Villodre ES, Migliavaca CB, Onzi GR, Felipe KB, Lenz G (2016) Ratiometric analysis of Acridine Orange staining in the study of acidic organelles and autophagy. J Cell Sci 129:4622–4632
39. Martí-Clua J (2022) Transmission electron microscopy in the study of autophagy: seeing is not always believing. https://doi.org/10.21203/rs.3.rs-2271893/v1
40. du Toit A, Hofmeyr J-HS, Gniadek TJ, Loos B (2018) Measuring autophagosome flux. Autophagy 14:1060–1071
41. Tooze SA, Yoshimori T (2010) The origin of the autophagosomal membrane. Nat Cell Biol 12:831–835
42. Haspel J, Shaik RS, Ifedigbo E, Nakahira K, Dolinay T, Englert JA, Choi AMK (2011) Characterization of macroautophagic flux in vivo using a leupeptin-based assay. Autophagy 7:629–642

Methods in Molecular Biology (2025) 2879: 219–224
DOI 10.1007/7651_2024_526

Published online: 07 March 2024

Measurement of Autophagic Activity in Cancer Cells with Flow Cytometric Analysis Using Cyto-ID Staining

Merve Şansaçar and Emel Başak Gencer Akçok

Abstract

Autophagy is an evolutionarily conserved process providing the energy that cells need to survive, especially in stress situations, through catabolic processes. Considering the dual role of autophagy in cancer cells depending on the cellular context, it is crucial to comprehend the effect of drug candidates put forward to prevent cancer through the autophagy pathway. The CYTO-ID® Autophagy Detection Kit allows a rapid, specific and quantitative measurement of autophagic activity at the cellular level using a 488 nm-excitable green fluorescent detection reagent via flow cytometer. In this chapter, we present the CYTO-ID® Autophagy Detection method with a stepwise protocol to monitor the autophagy flux after the application of any compound to suspension cancer cell lines with flow cytometric analysis.

Key words Autophagy, Flow cytometry, Autophagic flux, Autophagic compartments, Cyto-ID, Autophagosome, Cancer

1 Introduction

Autophagy is an evolutionary conserved catabolic process that is active in all cells. It is usually induced in response to starvation and stress [1]. The process involves the breakdown of dysfunctional organelles and misfolded or aggregated proteins and hence provides the necessary nutrients and supplies, such as proteins, lipids and carbohydrates, when needed [2]. This process also eradicates viral particles, bacteria, and parasites [3]. Therefore, autophagy maintains cellular or tissue homeostasis, provides energy for the cell, and has a cell-protective role [4]. Autophagy is known to play crucial roles in a variety of disorders, such as cancer, neurodegenerative diseases, aging, autoimmune diseases, and inflammation [5, 6]. The regulation of autophagy has a great impact on health and disease conditions. Knowing the crucial roles of autophagy, it is essential to precisely detect and measure the levels of autophagy in different types of physiological conditions [1, 7].

The role of autophagy in cancer cells could change depending on cellular context in different cell types. Various studies have implicated autophagy in regulating many oncogenes and tumor suppressor genes [8], while other studies show that autophagy plays a dual role in cancer progression or inhibition of cancer growth [9].

Three main types of autophagy direct the to-be-degraded contents to lysosomal degradation: macroautophagy, microautophagy, and chaperone-mediated autophagy [10]. The most common autophagic pathway is macroautophagy (hereafter referred to as autophagy), and it occurs via a specific double-membraned structure called autophagosome. The autophagosome then fuses with the lysosome to deliver the content for degradation [11]. Autophagic flux is referred to as the measure of the autophagic degradation activity and is described as the entire process of autophagy including the dynamic process of autophagosome formation, fusion of autophagosomes with lysosomes, and finally the degradation of the content in lysosomes [7]. Various methods measure the number of autophagosomes to detect the autophagic activity in cells. However, the detection of the number of autophagosomes could be misleading to estimate the autophagic flux because the number of autophagosomes could result from the formation of autophagosomes and the formation of autolysosomes, which forms as a result of the fusion of autophagosomes and lysosomes [10].

In this chapter, we describe a flow cytometry-based profiling of autophagy with CYTO-ID® Autophagy Detection Kit in cancer cells. Cyto-ID is an amphiphilic dye to monitor autophagic flux and it stains the membranes of autophagosomes and autolysosomes, with minimal staining of lysosomes or endosomes [12]. Given this feature of Cyto-ID staining, the researchers can distinguish the autophagosome formation from the inhibition of autolysosome formation, which will result in an increased autophagosome number.

2 Materials

2.1 Treatment of Cancer Cells with Compound of Interest

1. Compound of interest (the compound with unknown effect and to be tested on autophagy. This compound of interest will be called as "compound" hereafter).
2. Cell line of interest (adherent or suspension cells).
3. Required cell culture medium (RPMI 1640, DMEM, EMEM, etc.) with necessary complements (fetal bovine serum (FBS) and 100 U/100 μg/mL penicillin/streptomycin antibiotic (*see* **Note 1**)).
4. Dulbecco's phosphate buffered saline (DPBS).
5. Six-well cell culture dishes.

2.2 Detection of Autophagy Using CYTO-ID® Green Detection Kit with Flow Cytometry

1. CYTO-ID® Green Detection Reagent: Prepare the dye solution by diluting 1 μL of CYTO-ID® Green Detection Reagent with 1 mL of 1X Assay Buffer or Phenol Red Indicator-free cell culture medium supplemented with 5% FBS. (*see* **Note 2**).
2. 1X Assay Buffer: Warm the 10X Assay Buffer to room temperature. Make sure the assay buffer does not crystallize before diluting. Prepare enough 1X Assay Buffer for the number of samples to be assayed by diluting 10X Assay Buffer with 1:10 ratio.
3. Positive controls (as autophagy inhibitors: chloroquine (CQ), recommended starting concentration of 10 μM; as autophagy inducer: rapamycin, recommended starting concentration of 0.5 μM; PP242, recommended starting concentration of 10 μM) (*see* **Note 3**).
4. Negative controls (DMSO, deionized water, cell culture medium, or other solvents used to reconstitute or dilute the positive control).

3 Methods

3.1 Treatment of Cancer Cells with the Compound

1. Seed the cells in a six-well plate to a maximum density of 1×10^6 cells per 1 mL (*see* **Notes 4** and **5**).
2. Treat the cells with desired concentrations of the compound and incubate the cells for the desired duration under the standard cell culture conditions (at 37 °C, 5% CO_2) in a humidified incubator (*see* **Note 6**).
3. Negative and positive control wells should be included in the experimental setup, and cells should be treated with the desired controls (*see* Subheading 2.2, **steps 3** and **4**).
4. Incubate the cells with the chosen lysosomotropic agent (NH_4Cl (10–20 mmol/L) or chloroquine (5–10 μmol/L)) for the last 5 h of the incubation period that is decided in the 2nd step in this section (*see* **Note 5**).

3.2 Detection of Autophagy Using CYTO-ID® Green Detection Kit with Flow Cytometry

1. At the end of the incubation period, collect the cells via centrifugation. (This protocol is prepared for suspension cells. For attached cells, please *see* **Note** 7.)
2. Centrifuge the cells at 1000 rpm for 5 min, remove the supernatant, and then wash the cells with cell culture medium, DPBS, or 1X Assay Buffer and remove the supernatant.
3. Resuspend the cell pellet with 250 μL 1X Assay Buffer or indicator-free cell culture medium containing 5% FBS.

4. Add 250 μL of CYTO-ID® Green staining solution that is prepared in 1X Assay Buffer (*see* Subheading 2.2, **step 1**) to each sample (*see* **Note 8**).
5. Incubate for 30 min at room temperature or 37°C in the dark.
6. After 30 min of incubation, collect the cells by centrifugation at 1500 rpm for 5 min and discard the supernatant.
7. Wash the pellet with 1X Assay Buffer, centrifuge at 1500 rpm for 5 min and discard the supernatant.
8. Resuspend the cell pellets in 500 μL of fresh 1X Assay Buffer and transfer the samples to FACS tubes (*see* **Note 9**).
9. ***Optional:*** Incubate for 20 min with 4% formaldehyde (or 10% formalin) for the fixation of the samples and then wash three times with 1X Assay Buffer.
10. Analyze the samples by flow cytometry in the green (FL1) or orange (FL2) channel (*see* **Note 10).**

3.3 Analysis of Autophagic Flux

1. To calculate ΔMFI (mean fluorescence intensity) Cyto-ID, conditions both with +CQ/NH_4Cl and without -CQ/NH_4Cl are required for each sample (*see* **Notes 11** and **12**).
2. The following formula is used to evaluate the effect of a compound with unknown autophagic effect [13]:

$$\Delta\text{MFI Cyto} - \text{ID} = \text{MFI Cyto} - \text{ID}\ (+\text{CQ}/\text{NH}_4\text{Cl}) \\ - \text{MFI Cyto} - \text{ID}\ (-\ \text{CQ}/\text{NH}_4\text{Cl})$$

3. Values below zero show inhibition of autophagic flux, while values above zero point out activation of autophagy.

4 Notes

1. The percentage of FBS varies depending on the cell type and the needs of the cell that are used in the experimental setup. If the starvation condition is going to be applied for the induction of autophagy, the amount of FBS in the culture medium should be adjusted accordingly.
2. 250 μL of diluted CYTO-ID® Green dye staining solution will be used for each sample to be stained, the total amount of staining solution is dependent on the number of samples and should be calculated accordingly.
3. Autophagy can be modulated by using either activators or inhibitors of autophagic pathways in different stages. The recommended inducers or inhibitors could be selected by the researcher.

4. Since the results of the experiments will depend significantly on the general condition of the cells, the cells must be healthy and not overcrowded. Thus, the cell density should not exceed 1×10^6/mL. The recommended cell density for the staining step is between 10^5 and 10^6 cells/mL.
5. To calculate ΔMFI Cyto-ID, conditions both with +CQ or NH_4Cl and without -CQ or NH_4Cl are required for each compound treatment. Therefore, the researcher should seed the compound-treated cells in duplicate to include the lysosomotropic agents to calculate the ΔMFI Cyto-ID.
6. Prolonged analysis under conditions that could potentially affect cell number and viability, such as when the cell doubling time is exceeded or when the nutrients required for the cell are depleted/insufficient in the culture, may affect the results, as autophagic activity is highly correlated with the presence of nutrients. It is recommended not to exceed 48 h for drug administration.
7. For adherent cells, the trypsinization procedure is required as follows. Firstly, the cell medium is collected into the 15 mL falcon tube, and then the cells in wells are washed with 1 mL of 1X DPBS and the DPBS is removed. Later, 1 X 0.05% Trypsin-EDTA was added to the cells and left for some time for detachment (the amount of trypsin and the incubation time with trypsin are dependent on cell type and need to be optimized.). Following trypsinization, fresh cell medium is added to neutralize the trypsin. Lastly, the cells are collected in the same 15 mL falcon tube and centrifuged.
8. In this step, it is important to obtain a monodisperse cell suspension by gently pipetting up and down repeatedly.
9. Take care of the samples after the staining step is completed and always keep tubes containing the resuspended cells on ice and in the dark as much as possible.
10. A 488 nm laser source is recommended for the CYTO-ID® Autophagy Detection Kit.
11. Autophagic flux is measured via the accumulation of autophagic compartments after blockage of fusion of autophagosomes to lysosomes by altering the acidic environment of lysosomes using a lysosomotropic agent such as CQ or NH_4Cl (Fig. 1).
12. Autophagic flux can also be calculated by normalizing to untreated control.

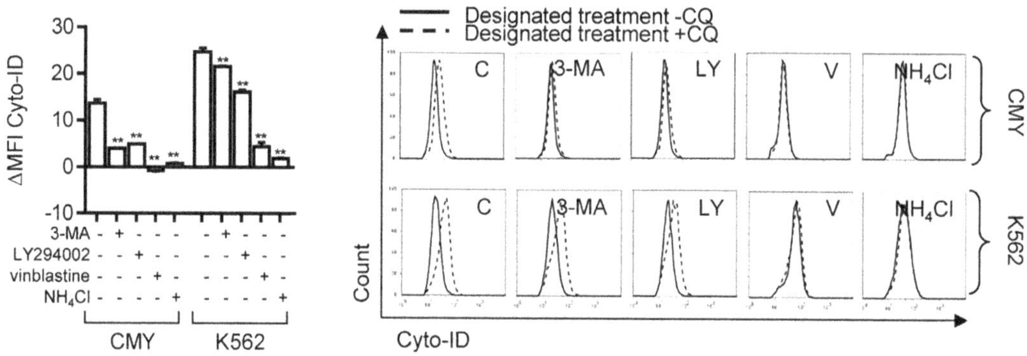

Fig. 1 Autophagic flux was assessed as accumulation of autophagic compartments (pre-autophagosomes, autophagosomes, and autophagolysosomes) after the addition of late autophagic inhibitors NH_4Cl or CQ for the last 5 h measured by changes in total cellular Cyto- ID® signal (ΔMFI Cyto-ID). Basic autophagic flux in CMY and K562 cells pretreated with autophagy inhibitors 3-MA (10mM), LY294002 (20μM), vinblastine (V) (12μM), and NH_4Cl (20mM) (left) and representative histograms (right). (Reproduced from Ref. [14] with permission from Springer Nature)

References

1. Levine B, Kroemer G (2008) Autophagy in the pathogenesis of disease. Cell 132:27–42
2. Dikic I, Elazar Z (2018) Mechanism and medical implications of mammalian autophagy. Nat Rev Mol Cell Biol 19:349–364
3. Levine B, Deretic V (2007) Unveiling the roles of autophagy in innate and adaptive immunity. Nat Rev Immunol 7:767–777
4. Anding AL, Baehrecke EH (2017) Cleaning house: selective autophagy of organelles. Dev Cell 41:10–22
5. Lei Y, Klionsky DJ (2021) The emerging roles of autophagy in human diseases. Biomedicines 9:1651
6. Choi AMK, Ryter SW, Levine B (2013) Autophagy in human health and disease. N Engl J Med 368:651–662
7. Klionsky DJ, Kamal Abdel-Aziz A, Abdelfatah S et al (2021) Guidelines for the use and interpretation of assays for monitoring autophagy (4th edition). Autophagy 17:1–382
8. Auberger P, Puissant A (2017) Autophagy, a key mechanism of oncogenesis and resistance in leukemia. Blood 129:547–552
9. Yun C, Lee S (2018) The roles of autophagy in cancer. Int J Mol Sci 19:3466
10. Mizushima N, Yoshimori T, Levine B (2010) Methods in mammalian autophagy research. Cell 140:313–326
11. Yamamoto H, Zhang S, Mizushima N (2023) Autophagy genes in biology and disease. Nat Rev Genet 24:382–400
12. Guo S, Liang Y, Murphy SF et al (2015) A rapid and high content assay that measures cyto-ID-stained autophagic compartments and estimates autophagy flux with potential clinical applications. Autophagy 11:560–572
13. Stankov M, Panayotova-Dimitrova D, Leverkus M et al (2014) Flow cytometric analysis of autophagic activity with cyto-ID staining in primary cells. Bio-Protocol. https://doi.org/10.21769/BioProtoc.1090
14. Stankov MV, El KM, Kumar Thakur B et al (2014) Histone deacetylase inhibitors induce apoptosis in myeloid leukemia by suppressing autophagy. Leukemia 28:577–588

Methods in Molecular Biology (2025) 2879: 225–238
DOI 10.1007/7651_2024_571

Published online: 28 September 2024

Assessing Autophagy Flux in Glioblastoma Temozolomide Resistant Cells

Courtney Clark, Amir Barzegar Behrooz, Marco Cordani, Shahla Shojaei, and Saeid Ghavami

Abstract

Autophagy is a critical cellular process involved in the degradation and recycling of cytoplasmic components, playing a dual role in cancer by either promoting cell survival or facilitating cell death. In glioblastoma (GB), autophagy has been implicated in resistance to the chemotherapeutic agent temozolomide (TMZ). This study presents a novel method to accurately measure autophagy flux in TMZ-resistant glioblastoma cells, combining advanced imaging techniques with biochemical assays. By quantifying key autophagy markers such as LC3-II and SQSTM1, our approach provides detailed insights into the dynamic processes of autophagosome formation and clearance under therapeutic stress. This method advances our understanding of autophagy in GB chemoresistance and has significant implications for the development of autophagy-targeted therapies. The ability to monitor and manipulate autophagy flux in real time offers a promising avenue for monitoring and understanding TMZ resistance and improving patient outcomes in glioblastoma treatment.

Key words Autophagy flux, Glioblastoma, LC3-II, SQSTM1 degradation, Temozolomide resistance

1 Introduction

Autophagy is a fundamental cellular process responsible for the degradation and recycling of cytoplasmic components, including damaged organelles and misfolded proteins, through lysosomal degradation [1–3]. This dynamic process, known as autophagy flux, involves the formation of autophagosomes that sequester cellular debris and their subsequent fusion with lysosomes for degradation [4–6]. Measuring autophagy flux provides insights into the dynamic nature of autophagy, distinguishing between increased autophagosome formation and impaired autophagosome clearance [7–9]. Accurate assessment of autophagy flux is crucial for understanding its role in various physiological and pathological conditions, including cancer [10, 11].

Authors Shahla Shojaei and Saeid Ghavami have co-senior authorship.

Glioblastoma (GB) is the most aggressive and common primary brain tumor in adults, characterized by rapid growth, extensive infiltration, and resistance to conventional therapies [4, 5, 12]. Despite advances in surgical resection, radiotherapy, and chemotherapy, the prognosis for GB patients remains dismal, with a median survival of 12–15 months [13, 14]. One of the standard chemotherapeutic agents for GB is temozolomide (TMZ), an oral alkylating agent that induces DNA damage, leading to cell death. However, the development of TMZ resistance is a significant challenge, limiting its therapeutic efficacy and contributing to GB's poor prognosis [2, 15, 16].

Emerging evidence suggests that autophagy plays a dual role in cancer, acting as a tumor suppressor by degrading damaged organelles and proteins and as a survival mechanism by providing nutrients by recycling cellular components under stress conditions [17–19]. Autophagy has promoted resistance to TMZ in GB, allowing cancer cells to survive and adapt to chemotherapy-induced stress [9, 20, 21]. Autophagy flux in GB TMZ-resistant cells is a dynamic process that contributes to maintaining cellular homeostasis and the survival of cancer cells under therapeutic pressure [22, 23].

Several studies have highlighted the role of autophagy flux in GB TMZ resistance, demonstrating that autophagy inhibitors can sensitize resistant GB cells to TMZ, thereby enhancing its cytotoxic effects [4, 5]. This indicates that autophagy flux is a critical adaptive response that enables GB cells to withstand TMZ-induced stress [10, 24]. Furthermore, autophagy-related genes (ATGs) and proteins involved in the autophagy pathway, such as Beclin-1, LC3, and SQSTM1, are often dysregulated in GB, contributing to the modulation of autophagy flux and chemoresistance [17]. Understanding the mechanisms underlying autophagy flux in GB TMZ-resistant cells is essential for developing novel therapeutic strategies to overcome resistance and improve patient outcomes [23].

Our novel method for measuring autophagy flux in GB TMZ-resistant cells provides a comprehensive approach to quantifying the dynamic process of autophagy. This method involves advanced imaging techniques, such as confocal and transmission electron microscopy, combined with biochemical assays to monitor the levels of autophagy markers, including LC3-II and SQSTM1, under various treatment conditions. By integrating these techniques, we can accurately assess the formation and clearance of autophagosomes, providing a detailed understanding of autophagy flux in response to TMZ treatment. This method applies to basic science research and has significant implications for clinical scientists seeking to develop autophagy-targeted therapies for GB.

In summary, the ability to measure autophagy flux in GB TMZ-resistant cells is crucial for elucidating the role of autophagy in chemoresistance and identifying potential therapeutic targets.

Our method offers a robust and reliable approach to quantify autophagy flux, contributing to the advancement of both basic and clinical research in GB. By providing detailed insights into the autophagic process, this method can facilitate the development of novel strategies to overcome TMZ resistance, ultimately improving the therapeutic outcomes for GB patients.

2 Materials

2.1 Measuring Autophagy Flux in U251 TMZR and U251 TMZ NR Cells Using Immunoblotting

- U251 TMZ-resistant (TMZR) and U251 TMZ non-resistant (TMZ NR) cells.
- DMEM high glucose medium (CORNING; Cat#: 50–003-PB).
- Fetal bovine serum (FBS) (Gibco™; Cat#: 16000044).
- Penicillin-streptomycin (Gibco™; Cat#: 41400045).
- Temozolomide (TMZ) (CAS Number: 85622-93-1).
- Phosphate buffered saline (PBS) (Sigma-Aldrich, Cat #D8537).
- Trypsin–EDTA (GIBCO® (Thermo Fisher Scientific, MA, USA).
- Lysis buffer: 20 mM Tris–HCl (pH 7.5), 0.5 mM PMSF, 0.5% Nonidet P-40, 100 μM β-glycerol 3-phosphate, 0.5% protease and phosphatase inhibitor cocktail.
- Lowry protein assay kit (Bio-Rad, Oakville, ON, CA).
- SDS-PAGE reagents (Bio-Rad, Oakville, ON, CA): acrylamide/bis-acrylamide solution: for gel polymerization; SDS (sodium dodecyl sulfate): denaturing agent; Tris buffer: maintains pH; APS (ammonium persulfate): polymerization initiator; TEMED (tetramethylethylenediamine): polymerization accelerator; running buffer: contains Tris, glycine, and SDS; loading buffer: contains SDS, bromophenol blue, glycerol, and β-mercaptoethanol or DTT; protein ladder/marker: for molecular weight reference; sample buffer: typically contains SDS, Tris, glycerol, and a reducing agent.
- Polyvinylidene difluoride (PVDF) membranes (Thermo Fisher Scientific, Waltham, MA, USA; Cat#: 88518.
- Blocking solution: 5% skim milk in TBST (1X Tris-buffered saline, 0.025% Tween 20) (Sigma, St. Louis, MO, USA).
- Primary antibodies: LC3 (Sigma Aldrich, Oakville, CA; Cat#: L7543, 1:2500 dilution), SQSTM1 (Cell Signaling Technology, Canada; Cat#: 5114, 1:1000 dilution).
- Secondary antibodies: Fluorochrome-conjugated secondary antibodies (Alexa Fluor 488 (Thermo Fisher Scientific, Waltham, A, USA; Cat#: A21206) and Alexa Fluor 594 (Thermo Fisher Scientific, Waltham, MA, USA; Cat#: R37119)).

- Enhanced chemiluminescence (ECL) detection kit (Amersham-Pharmacia Biotech).
- Bafilomycin A1 (Sigma Aldrich, Cat#B1793-10UG).

2.2 Transmission Electron Microscopy (TEM) and Evaluation of Autophagy Flux

- U251 TMZ-resistant (TMZR) and U251 TMZ non-resistant (TMZ NR) cells.
- DMEM high glucose medium (CORNING; Cat#: 50–003-PB).
- Fetal bovine serum (FBS) (Gibco™; Cat#: 16000044).
- Penicillin-streptomycin (Gibco™; Cat#: 41400045).
- Karnovsky fixative.
- 5% sucrose in 0.1 M Sorenson's phosphate buffer.
- 1% osmium tetroxide in 0.1 M Sorenson's buffer.
- Ethanol series for dehydration (50%, 70%, 90%, 100%) (UN 1170, Greenfield Global, # PO25EAAN).
- Embed 812 resin.
- Uranyl acetate.
- Lead nitrate.
- Copper grids (EMS, #G200-Cu).
- Philips CM10 electron microscope.

3 Methods

3.1 Cell Culture: Culturing U251 Cells

1. Prepare DMEM high glucose medium supplemented with 10% FBS and 1% penicillin-streptomycin (*see* **Note 1**).
2. Thaw U251 TMZR and U251 TMZ NR cells and transfer them to T75 culture flasks containing 20 mL of prepared DMEM medium.
3. Maintain U251 TMZR cells in DMEM medium with 250 μM TMZ. Incubate all flasks at 37 °C in a 5% CO_2 humidified incubator.
4. Change the medium every 2–3 days. For TMZR cells, remove TMZ-containing medium and replace it with TMZ-free DMEM medium 24 h before collection to mimic withdrawal conditions (*see* **Note 2**).

3.2 Cell Passaging and Collection

3.2.1 Cell Passaging

1. When cells reach 80–90% confluency, aspirate the medium and wash cells with 5 mL PBS (*see* **Note 3**).
2. Add 3 mL of trypsin–EDTA solution to the flask and incubate at 37 °C for 2–3 min until cells detach (*see* **Note 4**).
3. Neutralize trypsin by adding 7 mL of DMEM medium containing FBS.

4. Transfer cell suspension to a 15 mL centrifuge tube and centrifuge at 300× *g* for 5 min (*see* **Note 5**).
5. Aspirate the supernatant and resuspend the cell pellet in a fresh DMEM medium.
6. Seed cells at the appropriate density (e.g., 2 × 10^5 cells/mL) in new T75 flasks or culture dishes for subsequent experiments.

3.2.2 Cell Collection

- At the designated time point, aspirate the medium and wash cells with 5 mL PBS.
- Add 3 mL of trypsin–EDTA solution to the flask and incubate at 37 °C for 2–3 min until cells detach (*see* **Note 4**).
- Neutralize trypsin by adding 7 mL of DMEM medium containing FBS.
- Transfer cell suspension to a 15 mL centrifuge tube and centrifuge at 300× *g* for 5 min (*see* **Note 5**).
- Wash the cell pellet with 5 mL cold PBS and centrifuge again at 300× *g* for 5 min (*see* **Notes 6** and 7).
- Aspirate the supernatant and resuspend the cell pellet in 1 mL cold PBS (*see* **Note** 7).

3.3 Preparation of Cell Lysates

3.3.1 Cell Lysis

1. Transfer the resuspended cell pellet to a microcentrifuge tube.
2. Add 200 μL of lysis buffer (20 mM Tris–HCl, pH 7.5, 0.5 mM PMSF, 0.5% Nonidet P-40, 100 μM β-glycerol 3-phosphate, 0.5% protease and phosphatase inhibitor cocktail) to the tube.
3. Sonicate the lysate briefly using a sonicator set to 3 cycles of 5 s each, with 10-s intervals on ice, to ensure complete cell disruption (*see* **Note 8**).
4. Centrifuge the lysate at 10,000× *g* for 10 min at 4 °C to remove cell debris.
5. Transfer the clear supernatant (protein lysate) to a new microcentrifuge tube.

3.3.2 Protein Quantification

1. Use the Lowry protein assay kit to determine the protein concentration in the lysate (*see* **Note 9**).
2. Prepare a standard curve using BSA standards ranging from 0 to 1 mg/mL.
3. Add 5 μL of each protein sample or standard to a 96-well plate in triplicate.
4. Add 25 μL of Lowry reagent to each well, mix gently, and incubate for 10 min at room temperature.
5. Add 125 μL of Folin-Ciocalteu reagent, mix gently, and incubate for 30 min at room temperature.

6. Measure the absorbance at 750 nm using a microplate reader.
7. Calculate the protein concentration of each sample based on the standard curve and adjust the lysate concentration to 1 mg/mL using lysis buffer.

3.4 Immunoblotting

3.4.1 SDS-PAGE and Transfer

1. Prepare a 10–15% polyacrylamide gel for SDS-PAGE.
2. Load 30 μg of protein from each sample into the wells and run the gel at 100 V for 1.5 h.
3. Transfer the separated proteins to PVDF membranes at 100 V for 2 h under reducing conditions.

3.4.2 Blocking and Primary Antibody Incubation

1. Block the membranes with 5% skim milk in TBST for 1 h at room temperature (*see* **Note 10**).
2. Incubate membranes overnight at 4 °C with primary antibodies against LC3 (1:2500 dilution, Sigma) and SQSTM1 (1:1000 dilution, Cell Signaling) in 1% skim milk in TBST.

3.4.3 Secondary Antibody Incubation and Detection

1. Wash membranes with TBST (3 washes, 10 min each).
2. Incubate with HRP-conjugated secondary antibodies for 1 h at room temperature.
3. Wash membranes again with TBST (3 washes, 10 min each).
4. Develop the blots using an ECL detection kit and visualize bands using a BioRad imager.

3.5 Evaluation of Autophagy Flux

3.5.1 LC3 Lipidation and SQSTM1 Degradation

1. Assess LC3-II levels (lipidated form) and SQSTM1 degradation using immunoblotting. Increased LC3-II indicates autophagosome formation, while decreased SQSTM1 indicates autophagosome degradation.

3.5.2 Bafilomycin A1 Treatment for Flux Assessment

(a) *Treatment*

1. Treat cells with 100 nM bafilomycin A1 for 6 h to inhibit autophagosome-lysosome fusion.
2. Include a control group without bafilomycin A1 treatment for comparison.

(b) *Harvesting Cells Post-bafilomycin Treatment*

1. Wash cells with PBS and harvest by trypsinization (*see* **Note 6**).
2. Pellet cells by centrifugation at 300× *g* for 5 min and wash with cold PBS (*see* **Note 7**).

(c) *Cell Lysis*

1. Resuspend the cell pellet in lysis buffer as previously described.

2. Sonicate the lysate briefly (3 cycles of 5 s each).
3. Centrifuge the lysate at 10,000× *g* for 10 min at 4 °C.

(d) Protein Quantification

1. Use the Lowry protein assay kit to determine protein concentration and adjust to 1 mg/mL (*see* **Note 9**).

(e) *SDS-PAGE and Transfer*

1. Separate proteins on 10–15% SDS-PAGE and transfer to PVDF membranes.

(f) *Blocking and Antibody Incubation*

1. Block with 5% skim milk in TBST.
2. Incubate overnight with primary antibodies (LC3 and SQSTM1) at specified dilutions (*see* **Note 10**).

(g) *Secondary Antibody and Detection*

1. Wash membranes with TBST, incubate with HRP-conjugated secondary antibodies.
2. Develop blots using ECL detection and visualize bands with a BioRad imager.

3.6 Detailed Evaluation of Autophagy Flux

3.6.1 Baseline Autophagy Measurement

1. Measure LC3-II and SQSTM1 levels in U251 TMZR and TMZ NR cells under normal conditions.

3.6.2 Autophagy Inhibition with Bafilomycin A1

1. Treat cells with 100 nM Bafilomycin A1 for 6 h.
2. Harvest cells, lyse, and prepare protein samples as described.
3. Perform immunoblotting to detect LC3-II and SQSTM1 levels.

3.6.3 Comparison and Interpretation

1. Compare the levels of LC3-II and SQSTM1 between treated and untreated cells.
2. A significant increase in LC3-II and SQSTM1 levels in bafilomycin A1 treated cells compared to untreated controls was observed.

3.7 Fixation

3.7.1 Cell Fixation

1. Wash cells with PBS and detach using trypsin-EDTA (*see* **Note 6**).
2. Pellet cells by centrifugation at 300× *g* for 5 min (*see* **Note 5**).
3. Resuspend the cell pellet in 2 mL of Karnovsky fixative and incubate for 1 h at 4 °C.

3.7.2 Post-fixation

1. After fixation, wash cells with 0.1 M Sorenson's phosphate buffer.
2. Resuspend the pellet in 5% sucrose in 0.1 M Sorenson's phosphate buffer and store overnight at 4 °C.

3.7.3 Osmium Tetroxide Post-fixation

1. Pellet cells again and fix with 1% osmium tetroxide in 0.1 M Sorenson's buffer for 2 h at room temperature.

3.8 Dehydration and Embedding

3.8.1 Dehydration

1. Dehydrate cells through a graded ethanol series: 50%, 70%, 90%, and 100% ethanol, each for 10 min.
2. Perform final dehydration with 100% ethanol for 20 min.

3.8.2 Embedding

1. Infiltrate cells with Embed 812 resin: first, 1:1 mixture of resin and ethanol for 1 h, then pure resin overnight at room temperature.
2. Embed the cells in fresh resin and polymerize at 60 °C for 48 h.

3.9 Sectioning and Staining

3.9.1 Sectioning

1. Cut thin sections (70–90 nm) using an ultramicrotome.
2. Place sections on copper grids.

3.9.2 Staining

1. Stain sections with 2% uranyl acetate for 10 min.
2. Wash with distilled water and stain with lead citrate for 5 min.

3.10 Imaging

1. Examine the stained sections using a Philips CM10 electron microscope.
2. Capture images of representative fields at appropriate magnifications (e.g., 10,000× to 20,000×).

The overall experimental workflow for evaluation of autophagy flux in U251 TMZ NR and R cells is summarized in Scheme 1.

3.11 Methods for Evaluation of Autophagy Flux Using Expression of LC3-Beta II (LC3 Lipidation) and SQSTM1 Degradation in Immunoblotting

Autophagy flux can be assessed by measuring the expression levels of LC3-beta II (LC3 lipidation) and SQSTM1 degradation using immunoblotting. The microtubule-associated proteins 1A/1B light chain 3B (LC3) is essential in autophagy, undergoing lipidation to form LC3-II, which is associated with autophagosome membranes. SQSTM1 is a selective substrate of autophagy, degraded within autolysosomes, serving as a marker for autophagic degradation.

Without any treatments, we compared the levels of LC3-beta II and SQSTM1 in U251 TMZ-NR and TMZ-R cells:

1. *U251 TMZ-NR Cells*: Higher levels of LC3-beta II and lower levels of SQSTM1 indicate active autophagy flux.

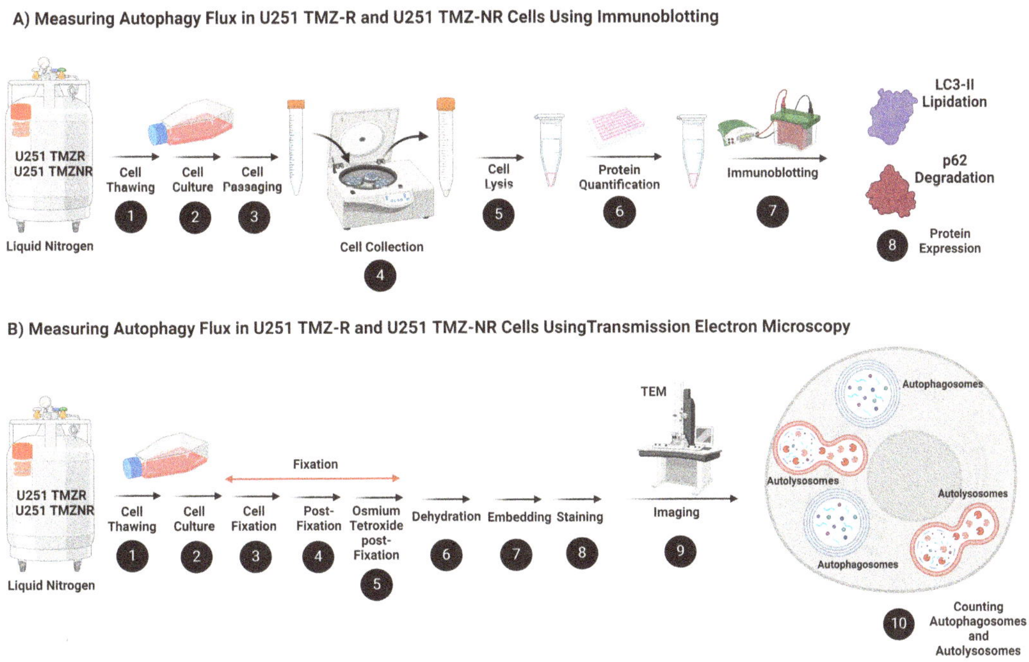

Scheme 1 Overview of the methodology used to measure autophagy flux in U251 TMZ-resistant (TMZR) and non-resistant (TMZ NR) cells. (**a**) Immunoblotting: The levels of autophagy markers LC3-II and SQSTM1 are quantified to assess autophagosome formation and degradation. (**b**) Electron microscopy: Ultrastructural analysis is performed to visualize and count autophagosomes and autolysosomes, providing detailed insights into the autophagic process at a cellular level. This combined approach offers a comprehensive evaluation of autophagy flux, which is crucial for understanding the mechanisms of TMZ resistance in glioblastoma cells

2. *U251 TMZ-R Cells*: Elevated levels of both LC3-beta II and SQSTM1 suggest impaired autophagy flux, potentially due to defective autophagosome-lysosome fusion or lysosomal function (Scheme 2).

3.12 Methods for Evaluation of Autophagy Flux Using TEM

3.12.1 Counting Autophagosomes and Autolysosomes

1. *Autophagosomes*: Identify and count double-membrane vesicles containing cytoplasmic material (autophagosomes) in the cytosol.
2. *Autolysosomes*: Identify and count single-membrane vesicles containing partially degraded material (autolysosomes).

3.12.2 Comparison Between Conditions

1. Compare the number of autophagosomes and autolysosomes in U251 TMZR and TMZ NR cells.
2. Analyze cells with and without treatment (e.g., bafilomycin A1) to determine the impact on autophagy flux.

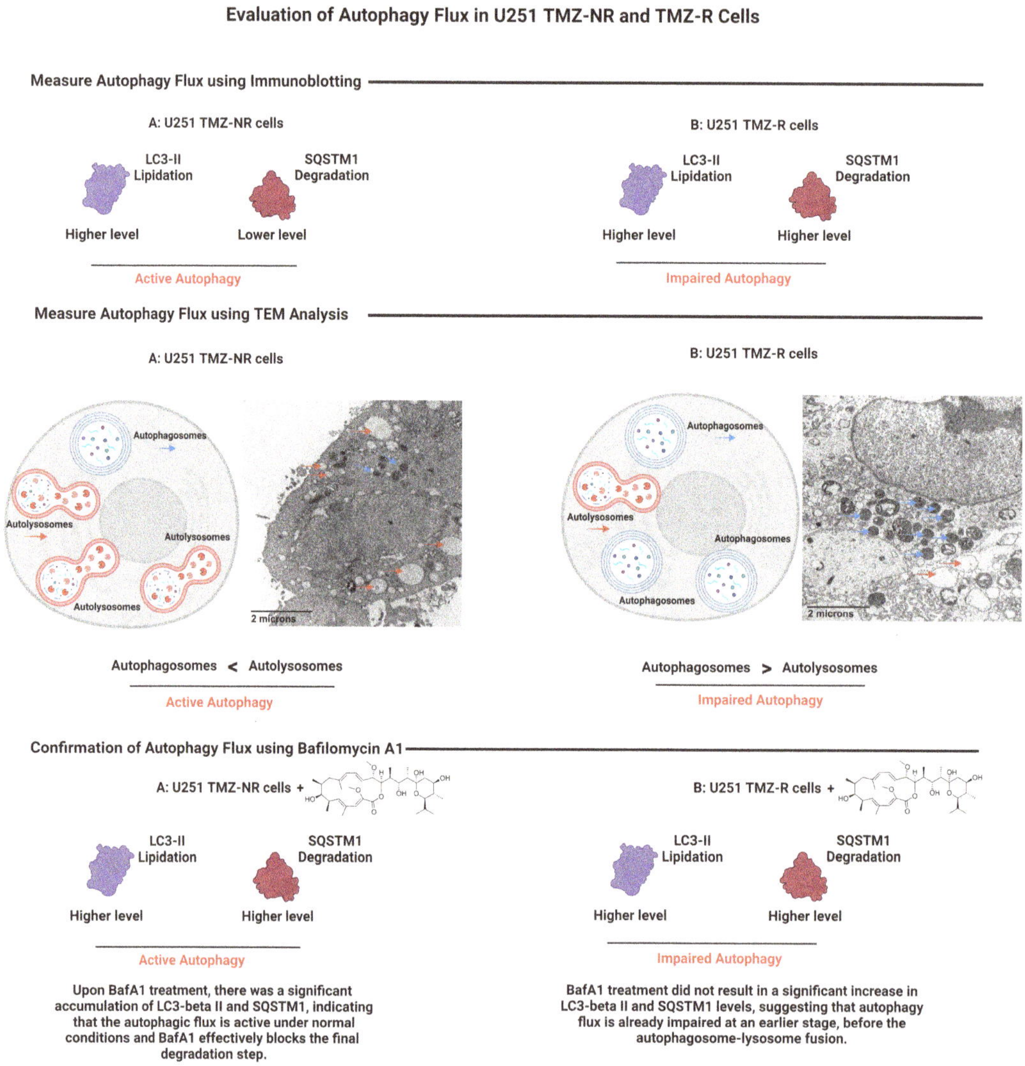

Scheme 2 Steps in the confirmation of autophagy flux. (**a**) Immunoblotting: The levels of LC3-II and SQSTM1 are measured to evaluate autophagosome degradation and autophagy flux. (**b**) Transmission electron microscopy (TEM): The numbers of autophagosomes and autolysosomes are quantified to assess autophagy flux at the ultrastructural level. (**c**) Bafilomycin A1 treatment: Final confirmation of autophagy flux is achieved by evaluating the accumulation of LC3-II and SQSTM1 after treatment, indicating blocked autophagosome-lysosome fusion. This multi-step approach provides a robust evaluation of autophagy flux in U251 cells

3.12.3 Data Interpretation

1. Increased number of autophagosomes in the presence of bafilomycin A1 indicates blocked autophagosome-lysosome fusion, reflecting active autophagosome formation.
2. A decrease in autophagosome numbers with a corresponding increase in autolysosomes indicates effective autophagosome-lysosome fusion and subsequent degradation.

3.12.4 Quantitative Analysis

1. Calculate the average number of autophagosomes and autolysosomes per cell.
2. Use statistical analysis to compare the differences between treated and untreated cells and between TMZR and TMZ NR cell lines.

This detailed TEM method provides a comprehensive approach to evaluate autophagy flux by analyzing the numbers of autophagosomes and autolysosomes, offering insights into the autophagic activity in glioblastoma cells (Scheme 2).

3.13 Methods for Evaluation of Autophagy Flux Using TEM

3.13.1 Counting Autophagosomes and Autolysosomes

1. *Autophagosomes*: Identify and count double-membrane vesicles containing cytoplasmic material (autophagosomes) in the cytosol.
2. *Autolysosomes*: Identify and count single-membrane vesicles containing partially degraded material (autolysosomes).

3.13.2 Comparison Between Conditions

1. Compare the number of autophagosomes and autolysosomes in U251 TMZR and TMZ NR cells.
2. Analyze cells with and without treatment (e.g., bafilomycin A1) to determine the impact on autophagy flux.

3.13.3 Data Interpretation

1. Increased number of autophagosomes in the presence of bafilomycin A1 indicates blocked autophagosome-lysosome fusion, reflecting active autophagosome formation.
2. A decrease in autophagosome numbers with a corresponding increase in autolysosomes indicates effective autophagosome-lysosome fusion and subsequent degradation.

3.13.4 Quantitative Analysis

1. Calculate the average number of autophagosomes and autolysosomes per cell.
2. Use statistical analysis to compare the differences between treated and untreated cells and between TMZR and TMZ NR cell lines.

This detailed TEM method provides a comprehensive approach to evaluate autophagy flux by analyzing the numbers of autophagosomes and autolysosomes, offering insights into the autophagic activity in glioblastoma cells (Scheme 2).

3.14 Methods for Evaluation of Autophagy Flux Using Bafilomycin A1

To further confirm the autophagy flux status, we used bafilomycin A1 (BafA1), an inhibitor of vacuolar-type H+-ATPase, which prevents the acidification of lysosomes and inhibits autophagosome-lysosome fusion.

3.14.1 Step III: Immunoblotting with Bafilomycin A1

We treated U251 TMZ-NR and TMZ-R cells with bafilomycin A1 and evaluated the changes in LC3-beta II and SQSTM1 levels.

1. *U251 TMZ-NR Cells*: Upon BafA1 treatment, there was a significant accumulation of LC3-beta II and SQSTM1, indicating that the autophagic flux is active under normal conditions and BafA1 effectively blocks the final degradation step.
2. *U251 TMZ-R Cells*: BafA1 treatment did not result in a significant increase in LC3-beta II and SQSTM1 levels, suggesting that autophagy flux is already impaired at an earlier stage, before the autophagosome-lysosome fusion.

4 Notes

1. *Consistent medium preparation*: Ensure the DMEM high glucose medium is consistently supplemented with 10% FBS and 1% penicillin-streptomycin. This is crucial for maintaining the health and growth of U251 cells.
2. *TMZ treatment specificity*: For U251 TMZR cells, always maintain them in DMEM medium with 250 μM TMZ. Remember to remove the TMZ-containing medium and replace it with TMZ-free DMEM medium 24 h before collection to accurately mimic withdrawal conditions.
3. *Cell confluency check*: Regularly monitor cell confluency, ensuring cells are passaged at 80–90% confluency. Over-confluent or under-confluent cultures can affect cell behavior and experimental results.
4. *Trypsinization timing*: Be precise with the trypsin–EDTA incubation time (2–3 min). Over-trypsinization can damage cells, while under-trypsinization may not detach cells effectively.
5. *Centrifugation consistency*: Maintain consistent centrifugation conditions (300× *g* for 5 min). Variations can result in incomplete cell pelleting or cell damage.
6. *PBS washing*: Thoroughly wash cells with PBS to remove any residual medium or trypsin before proceeding to the next steps, such as passaging or collection.
7. *Cold PBS for cell collection*: Use cold PBS for washing and resuspending cell pellets during cell collection to minimize cell metabolism and preserve cellular integrity for downstream applications.
8. *Sonication for complete disruption*: When sonicating the lysate, ensure the sonicator is set to 3 cycles of 5 s each with 10-s intervals on ice. This helps in complete cell disruption without overheating the samples.

9. *Accurate protein quantification*: Use the Lowry protein assay kit with a carefully prepared standard curve to accurately determine protein concentrations. Inaccurate quantification can lead to errors in downstream applications like immunoblotting.
10. *Controlled blocking and antibody incubation*: During immunoblotting, ensure that membranes are blocked with 5% skim milk in TBST for exactly 1 h and primary antibody incubation is done overnight at 4 °C. These steps are critical for reducing background noise and ensuring specific antibody binding.

Acknowledgments

SG conceptualized and designed the study. CC performed the experiments. SG, SS, MC, SS, and ABB analyzed the data. SG and MC wrote the manuscript. SG, CC, SS, ABB, and MC reviewed and edited the manuscript. SS developed GB TMR R cells. SG funded and supervised the experiments.

Funding

M.C. is supported by grant RYC2021-031003I funded by MICIU/AEI/https://doi.org/10.13039/501100011033 and by European Union NextGenerationEU/PRTR.

References

1. Eshraghi M, Ahmadi M, Afshar S, Lorzadeh S, Adlimoghaddam A, Rezvani Jalal N et al (2022) Enhancing autophagy in Alzheimer's disease through drug repositioning. Pharmacol Ther 237:108171
2. Dastghaib S, Shojaei S, Mostafavi-Pour Z, Sharma P, Patterson JB, Samali A et al (2020) Simvastatin induces unfolded protein response and enhances Temozolomide-induced cell death in glioblastoma cells. Cells 9(11):2339
3. Siri M, Behrouj H, Dastghaib S, Zamani M, Likus W, Rezaie S et al (2021) Casein Kinase-1-alpha inhibitor (D4476) sensitizes microsatellite instable colorectal cancer cells to 5-fluorouracil via Authophagy flux inhibition. Arch Immunol Ther Exp 69(1):26
4. Levy JM, Thorburn A (2011) Targeting autophagy during cancer therapy to improve clinical outcomes. Pharmacol Ther 131(1):130–141
5. Levy JMM, Towers CG, Thorburn A (2017) Targeting autophagy in cancer. Nat Rev Cancer 17(9):528–542
6. Singh N, Ulmer B, Medapati MR, Zhang C, Schroth RJ, Ghavami S et al (2024) Bitter taste receptor T2R14 and autophagy flux in gingival epithelial cells. Cells 13(6):531
7. Dossou AS, Basu A (2019) The emerging roles of mTORC1 in macromanaging autophagy. Cancers (Basel) 11(10):1422
8. Shojaei S, Barzegar Behrooz A, Cordani M, Aghaei M, Azarpira N, Klionsky DJ et al (2024) Unlocking a new path: an Autophagometer that measures flux using a non-fluorescent immunohistochemistry method. bioRxiv. https://doi.org/10.1101/2024.06.26.600741
9. Hajiahmadi S, Lorzadeh S, Iranpour R, Karima S, Rajabibazl M, Shahsavari Z et al (2023) Temozolomide, simvastatin and Acetylshikonin combination induces mitochondrial-dependent apoptosis in GBM cells, which is regulated by autophagy. Biology (Basel) 12(2):302

10. Saxton RA, Sabatini DM (2017) mTOR Signaling in growth, metabolism, and disease. Cell 169(2):361–371
11. Habibzadeh P, Dastsooz H, Eshraghi M, Los MJ, Klionsky DJ, Ghavami S (2021) Autophagy: the potential link between SARS-CoV-2 and cancer. Cancers (Basel) 13(22):5721
12. Jacobs J, Iranpour R, Behrooz AB, da Silva Rosa SC, Ghavami S (2024) The role of BCL2L13 in glioblastoma: turning a need into a target. Biochem Cell Biol 102(2): 127–134
13. Barzegar Behrooz A, Darzi Ramandi H, Latifi-Navid H, Peymani P, Tarharoudi R, Momeni N et al (2024) Genetic prognostic factors in adult diffuse gliomas: a 10-year experience at a single institution. Cancers (Basel) 16(11):2121
14. da Silva Rosa SC, Barzegar Behrooz A, Guedes S, Vitorino R, Ghavami S (2024) Prioritization of genes for translation: a computational approach. Expert Rev Proteomics 21(4): 125–147
15. Shojaei S, Koleini N, Samiei E, Aghaei M, Cole LK, Alizadeh J et al (2020) Simvastatin increases temozolomide-induced cell death by targeting the fusion of autophagosomes and lysosomes. FEBS J 287(5):1005–1034
16. Shojaei S, Alizadeh J, Thliveris J, Koleini N, Kardami E, Hatch GM et al (2018) Statins: a new approach to combat temozolomide chemoresistance in glioblastoma. J Investig Med 66(8):1083–1087
17. Wu WK, Coffelt SB, Cho CH, Wang XJ, Lee CW, Chan FK et al (2012) The autophagic paradox in cancer therapy. Oncogene 31(8): 939–953
18. Feng X, Chen L, Guo W, Zhang Y, Lai X, Shao L et al (2018) Graphene oxide induces p62/SQSTM-dependent apoptosis through the impairment of autophagic flux and lysosomal dysfunction in PC12 cells. Acta Biomater 81: 278–292
19. Samiei E, Seyfoori A, Toyota B, Ghavami S, Akbari M (2020) Investigating programmed cell death and tumor invasion in a three-dimensional (3D) microfluidic model of glioblastoma. Int J Mol Sci 21(9):3162
20. Palizkaran Yazdi M, Barjasteh A, Moghbeli M (2024) MicroRNAs as the pivotal regulators of Temozolomide resistance in glioblastoma. Mol Brain 17(1):42
21. Ou M, Cho HY, Fu J, Thein TZ, Wang W, Swenson SD et al (2023) Inhibition of autophagy and induction of glioblastoma cell death by NEO214, a perillyl alcohol-rolipram conjugate. Autophagy 19(12):3169–3188
22. Yang C, Li Y, Wu Q, Tang J, Chen M, Zhang B et al (2024) Unveiling the pharmacological role of human deubiquitinating enzymes in Temozolomide response of glioblastoma cells. Cell Biochem Biophys. https://doi.org/10.1007/s12013-024-01325-6
23. Xiang H, Zhang J, Lin C, Zhang L, Liu B, Ouyang L (2020) Targeting autophagy-related protein kinases for potential therapeutic purpose. Acta Pharm Sin B 10(4):569–581
24. Saxton RA, Sabatini DM (2017) mTOR Signaling in growth, metabolism, and disease. Cell 168(6):960–976

Methods in Molecular Biology (2025) 2879: 239–260
DOI 10.1007/7651_2024_524

Published online: 06 March 2024

A Lipidomics Approach to Determine the Role of Lipids and Its Crosstalk with Autophagy in Lung Cancer Metastasis

Simone C. da Silva Rosa, Javad Alizadeh, Rui Vitorino, Arun Surendran, Amir Ravandi, Biniam Kidane, and Saeid Ghavami

Abstract

Non-small cell lung cancer (NSCLC) is among the most malignant tumors with high propensity for metastasis and is the leading cause of cancer-related death globally. Most patients present with regional and distant metastasis, associated with poor prognosis. Lipids may play an essential role in either activating or inhibiting detachment-induced apoptosis (anoikis), where the latter is a crucial mechanism to prevent metastasis, and it may have a cross-talk with autophagy. Autophagy has been shown to be induced in various human cancer metastasis, modulating tumor cell motility and invasion, cancer cell differentiation, resistance to anoikis, and epithelial to mesenchymal transition. Hence, it may play a crucial role in the transition of benign to malignant phenotypes, the core of metastasis initiation. Here, we provide a method we have established in our laboratory for detecting lipids in attached and detached non-small lung cancer cells and show how to analyze lipidomics data to find its correlation with autophagy-related pathways.

Key words Anoikis, Autophagy, Lipidomics, Lung cancer, Mass spectrometry, MetaboAnalyst, Metastasis

1 Introduction

Non-small cell lung cancer (NSCLC) is among the most malignant tumors with high propensity for metastasis and is the leading cause of cancer-related death globally [1]. The NSCLC accounts for approximately 85% of all lung cancers worldwide, where the most common subtype is adenocarcinoma [2, 3]. Many patients present with regional and distant metastasis [4], associated with poor prognosis [5]. However, much yet must be elucidated on the precise mechanisms of NSCLC progression; hence, we need a deeper understanding of this issue to effectively identify effective diagnostic biomarkers and novel therapeutic targets.

Autophagy is an intracellular recycling mechanism that targets cytoplasmatic components to lysosomes for degradation and has been implicated in cancer [6]. After the autophagic breakdown of

Authors Simone C. da Silva Rosa and Saeid Ghavami have senior authorship.

encapsulated lipids, proteins, carbohydrates, or nucleic acids is completed, the product macromolecules return to the cytoplasm to support homeostatic cellular processes such as metabolism [7]. Although questionable, autophagy's role in cancer has been reported as a dual-edged sword, promoting both tumor suppression and proliferation, depending on the developmental stage of the tumor, regulating mitochondrial function and stress survival [7–9]. Studies have shown that autophagy is induced in various human cancer metastasis, modulating tumor cell motility and invasion, cancer cell differentiation, resistance to anoikis, and epithelial to mesenchymal transition [10, 11]. As a result, autophagy may play an essential role in the transition of benign to malignant phenotypes, the core of metastasis initiation [12].

Cancer cells have advanced mechanisms to escape cell death, hijacking the programmed cell death, the natural barrier in cancer progression initiated by apoptosis mechanisms, and additional programmed cell death types, such as autophagy [13, 14]. Lipid metabolism is another critical player in the regulation of cancer cell fate. The growing tumor cells heavily depend on lipid metabolic plasticity and reprogramming to sustain metabolic adaptations. As cancer cells develop a malignant phenotype, lipid metabolism changes, presenting increased lipid synthesis, storage and/or uptake, ultimately contributing to cancer progression [15].

Over the recent years, various studies have demonstrated the complex interplay between lipid metabolism and autophagy in regulating cell death and survival [16–18]. Lipids play an important role in stress-induced autophagic membrane structure formation [19]. Moreover, the association between lipid metabolism and autophagy is well demonstrated through commonly shared regulators such as AMPK and mTORC signaling molecules, among others [20–22]. Lipid metabolism has been suggested to modulate autophagy during cancer progression [23]; hence, therapeutic strategies targeting lipid metabolism and autophagy may potentially enhance treatment effectiveness, eventually improving cancer patient survival [14].

Guo et al. [24] showed that autophagy suppression in lung tumor maintains lipid homeostasis [24]. Previously, we have shown that Simvastatin, a widely used cholesterol-lowering agent, could further sensitize cancer cells (lung, breast, and brain) to temozolomide-induced cell death via autophagy inhibition [25, 26]. Following that first observation, we became particularly interested in understanding the role of lipids and lung cancer metastasis and its crosstalk with autophagy. Therefore, using a lipidomics approach in our studies, we hope to elucidate the molecular mechanisms of NSCLC metastasis to determine one or more reliable early metastasis biomarkers to improve outcomes for NSCLC patients. Mass spectrometry is a common method choice for qualitative and quantitative lipidomics analysis, allowing the identification of many individual lipids [27]. The recent advance in this field culminated in the rapid abundance of big data in cancer omics, which, combined with computational resources and bioinformatics, accelerate our fundamental understanding of cancer biology to translational advancements [28].

Here, we propose that lipids may play an important role in changing detachment-induced apoptosis (anoikis), where the latter is a crucial mechanism to prevent metastasis, and that it may have a cross-talk with autophagy. To further delineate the predominant lipids involved in metastasis, we cultured an adenocarcinoma human cell line (A549) in both attached and detached conditions, mimicking primary and metastatic tumor environments, and subjected these cells to a lipidomics screening through mass spectrometry (Fig. 1). We then quantified the lipid abundance and performed an enrichment analysis of possible altered pathways resulting from predominantly altered lipids in both conditions.

In the materials and methods section, we will list and describe, respectively, the following steps:

1. A549 cell culture
2. Lipid extraction and LC/MS
3. Lipidomics analysis

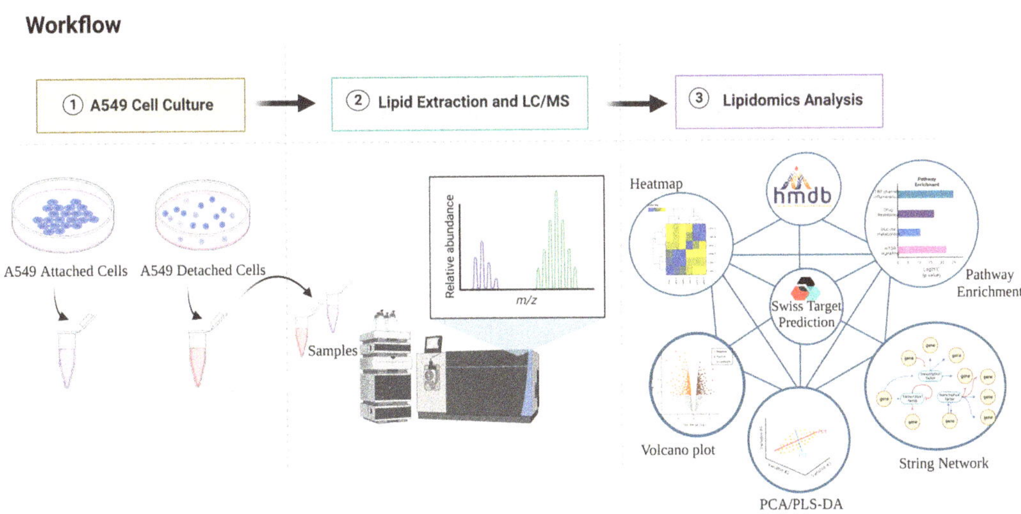

Fig. 1 Graphical abstract. Workflow of A549 attached and detached cell culture. Cells are collected, lipids are extracted and subjected to LC/MS processing. Next, lipidomics resulting data is analyzed through various steps involving utilization of MetaboAnlyst, HDMB, Swiss Target Prediction, and String

2 Materials

2.1 A549 Cell Culture

1. Lung adenocarcinoma cells (A549; ATCC).
2. Adherent and low-adherence plates culture plates.
3. Cell culture media: Dulbecco's Modified Eagle's Medium high glucose (4 mg/mL) (DMEM), 10% Fetal Bovine Albumin (FBS), 1% Penicillin-Streptomycin, and 1% insulin transferrin selenium (ITS).

2.2 Lipid Chromatography and Mass Spectrometry [29]

2.2.1 Lipid Extraction

1. Phosphate Buffer Saline 1x (PBS), 1 L: 800 mL of distilled water, 8 g of NaCl, 1.44 g of Na2HPO4, 0.2 g of KCl, 0.24 g of KH2PO4, pH 7.4 or 7.2, top up with additional water until reaches 1 L.
2. The lipids were quantified by comparing the lipid peak areas to those of the class-specific internal standards (ISTD). ISTD were added to the samples before the lipid extraction. The use of an ISTD is essential since they help to account for unavoidable variations in sample processing including suppression of ionization due to matrix interference and manual discrepancies in lipid extraction. The choice of an ISTD for a

Table 1
Lipid internal standards and quantification used in this study

Lipid class/subclass	Internal standard	μM/30 μL
SM	SM 12:0	0.5
PC	PC 13:0_13:0	0.5
PC(O)	PC 13:0_13:0	0.5
PC(P)	PC 13:0_13:0	0.5
LPC	LPC 13:0	0.5
LPC(O)	LPC 13:0	0.5
PE	PE 17:0_17:0	0.5
PE(O)	PE 17:0_17:0	0.5
PE(P)	PE 17:0_17:0	0.5
LPE	LPE 14:0	0.5
PI	PE 17:0_17:0	0.5
PS	PS 17:0_17:0	0.5
PG	PG 17:0_17:0	0.5
CE	CE 18:0 (*d6*)	20
DG	DG 15:0_15:0	5
TG	TG 17:0_17:0_17:0	5
FA	FA 15:0 (*d3*)	0.5

particular lipid class depends upon the structural similarity and likeness in MS/MS fragmentation pattern of the species present in that lipid class.

3. In this study, 30 μL of ISTD in chloroform/methanol (1:1, v/v) was added to 200 μL of chloroform/methanol (2:1, v/v) along with 100 μL of cell homogenate. The details of internal samples and their concentration are provided in Table 1.

2.2.2 Quality Assurance (QA) and Quality Control (QC) Measures

1. Two quality control (QC) samples, namely, pooled quality control sample (PQC), and a technical quality control sample (TQC, sample with all standards used).
2. Equal amounts from all the samples are pooled together to make the PQC sample.
3. To stabilize the LC/MS system, ten PQC samples are injected prior to injecting experimental samples.
4. PQC samples are injected at equal time intervals to monitor the analytical variability.
5. The lipid extracts without the matrix (here, cell homogenate) are called as TQC samples. They are evenly injected throughout the sample run to ensure the stability of the analytical platform by monitoring the variation in signal intensity over time.
6. Blank extracts are also run at the beginning of sample batch to filter out background ions. All solvents should be HPLC grade.

2.2.3 Lipid Separation

1. Reverse-phase liquid chromatography using Shimazu Prominence HPLC system.
2. HPLC column used: Zorbax C18, 1.8 μm, 50 × 2.1 mm column.
3. Flow rate: 300 μL/min using a linear gradient of mobile phase A and mobile phase B.
4. Mobile phase A and B in 10 mM ammonium formate: tetrahydrofuran, methanol, and water in the ratio 20:20:60 (v/v/v) and 75:20:5 (v/v/v), respectively.
5. Elution: begin with 0% solvent B; elevate to 100% B at 8.00 min; keep at 100% B for 2.5 min; and return to 0% B over 0.5 min.

2.2.4 Mass Spectral Analysis

1. The AbSciex 4000 QTRAP triple quadrupole linear hybrid mass spectrometer is used to introduce lipids eluted from the HPLC system. Operated in scheduled Multiple Reaction Monitoring (MRM) mode.
2. All lipid species are scanned in positive electrospray ionization mode [ESI^+] mode, except for fatty acids. Instrument settings optimized as curtain gas (psi), 26; collision gas (nitrogen), medium; ion spray voltage (V), 5500; temperature (°C),

500.0; ion source gas 1 (psi), 40.0; and ion and source gas 2 (psi), 30.0.

3. Flow injection analysis is used to fix collision energy and declustering potential for each lipid class. The MRM detection window is fixed between 45 and 90 s depending upon the chromatographic peak width of the lipid class.

2.3 Lipidomics Data Analysis

2.3.1 MetaboAnalyst v5.0

1. User-friendly, web-based interface metabolite data analysis platform.
2. Used to perform statistical and machine learning analysis, including univariate: t-test, volcano plot, ANOVA; multivariate: principal component analysis (PCA), partial least squares-discriminant analysis (PLS-DA); clustering: dendrogram and heatmap [30–32].
3. URL: https://www.metaboanalyst.ca/.

2.3.2 Human Metabolome Database v5.0 (HDMB)

1. It is a free electronic database sourcing detailed information on small molecule metabolites from the human body [33, 34].
2. This database contains or links chemical, clinical, and molecular biology or biochemistry data.
3. There are 220,945 metabolite entries up to November 2023, including water- and lipid-soluble metabolites. Furthermore, 8610 protein sequences are connected to these metabolite entries.
4. Lastly, HMDB provides unique IDs for each metabolite, such as the SMILES code.
5. URL: https://hmdb.ca/.

2.3.3 Swiss Target Prediction

1. It is a free web-based platform that allows the estimation of the most probable molecular targets of another molecule [35].
2. The estimation is possible based on 2D and 3D similarities with a library of 370,000 known actives for over 3000 proteins from three distinct species [36].
3. Given a lipid SMILES code, Swiss Target Prediction can predict possible interacting proteins with that metabolite (SMILES).
4. URL: http://www.swisstargetprediction.ch/.

2.3.4 String v12.0

1. This is a user-friendly free database of known and predicted protein-protein interactions, either physical (direct) or functional (indirect) associations [37].
2. The associations are originated from computational prediction coming from knowledge transfer and combination with other databases [38, 39].
3. URL: https://string-db.org/.

3 Methods

3.1 A549 Cell Culture

1. Carry out cell culture in biosafety cabinet in sterile condition.
2. Culture A549 lung adenocarcinoma cells in both adherent and non-adherent plates (100 mm); the former mimics primary tumor conditions (attached) and the latter metastatic tumors in circulation (detached). Use high glucose growth media. Three plates of cells for each condition (*see* **Note 1**).
3. Grow cells until reach 100% confluence, maintained in a humidified incubator with 95% air and 5% CO2 at 37 °C. Total yield per plate is approximately five million cells.
4. Once reached 100% confluence, collect cells using 1x PBS, in a 1.5 mL tube.
5. Sonicate samples (3 cycles, 5 s each).
6. Centrifuge samples (1000 g for 10 min).
7. Collect the supernatant for each sample and transfer to a newly labeled tube.
8. Measure protein concentration using Lowry Protein Assay method.
9. Ideally 1 mg/mL of sample is necessary for lipidomics. Therefore, normalize all proteins concentration to 1 mg/mL.
10. Proceed to lipidomics sample preparation and lipidomics analysis (*see* **Note 2**).

3.2 Lipid Chromatography and Mass Spectrometry [29]

3.2.1 Lipid Extraction

1. Lipid extraction was performed using chloroform and methanol [29, 40].
2. In a new 1.5 mL tube polypropylene tube, add 100 μL of cell homogenate and 30 μL of the internal standard (ISTD) plus 200 μL of chloroform/methanol (2:1, v/v) creating a new mixture (*see* **Note 3**).
3. Vortex mixture on a rotary mixer for 10 min and sonicate in a water bath at room temperature (RT) for 30 min.
4. Allow mixture to settle for 20 min and then centrifuge (20,000 g, 20 min, RT).
5. Transfer the upper lipid-containing phase into a clean polypropylene tube and dry under a stream of nitrogen gas at RT.
6. Reconstitute lipids in 50 μL water-saturated 1-butanol and sonicate for 10 min (water bath, RT).
7. Add 50 μL of 10 mM ammonium formate in methanol to final lipid extract.
8. Centrifuge the extract (10,000 g, 10 min, RT).

9. Transfer 80 μL of the supernatant into the micro insert in sample vials for lipid analysis.

3.2.2 Lipid Separation and Chromatography

1. Separate lipids on a reverse-phase liquid chromatography-electrospray ionization tandem mass spectrometry (LC-E-SI-MS/MS) platform using a Prominence chromatographic system.
2. Use Analyst v1.6 and Multi-Quant v2.1 software for instrument control and data processing.
3. Perform lipid separation on a Zorbax C18, 1.8 μm, 50 × 2.1 mm column at a flow rate set to 300 μL/min using a linear gradient of mobile phase A and mobile phase B (*see* **Note 4**).
4. Re-equilibrate the column to the starting conditions (0% mobile phase B) for 3 min prior the next sample injection.

3.2.3 Mass Spectral Analysis

1. Add to AbSciex 4000 QTRAP triple quadrupole linear hybrid mass spectrometer the eluted lipids from HPLC.
2. Mass spectrometer operation schedule mode: Multiple Reaction Monitoring (MRM).
3. Screen total of 322 unique lipids and 25 different lipid classes/subclasses for targeted semi-quantitation (Supplementary Table S3, Reference [29].
4. Scan all lipid species (except fatty acids) positive electrospray ionization mode [ESI^+] mode.
5. Identify the individual lipids in each lipid class by using lipid class-specific precursor ions or neutral losses (Supplementary Table S2, Reference [29].
6. Lipids are represented by the total carbon number of the fatty acids.
7. In the ESI^+ mode, optimize the instrument settings as follows: curtain gas (psi), 26; collision gas (nitrogen), medium; ion spray voltage (V), 5500; temperature (°C), 500.0; ion source gas 1 (psi), 40.0; ion and source gas 2 (psi), 30.0.
8. Fix the MRM detection window between 45 and 90 s depending upon the chromatographic peak width of the lipid class.
9. Scan M^{+1}/M^{+2} isotopes for high-intensity lipid species to prevent detection saturation.
10. Proceed to data analysis once lipid abundance has been detected.

3.3 Lipidomics Data Analysis

3.3.1 MetaboAnalyst 5.0

1. Create a matrix excel table with all lipids and labels and export to CSV format then upload it to MetaboAnalyst (*see* **Note 5**).

	A	B	C	D	E	F	G
1	Row Labels	A549 A_1	A549 A_2	A549 A_3	A549 D_1	A549 D_2	A549 D_3
2	Labels	A549 Attached	A549 Attached	A549 Attached	A549 Detached	A549 Detached	A549 Detached
3	GM3 16:0	2803.618328	3783.025922	3147.234799	8376.987476	6537.416486	6582.43898
4	GM3 18:0	173.3631857	239.3513903	329.0581878	613.991941	292.2301574	708.428204

Fig. 2 Screenshot of matrix formatting style to be uploaded at MetaboAnalyst

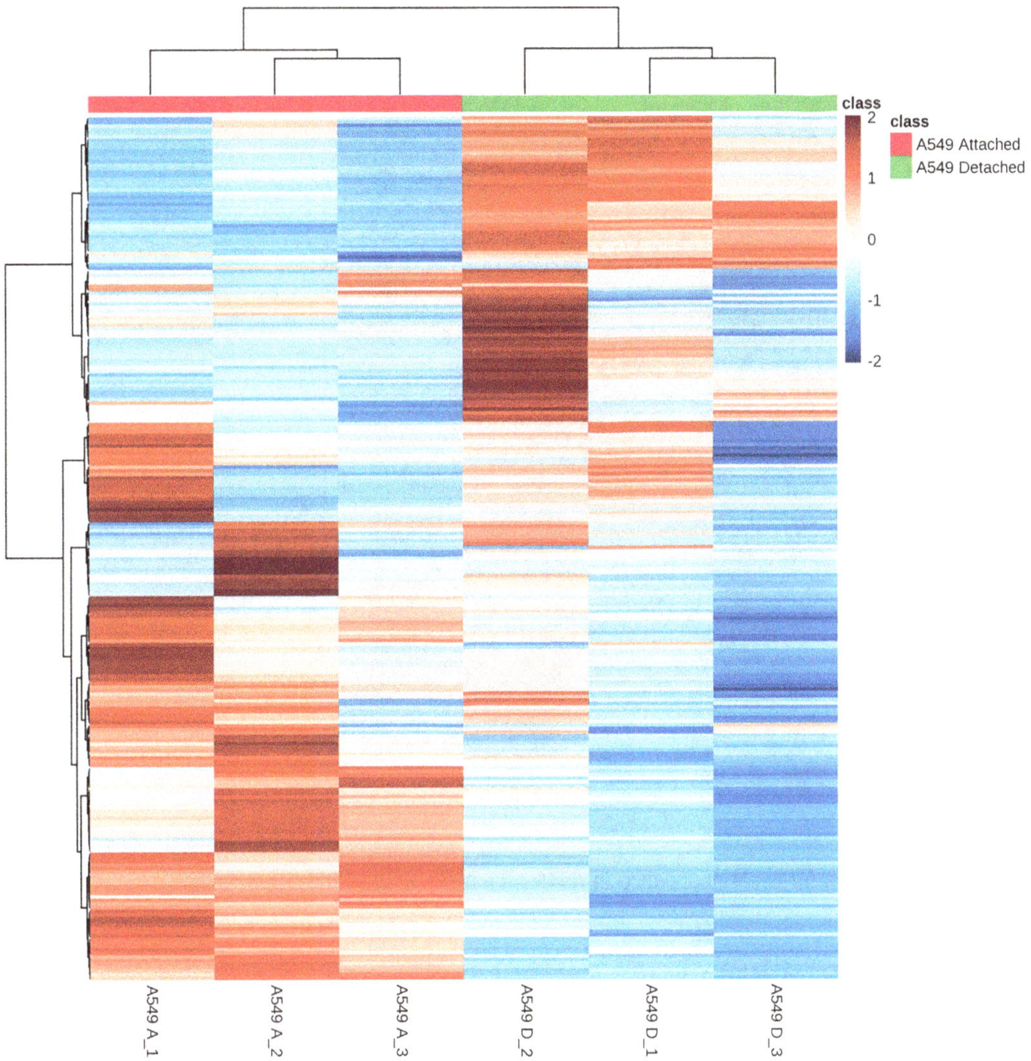

Fig. 3 Heatmap of all lipids abundance in A549 attached and A549 detached conditions. Generated in MetaboAnalyst

2. Normalize lipid abundance of the samples to 1 mg/mL to match protein determination normalization parameters.
3. Next, ensure there is no empty values (*see* **Note 6**).
4. First column should be listed all lipids, and following columns should be listed all groups, including a minimum of three replicates per group.

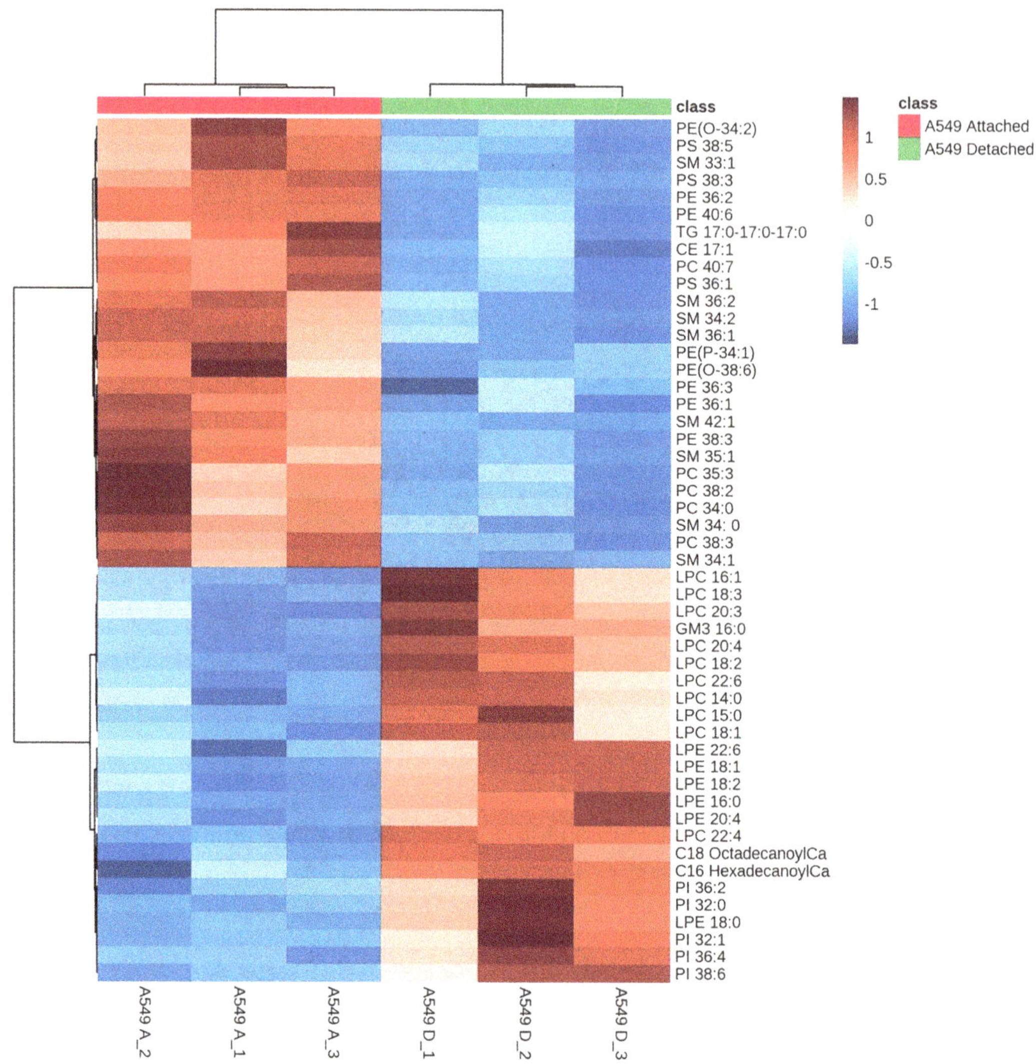

Fig. 4 Heatmap of top 50 lipids abundance in A549 attached and A549 detached conditions. Generated in MetaboAnalyst

5. Add two rows and name "A1" as "Row Labels, followed by group labels (A549 A_1, A549 A_2, A549 A_3, A549 D_1, A549 D_2, A549 D_3). On row "A2," name as "Labels", followed by group labels without replicate number (A549 Attached, A549 Attached, A549 Attached, A549 Detached, A549 Detached, A549 Detached), as shown in Fig. 2.
6. Once label formatting is done, export excel table as comma delimited format (CSV).
7. Open MetaboAnalyst website.
8. Click red button: "click here to start."

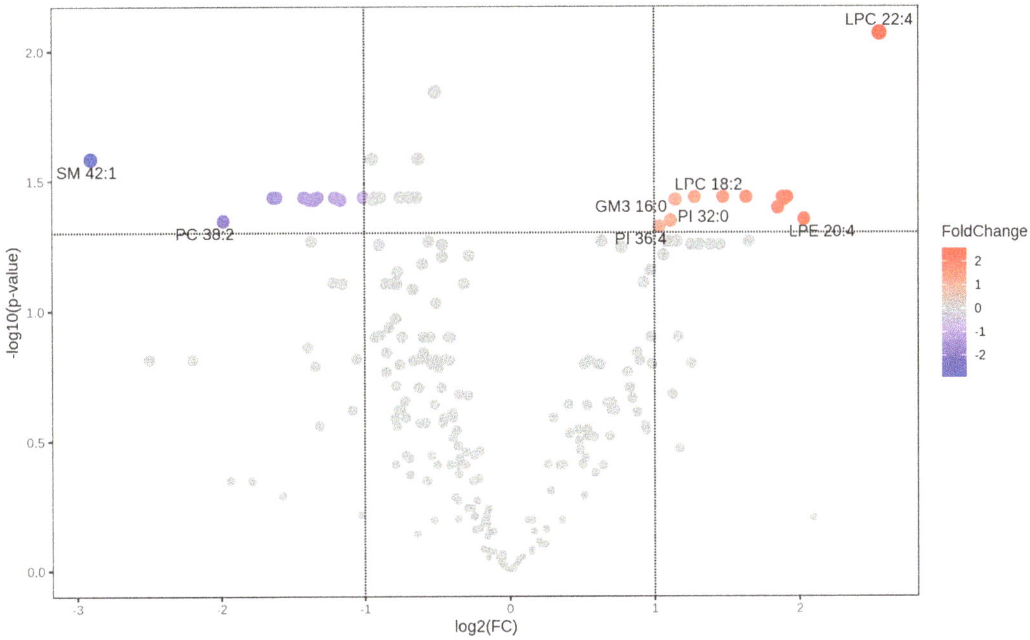

Fig. 5 Volcano plot of A549 detached condition versus A549 attached condition, displaying the most significant lipids either upregulated (red) or downregulated (blue), based on FC 2 and FDR 0.05

9. A new page will be opened with "module overview" of all available functions. Click "Statistical Analysis [one factor]", under impute data type "generic format .csv or .txt table files."
10. Next, upload data under the option "A plain text file (.txt or .csv)." And select the following options: Data Type "Concentrations"; Format "Samples in columns, unpaired"; Data File: "Choose" and upload csv formatted matrix/table.
11. "Data integrity check" page will be prompted. If matrix was well formatted, at this stage it will indicate the number of groups, samples, and compounds identified. Click on "Proceed".
12. Data Filtering: under "Statistical Filters", select "mean intensity value", then "Proceed".
13. A "Normalization overview" page will be prompted, select the following options: "Sample normalization – none", "Data transformation – none", "Data scaling – auto scaling". Click on "Normalize", next "View Results", next "Proceed".
14. Next, under "Select an analysis path to explore" new page, you can choose type of analysis to run and generate graphics.
15. MetaboAnalyst results (*see* **Note 7**).

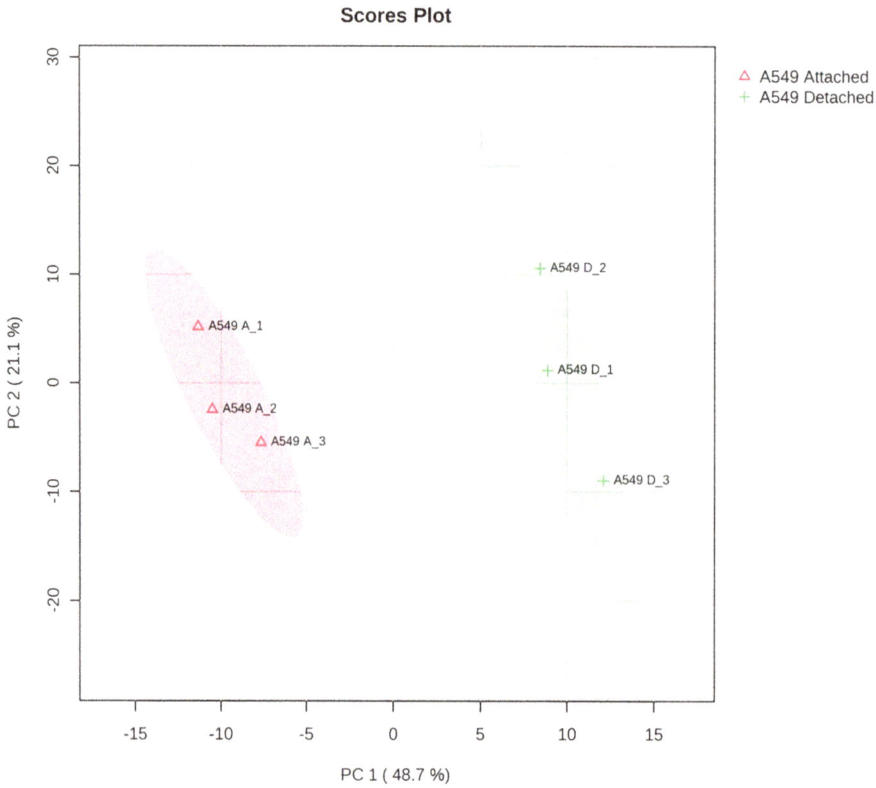

Fig. 6 Principal Component Analysis (PCA) plot of A549 attached and A549 detached conditions, unsupervised analysis, showing distinct biological variance between groups

16. On "Select an analysis path to explore" page, under drop down menu located on the left side of the page, click on "Statistics".
17. Next, click on "Heatmap" (Fig. 3) to visualize overview lipid abundance and select the following settings:
    1. Data source: normalized data
    2. Standardization: autoscale features
    3. Distance measure: eucledian
    4. Clustering: ward
    5. Color contrast: default
    6. Font size: 10
    7. View mode: overview
    8. Other view options: show heatmap color legend, samples/ T-test ANOVA

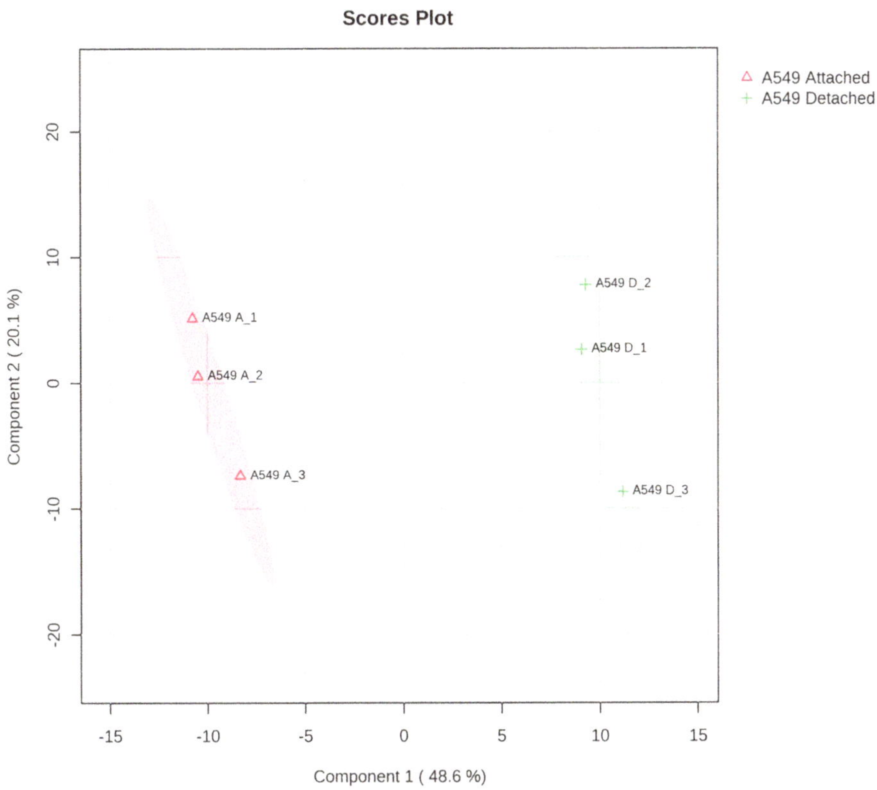

Fig. 7 Partial Square Least Discriminant Analysis (PLSDA) plot of A549 attached and A549 detached conditions, supervised analysis, showing distinct biological variance between groups

9. To visualize most significant lipid abundance and lipid names, under "Other view options", select "Use top", then enter the number of lipids to be displayed, e.g., 50, as shown in Fig. 4
10. Next, click "Show row names"

18. Back to "Statistics" menu, select "Volcano Plot" (*see* **Note 8**).
19. Select the following settings to generate Volcano Plot graph (Fig. 5):
    1. Analysis: unpaired
    2. Plot style: show label
    3. Theme: blackwhite
    4. X-axis: Fold Change threshold 2.0; Direction of comparison: A5459 Detached/A549 Attached
    5. Y-axis: p-Value threshold FDR 0.05; group variance equal (*see* **Note 9**)
20. Back to "Statistics" menu, select "PCA" for unsupervised analysis (Fig. 6).

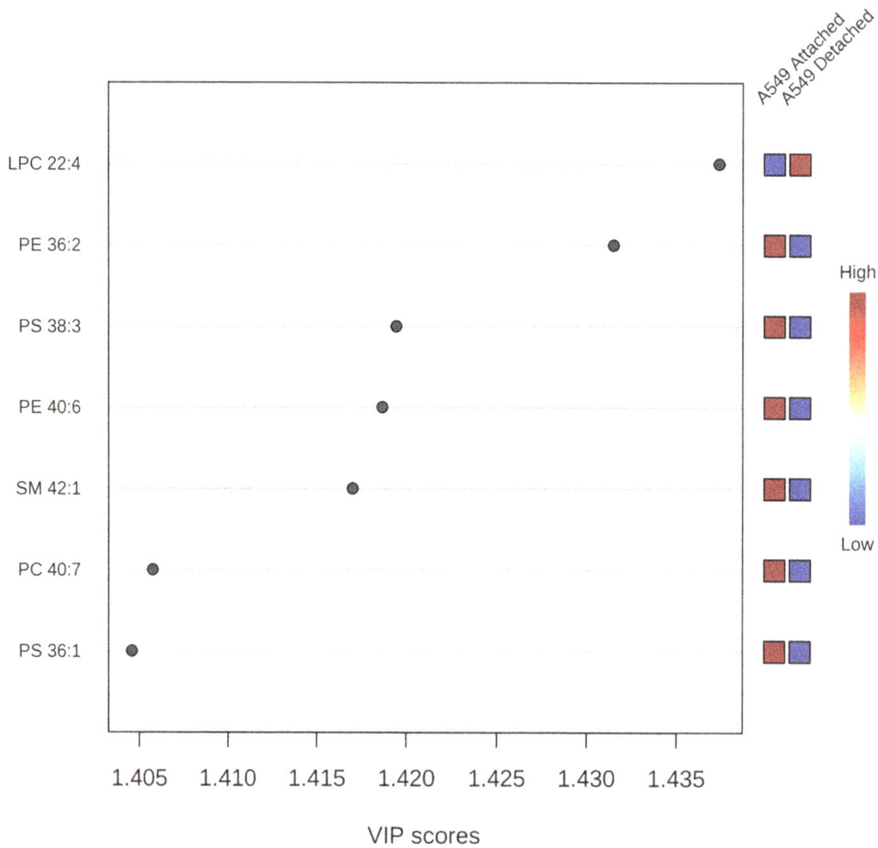

Fig. 8 Partial Square Least Discriminant Analysis (PLSDA) variable importance in projection (VIP) score plot of A549 attached and A549 detached conditions (VIP >1.4)

21. A new page "Principal Component Analysis (PCA)" will be prompted. Choose the following parameters:
    1. Select "2D Scores Plot"
    2. Specify PC on X-axis: 1
    3. Specify PC on Y-axis: 2
    4. Display 95% confidence regions: check
    5. Display sample names: check
22. Back to "Statistics" menu, select "PLSDA", supervised analysis (Fig. 7).
    1. A new page "Partial Least Squares Discriminant Analysis (PLS-DA)" will be prompted.
    2. Under 2D scores plot, apply same settings as indicated for "PCA."
23. Still within PLSDA menu, click on "Imp. Features" to measure the PLS-DA variable importance in projection (VIP) score (Fig. 8), and apply the following settings:

1. Importance measure: VIP Score, comp 1.
2. Show top features (display on graph should be min 1.2 VIP score).
3. Download excel sheet with all lipids with VIP score >1.4 and proceed to next steps of analysis to predict lipid-protein interaction.

3.3.2 Human Metabolome Database (HMDB)

1. At this step, all significant lipids selected from VIP score >1.4 will be searched at HMDB to identity their unique ID and smiles code. This smiles code will be used to predict possible interacting proteins with each metabolite.
2. First, create a new excel table with PC1 exported from VIP score tab on MetaboAnalyst (Fig. 9). In front of each lipid, add columns for corresponding VIP-PC1 value, lipid common name, and smiles code.
3. Enter lipid name in HMDB search engine and look for exact lipid. Click on the correct lipid and copy "lipid common name" and "smiles code" and past to excel table. Repeat this step for all significant lipids (VIP >1.4 in this case).
4. Proceed to Swiss Target Prediction to identify interacting proteins with each metabolite/lipid in query.

3.3.3 Swiss Target Prediction

1. First, add a new column to excel sheet and label as "Uniprot ID" where top 15 predicted interacting proteins will be pasted (Fig. 10).
2. On the Swiss Target Prediction main webpage, follow the following steps:
    1. Select a species: homo sapiens.
    2. Paste a smile code in the allocated box.
    3. Click red button "Predict targets", it may take a few minutes to load.
    4. A new page will be prompted with a list of predicted interacting proteins.
    5. Click under "show entries" and select 15.
    6. Lastly, under "Export Results" tab, select icon "CSV" and choose folder location for download.
    7. Repeat this step for all lipids in query.
    8. After all saved all CSV files containing predicted interacting proteins for all lipids, now open each one individually, copy the protein ID of top 15 targets, and paste in the new column created in excel sheet (within the same cell), labeled "Uniprot ID".

	A	B	C	D
1	Lipid	VIP - PC1	Lipid Common Name	Smiles Code
	PS 38:3	1.4194	PS(18:3(6Z,9Z,12Z)/20:0)	[H][C@](N)(COP(O)(=O)OC[C@@]([H])(COC(=O)CCCC\C=C/C\C=C/C\C=C/CCCCC)OC(=O)CCCCCCCCCCCCCCCCCCC)C(O)=O

Fig. 9 Screenshot of template table to annotate significant lipids details, including VIP score >1.4, lipid common name and smiles code obtained from HMDB

	A	B	C	D	E
1	Lipid	VIP - PC1	Lipid Common Name	Smiles Code	Uniprot ID
	PS 38:3	1.4194	PS(18:3(6Z,9Z,12Z)/20:0)	[H][C@](N)(COP(O)(=O)OC[C@@]([H])(COC(=O)CCCC\C=C/C\C=C/C\C=C/CCCCC)OC(=O)CCCCCCCCCCCCCCCCCCC)C(O)=O	Q9UPC5 O00398 Q9BXC1 P16109 Q13822 P05413 Q9Y2T6 P43657 Q9UBY5 Q9HBW0 Q92633 Q8NER1 Q9H1C0 Q99677 P25105

Fig. 10 Screenshot of template table to annotate significant lipids details, including VIP score >1.4, lipid common name, smiles code obtained from HMDB, and Uniprot IDs obtained from SwissTargetPrediction

9. Once excel table is complete with all corresponding protein targets, proceed to pathway enrichment analysis using String.

3.3.4 String

On the String main webpage, follow the following steps:

1. Left menu tab: click on "Multiple proteins".
2. Paste list of all protein ID from excel sheet, in a column, one protein per line (*see* **Note 10**).
3. Select organism: Homo sapiens, and click "search".
4. A list of matches will prompt, click "continue".
5. The network will be now generated, and a new panel of commands are available to adjust and set statistical parameters for the analysis.
6. Click on "Settings" and set the following parameters for "Basic Settings":
   1. Network type: full string network
   2. Meaning of network edges: confidence
   3. Active interaction sources: check all
   4. Minimum required interaction score: high confidence 0.700

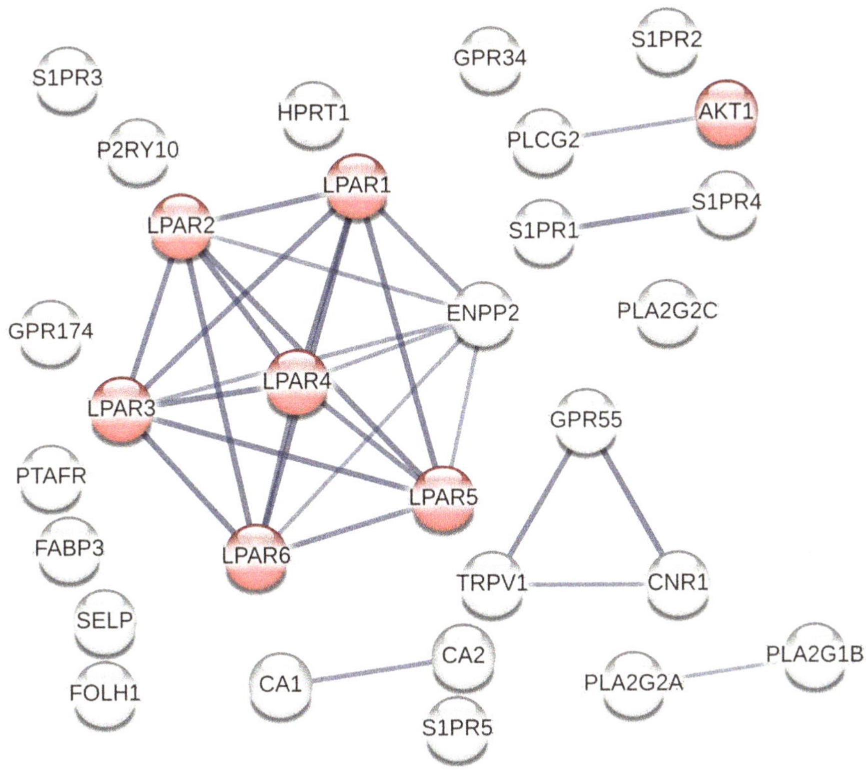

Fig. 11 String network analysis highlighting PI3K-AKT signaling pathway (KEGG) in red nodes (FDR 5.44e-05). Network was generated based on significant lipids VIP >1.4 and respective predicted interacting proteins. Minimum 15 possible interacting proteins were entered on String, per lipid. Confidence interaction min 0.700

5. Max number of interactions to show: 1st shell (none/query proteins only)

7. Still under "Settings", set the following parameters for "Advanced Settings":
   1. Network display mode: interactive
   2. Network display options: check boxes for "enable 3D bubble design", "disable structure previews inside network bubbles", "center protein names on nodes", "protein font size 14"
8. Click on "Update" to save new settings.
9. Next, click on "Analysis" tab.
10. Look for "KEGG" pathways, then click on "FDR" so pathways will be listed based on most significant FDR values, with arrow pointing up.
11. Click on KEGG pathway of interest listed, such as the "PI3K-AKT signaling pathway" (FDR 5.44e-05). It will be

Table 2
List of significant enriched KEGG pathways, highlighting PI3K-Akt signaling as one of the main correlation pathways to autophagy changes in A549 adenocarcinoma primary and metastatic tumors. KEGG) in red nodes (FDR 5.44e-05). Network was generated based on significant lipids VIP >1.4 and respective predicted interacting proteins. Minimum 15 possible interacting proteins were entered on String, per lipid

#term ID	Term description	Observed gene count	Background gene count	Strength	False discovery rate
hsa04080	Neuroactive ligand-receptor interaction	14	329	1.45	6.73E-15
hsa04072	Phospholipase D signaling pathway	8	147	1.55	1.03E-08
hsa04015	Rap1 signaling pathway	7	201	1.36	2.43E-06
hsa04071	Sphingolipid signaling pathway	6	116	1.53	2.43E-06
hsa04151	PI3K-Akt signaling pathway	7	349	1.12	5.44E-05
hsa05200	Pathways in cancer	8	515	1.01	5.44E-05
hsa00592	alpha-Linolenic acid metabolism	3	25	1.9	0.00049
hsa00591	Linoleic acid metabolism	3	29	1.83	0.00064
hsa04972	Pancreatic secretion	4	97	1.43	0.00064
hsa04014	Ras signaling pathway	5	225	1.16	0.00078
hsa04975	Fat digestion and absorption	3	41	1.68	0.0012
hsa00565	Ether lipid metabolism	3	47	1.62	0.0017
hsa00590	Arachidonic acid metabolism	3	60	1.52	0.0031
hsa05130	Pathogenic Escherichia coli infection	4	187	1.15	0.0046
hsa04810	Regulation of actin cytoskeleton	4	209	1.1	0.0065
hsa00910	Nitrogen metabolism	2	17	1.89	0.0079
hsa00564	Glycerophospholipid metabolism	3	97	1.31	0.0092
hsa04068	FoxO signaling pathway	3	126	1.19	0.0182
hsa01100	Metabolic pathways	8	1435	0.56	0.0194
hsa04270	Vascular smooth muscle contraction	3	132	1.17	0.0194

highlighted in the network the proteins from your query involved in this pathway, shown in red (Fig. 11).

12. Still under "Analysis" tab, go to the bottom of the page and click on "download" "KEGG pathways" (Table 2).
13. Next, click on "Exports" tab.
14. Under "Export your current network", download following files "high- resolution bitmap" network image and "functional annotations" table containing all analysis results, including KEGG pathways.

3.4 Conclusion

Our results showed a distinct biological variance among A549 adenocarcinoma attached and detached cells. Based on MetaboAnlyst lipidomics analysis, we could observe a significant increase in lysophosphatidylcholines (LPC), phosphatidylcholines (PC), phosphatidylethanolamine (PE), phosphatidylserine (PS), phosphatidylethanolamine (PE), and sphingomyelin (SM) lipids in A549 attached cell group, while there is a decrease in detached group. We further predicted the possible interacting proteins for each significant lipid and performed an enrichment analysis based on the Kyoto Encyclopedia of Genes and Genome (KEGG). The KEGG results indicated that the PI3K-AKT pathway is among the most significant enriched pathways. Furthermore, PI3K-AKT are at the core of mTOR and autophagy signaling pathways regulation. Hence, this method may serve as an advanced approach for exploratory research of the role of lipids in autophagy and cancer, and further experimental data analysis are necessary to validate the predicted biological findings (*see* **Note 11**).

4 Notes

1. Make sure to have minimum triplicates of each condition. Ideally, it is best if there are four or more replicates in each group/condition. The main reason is ensuring repeatability, consistency, and elimination of any variability or possible outliers. Moreover, a minimum of three replicates is required to perform statistical analysis. Also, make sure cells are serum starved and grown in 1% ITS as indicated in material section. The reason is that FBS may interfere in the circulating levels of chemokines and cytokines, while ITS does not impact it.
2. A typical workflow of lipidomics analysis of biological samples consists of sample collection and storage, homogenization, internal standard addition for calibration, analysis of lipids with and without separation, data processing and lipid identification, and biomedical application.
3. Lipid internal standards are either odd-chain or deuterated and not present endogenously.

4. Maintain analytical column and autosampler at 50 °C and 25 °C, respectively, during analysis. The injection volume is 5 μL.
5. All lipids should be abbreviated with character free. Ensure that all text is exactly the same within replicates nomenclature. If a tiny space is added extra in one replicate, it will be indicated as a new group and will be misinterpreted by MetaboAnalyst, which is an R-based web analysis tool.
6. It is important to note that this procedure does not account for all missing values, which must be eliminated for subsequent data analysis procedures such as principal component analysis (PCA). Rather than relying solely on the imputation of zero values, it is advantageous to use data-driven imputation methods such as Random Forest (RF), k- nearest neighbors (KNN), and Singular Value Decomposition (SVD) or use the Limit of Detection (LOD) value. The sample size generally had only a minimal influence on the results. However, with a sample size of $N = 10$, even a moderate amount of missing data significantly reduced the statistical power. In cases where more data is missing, KNN and RF prove to be effective means of statistical testing, causing minimal delta and standard deviation (SD) bias [41].
7. All generated graphs in MetaboAnalyst can be saved by clicking on a "paint palette" icon, usually displayed besides each graphic. When available, "Excel" icons will be displayed, indicating that a table is available for download.
8. Volcano plots are generated with two-group comparison at a time. If your matrix contains three groups or more, you will have to select only two groups at a time to perform analysis. Groups can be edited under left menu tab on MetaboAnlyst by clicking on "Data processing", then select "Data Editor", manually select chosen groups, and start data normalization steps once again.
9. Ideally, false discovery rates (FDR) should be set between 0.01 and 0.05 for the identification of significant statistical changes.
10. Make sure no special character is added before or after any proteins. Sometimes it may occur to automatically be added quotations (" ") when pasting from excel. Quickly scan over the list of proteins and remove any special characters that have been automatically added. If this step is not done, there will be errors.
11. The lipid-protein interaction is a bioinformatic prediction approach used, which may not be as accurate as if performed a sample proteomics analysis. However, in the absence of proteomics cellular data, the in silico screening is a great start and indication of possible target proteins to be considered.

Acknowledgments

Simone C. da Silva Rosa was funded by the Canadian Institutes of Health Research (CIHR) Postdoctoral Fellowship FRN 176508.

References

1. Brenner DR, Poirier A, Woods RR et al (2022) Projected estimates of cancer in Canada in 2022. CMAJ 194(17):E601–E607
2. Lung Cancer Statistics| How common is lung cancer? [Internet]. [cited 2023 Jul 13]; Available from: https://www.cancer.org/cancer/types/lung-cancer/about/key-statistics.html
3. Expert Consensus Panel, Kidane B, Bott M et al (2023) The American Association for Thoracic Surgery (AATS) 2023 Expert Consensus Document: staging and multidisciplinary management of patients with early-stage non-small cell lung cancer. J Thorac Cardiovasc Surg:S0022-5223(23) 00363-X. https://doi.org/10.1016/j.jtcvs.2023.04.039
4. Lung Cancer – Non-small cell – Statistics [Internet]. Cancer Net 2012 [cited 2023 Jul 13]; Available from: https://www.cancer.net/cancer-types/lung-cancer-non-small-cell/statistics
5. Lung Cancer Survival Rates | 5-Year Survival Rates for Lung Cancer [Internet]. [cited 2023 Jul 13]; Available from: https://www.cancer.org/cancer/types/lung-cancer/detection-diagnosis-staging/survival-rates.html
6. Debnath J, Gammoh N, Ryan KM (2023) Autophagy and autophagy-related pathways in cancer. Nat Rev Mol Cell Biol 24(8):560–575
7. Russell RC, Guan K (2022) The multifaceted role of autophagy in cancer. EMBO J 41(13): e110031
8. Chavez-Dominguez R, Perez-Medina M, Lopez-Gonzalez JS, Galicia-Velasco M, Aguilar-Cazares D (2020) The double-edge sword of autophagy in Cancer: from tumor suppression to pro-tumor activity. Front Oncol [Internet] [cited 2023 Dec 1]; 10. Available from: https://www.frontiersin.org/articles/10.3389/fonc.2020.578418
9. Panda PK, Mukhopadhyay S, Das DN, Sinha N, Naik PP, Bhutia SK (2015) Mechanism of autophagic regulation in carcinogenesis and cancer therapeutics. Semin Cell Dev Biol 39:43–55
10. Mowers EE, Sharifi MN, Macleod KF (2017) Autophagy in cancer metastasis. Oncogene 36(12):1619–1630
11. Fares J, Fares MY, Khachfe HH, Salhab HA, Fares Y (2020) Molecular principles of metastasis: a hallmark of cancer revisited. Sig Transduct Target Ther 5(1):1–17
12. Macintosh RL, Timpson P, Thorburn J, Anderson KI, Thorburn A, Ryan KM (2012) Inhibition of autophagy impairs tumor cell invasion in an organotypic model. Cell Cycle 11(10): 2022–2029
13. Peng F, Liao M, Qin R et al (2022) Regulated cell death (RCD) in cancer: key pathways and targeted therapies. Sig Transduct Target Ther 7(1):1–66
14. Zhang C, Zhu N, Li H et al (2022) New dawn for cancer cell death: emerging role of lipid metabolism. Mol Metab 63:101529
15. Fu Y, Zou T, Shen X et al (2020) Lipid metabolism in cancer progression and therapeutic strategies. MedComm 2(1):27–59
16. Thelen AM, Zoncu R (2017) Emerging roles for the lysosome in lipid metabolism. Trends Cell Biol 27(11):833–850
17. Jaishy B, Abel ED (2016) Lipids, lysosomes, and autophagy. J Lipid Res 57(9):1619–1635
18. Caron A, Richard D, Laplante M (2015) The roles of mTOR complexes in lipid metabolism. Annu Rev Nutr 35:321–348
19. Xie Y, Li J, Kang R, Tang D (2020) Interplay between lipid metabolism and autophagy. Front Cell Dev Biol [Internet] [cited 2023 Nov 17]; 8. Available from: https://www.frontiersin.org/articles/10.3389/fcell.2020.00431
20. Ward C, Martinez-Lopez N, Otten EG et al (2016) Autophagy, lipophagy and lysosomal lipid storage disorders. Biochim Biophys Acta (BBA) 1861(4):269–284
21. Bhatt V, Khayati K, Hu ZS et al (2019) Autophagy modulates lipid metabolism to maintain metabolic flexibility for Lkb1-deficient Kras-driven lung tumorigenesis. Genes Dev 33(3–4):150–165
22. Efeyan A, Comb WC, Sabatini DM (2015) Nutrient-sensing mechanisms and pathways. Nature 517(7534):302–310
23. Alizadeh J, Kavoosi M, Singh N et al (2023) Regulation of autophagy via carbohydrate and

lipid metabolism in Cancer. Cancers (Basel) 15(8):2195

24. Guo JY, Karsli-Uzunbas G, Mathew R et al (2013) Autophagy suppresses progression of K-ras-induced lung tumors to oncocytomas and maintains lipid homeostasis. Genes Dev 27(13):1447–1461
25. Shojaei S, Koleini N, Samiei E et al (2020) Simvastatin increases temozolomide-induced cell death by targeting the fusion of autophagosomes and lysosomes. FEBS J 287(5): 1005–1034
26. Alizadeh J, Zeki AA, Mirzaei N et al (2017) Mevalonate Cascade inhibition by simvastatin induces the intrinsic apoptosis pathway via depletion of isoprenoids in tumor cells. Sci Rep 7:44841
27. Köfeler HC, Fauland A, Rechberger GN, Trötzmüller M (2012) Mass spectrometry based Lipidomics: an overview of technological platforms. Metabolites 2(1):19–38
28. Jiang P, Sinha S, Aldape K, Hannenhalli S, Sahinalp C, Ruppin E (2022) Big data in basic and translational cancer research. Nat Rev Cancer 22(11):625–639
29. Surendran A, Atefi N, Ismail U, Shah A, Ravandi A (2022) Impact of myocardial reperfusion on human plasma lipidome. iScience 25(2):103828
30. Xia J, Psychogios N, Young N, Wishart DS (2009) MetaboAnalyst: a web server for metabolomic data analysis and interpretation. Nucleic Acids Res 37(suppl_2):W652–W660
31. Pang Z, Zhou G, Ewald J et al (2022) Using MetaboAnalyst 5.0 for LC–HRMS spectra processing, multi-omics integration and covariate adjustment of global metabolomics data. Nat Protoc 17(8):1735–1761
32. Lu Y, Pang Z, Xia J (2023) Comprehensive investigation of pathway enrichment methods for functional interpretation of LC–MS global metabolomics data. Brief Bioinform 24(1): bbac553
33. Wishart DS, Tzur D, Knox C et al (2007) HMDB: the human metabolome database. Nucleic Acids Res 35(Database issue): D521–D526
34. Wishart DS, Guo A, Oler E et al (2022) HMDB 5.0: the human metabolome database for 2022. Nucleic Acids Res 50(D1):D622–D631
35. Gfeller D, Michielin O, Zoete V (2013) Shaping the interaction landscape of bioactive molecules. Bioinformatics 29(23):3073–3079
36. Daina A, Michielin O, Zoete V (2019) SwissTargetPrediction: updated data and new features for efficient prediction of protein targets of small molecules. Nucleic Acids Res 47(W1): W357–W364
37. Snel B, Lehmann G, Bork P, Huynen MA (2000) STRING: a web-server to retrieve and display the repeatedly occurring neighbourhood of a gene. Nucleic Acids Res 28(18): 3442–3444
38. Szklarczyk D, Gable AL, Nastou KC et al (2021) The STRING database in 2021: customizable protein-protein networks, and functional characterization of user-uploaded gene/measurement sets. Nucleic Acids Res 49(D1): D605–D612
39. Szklarczyk D, Kirsch R, Koutrouli M et al (2023) The STRING database in 2023: protein-protein association networks and functional enrichment analyses for any sequenced genome of interest. Nucleic Acids Res 51(D1): D638–D646
40. Weir JM, Wong G, Barlow CK et al (2013) Plasma lipid profiling in a large population-based cohort. J Lipid Res 54(10):2898–2908
41. Taylor S, Ponzini M, Wilson M, Kim K (2022) Comparison of imputation and imputation-free methods for statistical analysis of mass spectrometry data with missing data. Brief Bioinform 23(1):bbab353

Methods in Molecular Biology (2025) 2879: 261–273
DOI 10.1007/7651_2024_532

Published online: 13 April 2024

Evaluation of Mitochondrial Phagy (Mitophagy) in Human Non-small Adenocarcinoma Tumor Cells

Javad Alizadeh, Simone C. da Silva Rosa, Marco Cordani, and Saeid Ghavami

Abstract

Non–small cell lung cancer (NSCLC) is a predominant form of lung cancer characterized by its aggressive nature and high mortality rate, primarily due to late-stage diagnosis and metastatic spread. Recent studies underscore the pivotal role of mitophagy, a selective form of autophagy targeting damaged or superfluous mitochondria, in cancer biology, including NSCLC. Mitophagy regulation may influence cancer cell survival, proliferation, and metastasis by modulating mitochondrial quality and cellular energy homeostasis. Herein, we present a comprehensive methodology developed in our laboratory for the evaluation of mitophagy in NSCLC tumor cells. Utilizing a combination of immunoblotting, immunocytochemistry, and fluorescent microscopy, we detail the steps to quantify early and late mitophagy markers and mitochondrial dynamics. Our findings highlight the potential of targeting mitophagy pathways as a novel therapeutic strategy in NSCLC, offering insights into the complex interplay between mitochondrial dysfunction and tumor progression. This study not only sheds light on the significance of mitophagy in NSCLC but also establishes a foundational approach for its investigation, paving way for future research in this critical area of cancer biology.

Key words Autophagy, Cellular homeostasis, Mitochondrial dynamics, Mitophagy, Non-small cell adenocarcinoma

1 Introduction

Non–small cell lung cancer (NSCLC) is recognized as one of the deadliest malignancies, notorious for its high metastatic potential and status as the principal contributor to cancer-related mortalities worldwide [1, 2]. Representing roughly 85% of lung cancer cases, NSCLC predominantly manifests in the form of adenocarcinoma, the most prevalent subtype of this disease [3, 4]. The presentation of NSCLC often includes regional and distant metastases, which are associated with a severe prognosis [5, 6]. Despite considerable research efforts, the complex mechanisms driving NSCLC progression are not yet fully comprehended. Therefore, it is crucial to deepen our understanding to unearth dependable diagnostic markers and pioneer new treatments.

Autophagy is a cellular process that directs components within the cytoplasm to lysosomes for breakdown and has been associated with the development of cancer [7, 8]. Following the degradation of sequestered lipids, proteins, carbohydrates, or nucleic acids, the resultant biomolecules are recycled back into the cytoplasm, contributing to the maintenance of metabolic processes and cellular homeostasis [9, 10]. Autophagy selectively targets different cellular components using specialized adaptors, leading to the creation of specific terms such as pexophagy for peroxisome degradation, mitophagy for mitochondrial clearance, and aggrephagy for the removal of protein aggregates [11, 12].

Mitophagy, a form of autophagy dedicated to the elimination of mitochondria, is essential for maintaining a proper mitochondrial cellular fitness [13]. The removal of defective mitochondria through mitophagy is predominantly facilitated by proteins like NIP3-like protein X (NIX or BNIP3L), BNIP3, FUNDC, and the PINK1/PARKIN complex, as well as BCL2L13, all of which are situated on the mitochondrial outer membrane [14–18]. The autophagy adapter protein p62/sequestosome-1 has been identified to enhance mitochondrial ubiquitination, a process essential for the degradation and recycling of mitochondria [19]. All these mitophagy mediators bind to LC3β-II via a cytoplasmic LC3 β-interacting region (LIR) motif to mediate the autophagosome formation and the selective targeting of mitochondria [20–22]. TOMM20 functions within the TOM complex as a receptor on the mitochondrial outer membrane [23] and is degraded during mitophagy, independently of autophagic pathways [24]. Enhancing TOMM20 expression can mitigate mitochondrial dysfunction and avoid cell death, showing its protective role in cellular health [25]. LAMP1 (lysosomal associated membrane protein 1) plays a pivotal role in the late stages of mitophagy, serving as a marker for the lysosomal degradation phase. The presence of LAMP1 marks the lysosomal activity in dismantling mitochondria, underscoring its importance in evaluating mitophagy's efficiency and completion. Thus, both TOMM20 and LAMP1 are crucial markers for the identification and study of mitophagy, offering insights into the dynamics and regulation of this process within cells [22]. Deregulation of mitophagy is linked to the impairment of mitochondrial functions [26].

Numerous studies have shown a key role for mitophagy in the progression of different cancers, including colorectal [27], breast [28], and lung [29, 30], as well as in the chemosensitivity of pancreatic cancer [31]. Receptor-interacting protein kinase-1 (RIPK1) has been shown to induce mitophagy and reactive oxygen species (ROS) production, leading to anoikis —the elimination of cells detached from the extracellular matrix [32]. This underscores the potential of mitophagy in NSCLC where detachment and reattachment processes are critical for metastasis. Mitochondrial dysfunction in head and neck cancer (HNC) cell lines with specific inhibitors leads to increased expression of EMT markers in some cell lines, suggesting a potential link between mitochondrial dysfunction and EMT [33]. The detection of mitophagy in cancer cells, including NSCLC, is paramount for unraveling the complex interplay between mitochondrial dysfunction and tumor progression. It offers a promising avenue for the development of targeted therapies aimed at mitigating cancer aggressiveness and enhancing treatment efficacy. Hence, in this study, we unveil a comprehensive framework for probing the intricacies of mitophagy within NSCLC cell lines, underpinning the critical role of this autophagic process in cancer cell survival and proliferation. By integrating methodologies such as immunocytochemistry, subcellular fractionation, and protein interaction assays, we delineate the pivotal contributions of key mitophagy mediators—p62, LC3, TOMM20, and LAMP1—in dictating the fate of mitochondria, thus offering a nuanced understanding of their involvement in NSCLC pathology. This approach not only sheds light on the fundamental mechanisms underlying mitochondrial dynamics in cancer but also propels us toward identifying novel therapeutic targets, poised to revolutionize the management of NSCLC by harnessing the potential of mitophagy modulation.

In the materials and methods section, we will list and describe respectively the following steps:

1. NSCLC cell culture.
2. Subcellular fractionation.
3. Immunoblotting for mitophagy markers (LC3, p62, and TOMM20).
4. Immunocytochemistry for mitophagy markers.

2 Materials

2.1 NSCLC Cell Culture

1. Lung adenocarcinoma cells (A549; ATCC-CCL-185™), Mouse Lewis Lung Carcinoma cells (LL/2-LLC1; ATCC® CRL1642).
2. Adherent plates culture plates (VWR, Toronto, ON, Canada).
3. Dulbecco's modified Eagle's medium (DMEM) high glucose (4 mg/mL) (CORNING; Cat#: 50–003-PB), 10% fetal bovine serum (FBS) (Gibco™; Cat#: 16000044), 1% penicillin-streptomycin, 1% insulin transferrin selenium (ITS) (Gibco™; Cat#: 41400045).

2.2 Subcellular Fractionation

1. *Lysis buffer*: Protease and phosphatase inhibitor-containing lysis buffer system (Sigma-Aldrich, Oakville, CA).
2. *Mitochondrial isolation*: Qiagen Qproteome mitochondrial isolation kit.
3. *Nuclear and cytoplasmic extraction*: Thermo Fisher Scientific nuclear and cytoplasmic extraction kit (Waltham, MA, USA).
4. *Protein quantification*: Lowry protein assay (Bio-Rad, Oakville, ON, CA).
5. *Primary antibodies*: Used as per manufacturer's protocol at a 1: 1000 dilution.
6. *Secondary antibodies*: HRP-conjugated goat anti-rabbit (1: 5000 dilution), HRP-conjugated goat anti-mouse (1:3000 dilution).
7. *Visualization system*: Bio-Rad imager (Mississauga, ON, Canada).
8. *Fractionation purity markers*: monoclonal COX IV (Cell Signaling Technology, Canada; Cat#: 4850) and Lamin A/C (Santa Cruz Technology, Dallas, TX, USA; Cat#: sc-517579).
9. *Analysis software*: Image Lab Software, Bio-Rad (Mississauga, ON, Canada).

2.3 Immunoblotting for SQSTM1/p62, LC3B, and TOMM20

1. *Lysis buffer for protein extraction*: 20 mM Tris–HCl (pH 7.5), 0.5 mM PMSF, 0.5% Nonidet P-40, 100 μM β-glycerol 3-phosphate, and 0.5% protease and phosphatase inhibitor cocktail.
2. *Protein quantification*: Lowry protein assay (Bio-Rad, Oakville, ON, CA).
3. *SDS-PAGE supplies*: Gels, running buffer, protein loading marker, power supply (100 V, 1.5 h).
4. *Transfer materials*: PVDF transfer membranes (Thermo Fisher Scientific, Waltham, MA, USA; Cat#: 88518), transfer buffer, chromatography paper, power supply (100 V, 2 h).
5. *Blocking solution*: Non-fat dried milk (5%) (AppliChem, Darmstadt, Germany) and Tween-20 (0.5%) (Sigma, St. Louis, MO, USA).
6. *Primary antibodies*: polyclonal LC3B (Sigma Aldrich, Oakville, CA; Cat#: L7543); polyclonal SQSTM1/p62 (Cell Signaling Technology, Canada; Cat#: 5114); monoclonal β-Actin (Sigma Aldrich, Oakville, CA; Cat#: A5441) and monoclonal GAPDH (Santa Cruz technology, Dallas, TX, USA; Cat#: sc-32,233) as loading controls.

7. *Secondary antibodies*: Anti-Rabbit IgG (whole molecule)—HRP-conjugated antibody produced in goat (Sigma Aldrich, Oakville, CA, Cat#: A0545); anti-Mouse IgG (whole molecule)—HRP-conjugated antibody produced in goat (Sigma Aldrich, Oakville, CA; Cat#: A4416).
8. *Detection system*: Enhanced chemiluminescence (ECL) detection kit (Amersham-Pharmacia Biotech).
9. *Imaging and analysis*: Bio-Rad imager (Mississauga, ON, Canada). For signal quantification utilize Image Lab Software suite, which includes the Alpha Ease FC functionality, provided by Bio-Rad.

2.4 Immunocytochemistry (ICC) for Mitophagy Markers

1. *Culture substrates*: Cover slips for cell culture (VWR, Toronto, ON, Canada).
2. *Culture media and supplements*: DMEM (high glucose) with 10% FBS or 1% ITS for starvation.
3. *Mitophagy inducers*: TGFβ1 (5 ng/mL) (VWR, Toronto, ON, Canada) for treatment.
4. *Mitochondrial and nucleus staining*: MitoTracker™ Red CMXRos (200 nM) (Thermo Fisher Scientific, Waltham, MA, USA; Cat#: M7512) and Hoechst (10 μM) (Sigma, St. Louis, MO, USA).
5. *Fixative*: 3.7% formaldehyde (VWR International, Milan, Italy).
6. *Permeabilization reagent*: 0.25% Triton X-100 (Sigma, St. Louis, MO, USA).
7. *Blocking solution*: 5% normal donkey serum (NDS) (Jackson lab).
8. *Primary antibodies*: Rabbit anti-LC3 (Sigma-Aldrich, Canada; Cat#: L7543), anti-p62 (Cell Signaling Technologies, Canada; Cat#: #5114), anti-TOMM20 (Cell Signaling Technologies, Canada; Cat#: 42406), and anti-LAMP1 IgG-Lysosome Marker (Abcam; ab24170).
9. *Secondary antibodies*: Fluorochrome-conjugated secondary antibodies (Alexa Fluor 488 (Thermo Fisher Scientific, Waltham, MA, USA; Cat#: A21206) and Alexa Fluor 594 (Thermo Fisher Scientific, Waltham, MA, USA; Cat#: R37119)).
10. *Mounting medium*: Nuclear counterstain (Thermo Fisher Scientific).
11. *Imaging software*: ImageJ (NIH, Bethesda, MD, USA) and Zen 2.3 Pro software (Zeiss, Heidelberg, Germany) for image processing and quantification.

3 Methods

3.1 NSCLC Cell Culture

3.1.1 Cell Culture

1. Conduct all cell culture activities within a biosafety cabinet under sterile conditions to prevent contamination.
2. Utilize both A549 and LLC NSCLC cell lines for the experiments.

3.1.2 Culture Conditions

1. Culture A549 and LLC NSCLC cells on cover slips and in adherent (100 mm) plates. Count and seed 100 K cells per plate (*see* **Note 1**).
2. Employ high-glucose DMEM supplemented with 10% FBS for optimal cell growth and maintenance. Change the media once a day till the confluency of cells reaches about 40% (*see* **Note 2**).
3. Allocate three plates for each condition to ensure replicability.

3.1.3 Starvation and Treatment Protocol

1. Induce a starvation state by treating cells with 1% ITS for 24 h, creating conditions conducive to autophagy/mitophagy (*see* **Note 3**).
2. Administer TGFβ1 (5 ng/mL) for 36 h to further promote mitophagy. This process mimics physiological stress that cancer cells might encounter.

3.1.4 Cell Harvesting

1. After 36 h of treatment, gently wash the cells with 1X PBS and then detach cells using 1 mL of 1X PBS.

3.1.5 Cell Lysis and Protein Extraction

1. Proceed with cell lysis by sonication (3 cycles, 5 s each) to ensure complete disruption of cellular membranes.
2. Centrifugate at 10,000 g for 10 min to separate the soluble protein fraction from cell debris.
3. Transfer the supernatant (containing the soluble fraction) into another tube and protein quantification.

3.1.6 Protein Quantification

1. Utilize Lowry protein assay to measure the protein concentration of the supernatant.
2. Adjust the protein concentration to an ideal level of 1 mg/mL for uniformity across samples.

3.1.7 Preparation for Analysis

Prepare samples according to specified requirements for subsequent immunoblotting or immunocytochemistry analysis to assess mitophagy markers.

3.2 Subcellular Fractionation

3.2.1 Preparation of Protein Samples

Extract proteins using a lysis buffer containing protease and phosphatase inhibitors (Sigma-Aldrich, Oakville, CA) to preserve protein integrity and prevent degradation.

3.2.2 Subcellular Organelle Fractionation

1. Utilize the Mitochondrial Isolation Kit (Qiagen Qproteome) for mitochondrial fractionation.
2. Employ the Nuclear and Cytoplasmic Extraction Kit (Thermo Fisher Scientific) to separate nuclear and cytoplasmic components (*see* **Note 4**).

3.2.3 Protein Quantification

Determine protein concentration using the Lowry protein assay, which provides a reliable measurement of protein levels for subsequent analyses.

3.2.4 Immunoblotting

1. Separate proteins by SDS-PAGE (load 20 μg).
2. Transfer the proteins to PVDF membranes under reducing conditions.
3. Block membranes with non-fat dried milk and Tween 20 to prevent non-specific binding.
4. Incubate membranes overnight at 4 °C with primary antibodies (diluted 1:1000 as per manufacturer's protocol) targeting specific proteins of interest present in the mitochondrial or nuclear fractions.
5. Use HRP-conjugated secondary antibodies for detection: Goat Anti-Rabbit (1:5000 dilution) and Goat Anti-Mouse (1:3000 dilution).
6. Develop blots using enhanced chemiluminescence (ECL) and visualize bands with a Bio-Rad imager.

3.2.5 Confirmation of Fractionation Purity

Verify the purity of mitochondrial and nuclear fractions by immunoblotting for COX IV (mitochondrial marker) and Lamin A/C (nuclear marker), respectively.

3.2.6 Imaging and Analysis

Ensure linear exposure of blots using Image Lab Software provided by the Bio-Rad imager for accurate quantification and analysis of protein bands.

3.3 Immunoblotting for SQSTM1/p62 and LC3β

3.3.1 Preparation of Protein Extracts

Use lysis buffer containing 20 mM Tris–HCl (pH 7.5), 0.5 mM PMSF, 0.5% Nonidet P-40, 100 μM β-glycerol 3-phosphate, and 0.5% protease and phosphatase inhibitor cocktail to prepare protein extracts (*see* **Note 5**).

3.3.2 Quantification of Protein Content

1. Centrifugation at 10,000 g for 10 min.
2. Transfer the supernatant (containing the soluble fraction) into another tube and do protein quantification.
3. Measure the supernatant protein content with the Lowry protein assay.

3.3.3 SDS-PAGE and Transfer

1. Separate proteins by SDS-PAGE (20 μg) (*see* **Note 6**).
2. Transfer them onto nylon membranes under reducing conditions.

3.3.4 Blocking and Primary Antibody Incubation

1. Block membranes with non-fat dried milk (5%) and Tween 20 (0.5%).
2. Incubate overnight at 4 °C with primary antibodies against LC3B, SQSTM1/p62, and TOMM20.

3.3.5 Washing and Secondary Antibody Incubation

1. Wash the membrane three times for 5 min with 1X PBS.
2. Incubate for 2 h at room temperature with HRP-conjugated secondary antibody: Goat Anti-Rabbit (1:5000 dilution) and Goat Anti-Mouse (1:3000 dilution).

3.3.6 Detection

1. Develop blots using enhanced chemiluminescence (ECL).
2. Visualize bands with a Bio-Rad imager.

3.3.7 Loading Controls

Use Ponceau S as loading control to ensure equal protein loading and transfer efficiency.

3.3.8 Imaging and Quantification

1. Ensure linear exposure using Image Lab Software provided by the Bio-Rad imager.
2. Quantify the signals using dosimetry software Alpha Ease FC.

3.4 Immuno-cytochemistry (ICC) for Mitophagy Markers

3.4.1 Cell Seeding and Preparation

1. Culture A549 and LLC cells on cover slips.
2. Seeding approximately 5000 cells per spot in high-glucose DMEM supplemented with 10% FBS for 24 h.

3.4.2 Starvation and Induction

1. Subject cells to starvation using 1% ITS for 24 h.
2. TGFβ1 treatment (5 ng/mL) for 36 h to induce mitophagy.

3.4.3 Staining

1. Incubate cells with MitoTracker Red (200 nM) for mitochondrial staining for 30 min at 37 °C.
2. Incubate cells with Hoechst (10 μM) for nuclear staining for 30 min at 37 °C.

3.4.4 Fixation

1. Fix cells with 3.7% formaldehyde for 20 min at room temperature (RT).
2. Washing the cells three times with 1x PBS.

3.4.5 Permeabilization and Blocking

1. Permeabilize with 0.25% Triton X-100 for 15 min.
2. Block with 5% normal donkey serum for 1 h at RT to prevent non-specific antibody binding.

3.4.6 Primary Antibody Incubation

1. Incubate overnight at 4 °C with rabbit anti-LC3, anti-TOMM20, and anti-LAMP1 IgG primary antibodies.
2. The antibodies are diluted in 5% NDS to specifically target autophagy and mitophagy markers.

3.4.7 Secondary Antibody Application

1. Wash slides very gently three times for 5 min with 1x PBS (LLC cells are loosely attached to the cover slip, so practice extra care while washing).
2. Incubate with fluorochrome-conjugated secondary antibodies (Alexa Fluor 488 and Alexa Fluor 594) for 1 h at RT to enable fluorescence detection.

3.4.8 Mounting and Storage

1. Mount cover slips on slides using a nuclear counterstain.
2. Store at −20 °C until ready for imaging.

3.4.9 Imaging and Analysis

1. Perform imaging with appropriate fluorescence microscopy.
2. Quantification, scale bar adjustment, background subtraction, and image processing should be conducted using ImageJ and Zen 2.3 Pro software for accurate representation and analysis of mitophagy markers.

3.5 Results

TGFβ1 plays a pivotal role in promoting the metastasis of NSCLC by inducing EMT [4]. Previous studies have indicated that TGFβ1-induced EMT relies on general autophagy, and inhibiting autophagy enhances the epithelial phenotype in NSCLC adenocarcinoma cells [34]. Our recent preliminary research revealed structures resembling mitochondria within the cytosolic double membrane structure of NSCLC adenocarcinoma cells treated with TGFβ1. Recent studies have suggested a potential involvement of mitophagy in the mechanisms underlying tumor metastasis [35].

Therefore, we developed a method to measure TGFβ1-inuduced mitophagy in human and mouse NSCLC adenocarcinoma cells (A549 and LLC). Our immunocytochemistry results showed a significant increase in the colocalization of LC3β puncta and MitoTracker (indicative of early mitophagy) as well as the colocalization of TOMM20 and LAMP1 (indicative of late mitophagy) in both cell lines upon TGFβ1 treatment (Fig. 1 *for A549*). Additionally, immunoblotting analysis of subcellular fractions revealed that TGFβ1 treatment induced the lipidation of LC3β-II, degradation of p62, and degradation of TOMM20 in the mitochondrial fraction compared to baseline in both A549 and LLC cells (Fig. 2 *for A549*). This data suggests that TGFβ1 induces mitophagy in both NSCLC adenocarcinoma cell lines.

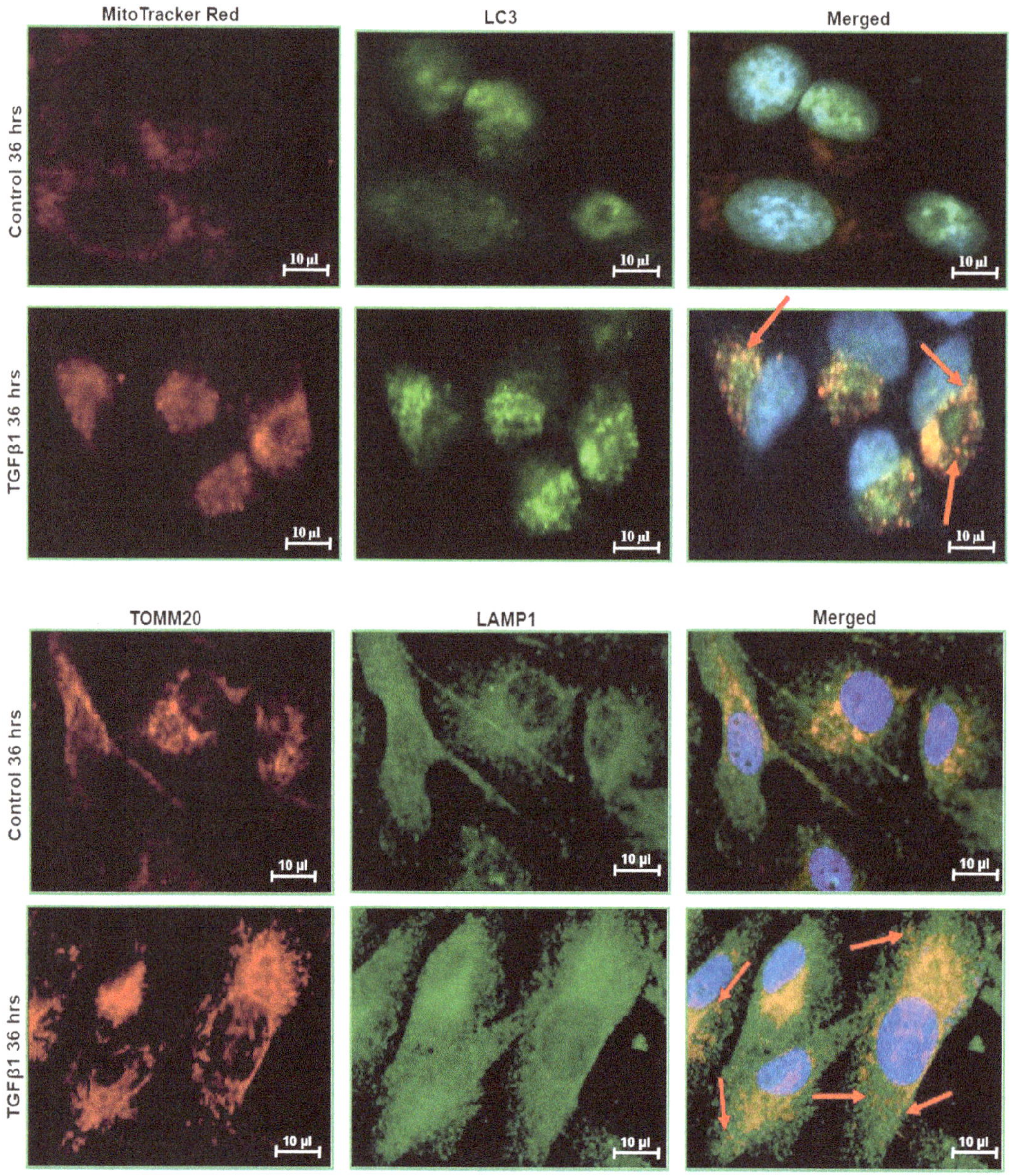

Fig. 1 Immunocytochemistry analysis of A549 cells after treatment with TGFβ1 to measure early and late mitophagy. A549 cells were cultured on sterile coverslips until reaching 30–40% confluency, followed by exposure to recombinant human TGFβ1 at a concentration of 5 ng/mL for durations of 36 h to induce early and late mitophagy, respectively. Immunocytochemistry (ICC) was then conducted on the coverslips to observe alterations in various mitophagy markers. Specifically, cells were stained with MitoTracker (appearing red) and incubated with LC3β antibody (appearing green) for early mitophagy (upper panel) and with TOM20 antibody (appearing red) and LAMP1 antibody (appearing green) (lower panel) for late mitophagy detection. These samples were then visualized using standard fluorescence microscopy, with scale bars indicating a size of 10 μm. Colocalization of mitotracker/LC3 and TOMM20/LAMP1 shows early and late mitophagy events

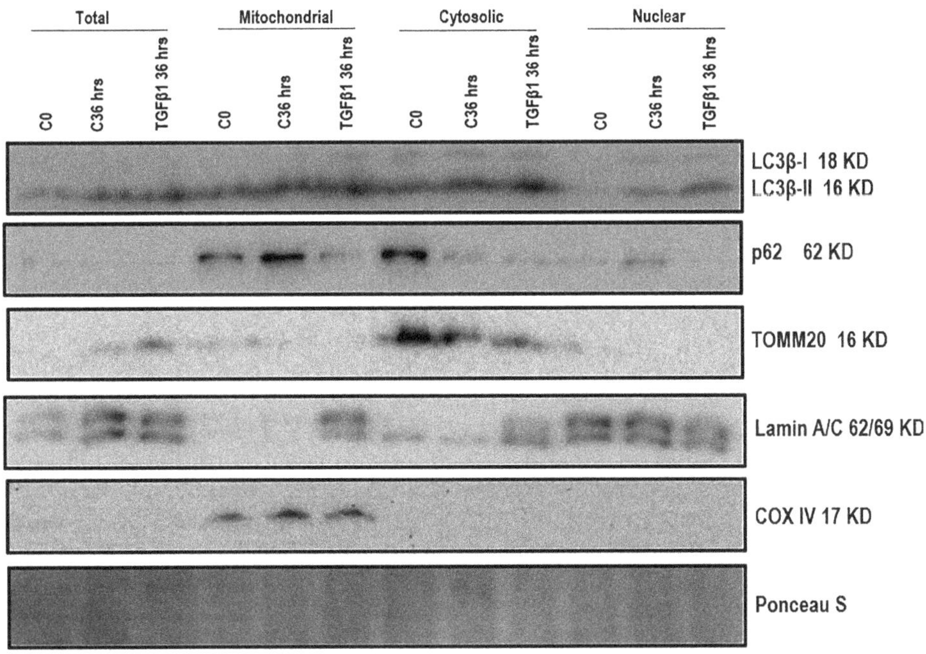

Fig. 2 Immunoblotting of LC3β lipidation and p62 and TOMM20 degradation in cytosolic, mitochondria, and nuclear fraction of A549 cells after treatment with TGFβ1 to measure mitophagy. A549 cells ($n = 3$) were treated in the presence and absence of TGFβ1 (5 ng/mL, 36 h), and protein extracts were immunoblotted ($n = 3$) as indicated. Equal protein loading was confirmed using Ponceau S staining and Beta actin. Lipidation of LC3β and degradation of p62 and TOMM20 in mitochondrial fraction is a confirmation of mitophagy and mitochondrial degradation

4 Notes

1. Ensure that no cell clumps are present in the initial cell culturing.
2. The confluency of the cells for starting the experiment is very important. Higher confluency will increase the chance of false positive mitophagy results.
3. ITS starvation is a key step for getting accurate mitophagy detection. Avoid starvation in pure cell culture media without ITS.
4. Use fresh prepared solution of the kits (prepare on the same day).
5. pH of the lysis buffer is a key step for proper detection of LC3 lipidation.
6. Use 15% gel for the best separation of LC3 bands.

Acknowledgments

SG conceptualized and designed the study. JA performed the experiments. SG and JA SCDR analyzed the data. JA and MC wrote the manuscript. SG, JA, SCDR, and MC reviewed and edited the manuscript. SG funded and supervised the experiments.

Funding

This study was funded by Canercare Manitoba Foundation. JA was supported by CIHR Vaneir PhD studentship. SCDR was supported with a Postdoctoral Fellowship sponsored by the Canadian Institutes of Health and Research (CIHR), FRN 176508. MC was supported by grant RYC2021-031003I funded by MICIU/AEI/10.13039/501100011033 and, by European Union NextGenerationEU/PRTR.

References

1. Brenner DR, Poirier A, Woods RR, Ellison LF, Billette JM, Demers AA et al (2022) Projected estimates of cancer in Canada in 2022. CMAJ 194(17):E601–E6E7
2. Alizadeh J, Kavoosi M, Singh N, Lorzadeh S, Ravandi A, Kidane B et al (2023) Regulation of autophagy via carbohydrate and lipid metabolism in cancer. Cancers (Basel) 15(8):2195
3. Expert Consensus P, Kidane B, Bott M, Spicer J, Backhus L, Chaft J et al (2023) The American Association for Thoracic Surgery (AATS) 2023 Expert Consensus Document: staging and multidisciplinary management of patients with early-stage non-small cell lung cancer. J Thorac Cardiovasc Surg 166(3):637–654
4. Alizadeh J, Shojaei S, Sepanjnia A, Hashemi M, Eftekharpour E, Ghavami S (2019) Simultaneous detection of autophagy and epithelial to mesenchymal transition in the non-small cell lung cancer cells. Methods Mol Biol 1854:87–103
5. Araghi M, Mannani R, Heidarnejad Maleki A, Hamidi A, Rostami S, Safa SH et al (2023) Recent advances in non-small cell lung cancer targeted therapy; an update review. Cancer Cell Int 23(1):162
6. Nikseresht M, Shahverdi M, Dehghani M, Abidi H, Mahmoudi R, Ghalamfarsa G et al (2019) Association of single nucleotide autophagy-related protein 5 gene polymorphism rs2245214 with susceptibility to non-small cell lung cancer. J Cell Biochem 120(2):1924–1931
7. Debnath J, Gammoh N, Ryan KM (2023) Autophagy and autophagy-related pathways in cancer. Nat Rev Mol Cell Biol 24(8):560–575
8. Cordani M, Strippoli R, Trionfetti F, Barzegar Behrooz A, Rumio C, Velasco G et al (2024) Immune checkpoints between epithelial-mesenchymal transition and autophagy: a conflicting triangle. Cancer Lett 585:216661
9. Behrooz AB, Cordani M, Fiore A, Donadelli M, Gordon JW, Klionsky DJ, Ghavami S (2024) The obesity-autophagy-cancer axis: mechanistic insights and therapeutic perspectives. Semin Cancer Biol 99:24–44
10. Vargas JNS, Hamasaki M, Kawabata T, Youle RJ, Yoshimori T (2023) The mechanisms and roles of selective autophagy in mammals. Nat Rev Mol Cell Biol 24(3):167–185
11. Behrooz AB, Cordani M, Donadelli M, Ghavami S (2024) Metastatic outgrowth via the two-way interplay of autophagy and metabolism. Biochim Biophys Acta Mol basis Dis 1870(3):166824
12. Kirkin V, Rogov VV (2019) A diversity of selective autophagy receptors determines the specificity of the autophagy pathway. Mol Cell 76(2):268–285
13. Pickles S, Vigie P, Youle RJ (2018) Mitophagy and quality control mechanisms in mitochondrial maintenance. Curr Biol 28(4):R170–R185
14. Alizadeh J, da Silva Rosa SC, Weng X, Jacobs J, Lorzadeh S, Ravandi A et al (2023) Ceramides and ceramide synthases in cancer: focus on

apoptosis and autophagy. Eur J Cell Biol 102(3):151337

15. Huang CF, Hsieh YH, Yang SF, Kuo CH, Wang PH, Liu CJ, Lin RC (2023) Mitophagy effects of Protodioscin on human osteosarcoma cells by inhibition of p38MAPK targeting NIX/LC3 Axis. Cells 12(3):395
16. Qin S, Liu C, Chen Y, Yao M, Liao S, Xin W et al (2023) Cobaltosic oxide-polyethylene glycol-triphenylphosphine nanoparticles ameliorate the acute-to-chronic kidney disease transition by inducing BNIP3-mediated mitophagy. Kidney Int 103(5):903–916
17. Bi Y, Liu S, Qin X, Abudureyimu M, Wang L, Zou R et al (2024) FUNDC1 interacts with GPx4 to govern hepatic ferroptosis and fibrotic injury through a mitophagy-dependent manner. J Adv Res 55:45–60
18. Fang Q, Zheng S, Chen Q, Chen L, Yang Y, Wang Y et al (2023) The protective effect of inhibiting mitochondrial fission on the juvenile rat brain following PTZ kindling through inhibiting the BCL2L13/LC3 mitophagy pathway. Metab Brain Dis 38(2):453–466
19. Yamada T, Murata D, Adachi Y, Itoh K, Kameoka S, Igarashi A et al (2018) Mitochondrial stasis reveals p62-mediated ubiquitination in Parkin-independent Mitophagy and mitigates nonalcoholic fatty liver disease. Cell Metab 28(4):588–604 e5
20. Wirth M, Zhang W, Razi M, Nyoni L, Joshi D, O'Reilly N et al (2019) Molecular determinants regulating selective binding of autophagy adapters and receptors to ATG8 proteins. Nat Commun 10(1):2055
21. Villa E, Marchetti S, Ricci JE (2018) No Parkin zone: Mitophagy without Parkin. Trends Cell Biol 28(11):882–895
22. Johansen T, Lamark T (2020) Selective autophagy: ATG8 family proteins, LIR motifs and cargo receptors. J Mol Biol 432(1):80–103
23. Su J, Liu D, Yang F, Zuo MQ, Li C, Dong MQ et al (2022) Structural basis of Tom20 and Tom22 cytosolic domains as the human TOM complex receptors. Proc Natl Acad Sci USA 119(26):e2200158119
24. Chan NC, Salazar AM, Pham AH, Sweredoski MJ, Kolawa NJ, Graham RL et al (2011) Broad activation of the ubiquitin-proteasome system by Parkin is critical for mitophagy. Hum Mol Genet 20(9):1726–1737
25. Franco-Iborra S, Cuadros T, Parent A, Romero-Gimenez J, Vila M, Perier C (2018) Defective mitochondrial protein import contributes to complex I-induced mitochondrial dysfunction and neurodegeneration in Parkinson's disease. Cell Death Dis 9(11): 1122
26. Luciani A, Schumann A, Berquez M, Chen Z, Nieri D, Failli M et al (2020) Impaired mitophagy links mitochondrial disease to epithelial stress in methylmalonyl-CoA mutase deficiency. Nat Commun 11(1):970
27. Zhang C, Zeng C, Xiong S, Zhao Z, Wu G (2022) A mitophagy-related gene signature associated with prognosis and immune microenvironment in colorectal cancer. Sci Rep 12(1):18688
28. Li Q, Chu Y, Li S, Yu L, Deng H, Liao C et al (2022) The oncoprotein MUC1 facilitates breast cancer progression by promoting Pink1-dependent mitophagy via ATAD3A destabilization. Cell Death Dis 13(10):899
29. Yu H, Liu Q, Jin M, Huang G, Cai Q (2023) Comprehensive analysis of mitophagy-related genes in NSCLC diagnosis and immune scenery: based on bulk and single-cell RNA sequencing data. Front Immunol 14:1276074
30. Liu Z, Shan S, Yuan Z, Wu F, Zheng M, Wang Y et al (2023) Mitophagy bridges DNA sensing with metabolic adaption to expand lung cancer stem-like cells. EMBO Rep 24(2):e54006
31. Qin C, Wang Y, Zhao B, Li Z, Li T, Yang X et al (2023) STOML2 restricts mitophagy and increases chemosensitivity in pancreatic cancer through stabilizing PARL-induced PINK1 degradation. Cell Death Dis 14(3):191
32. Hawk MA, Schafer ZT (2018) RIPK1-dependent mitophagy: a novel mechanism to eliminate cells detached from the extracellular matrix. Mol Cell Oncol 5(4):e1465015
33. Greier MC, Runge A, Dudas J, Pider V, Skvortsova II, Savic D, Riechelmann H (2022) Mitochondrial dysfunction and epithelial to mesenchymal transition in head neck cancer cell lines. Sci Rep 12(1):13255
34. Alizadeh J, Glogowska A, Thliveris J, Kalantari F, Shojaei S, Hombach-Klonisch S et al (2018) Autophagy modulates transforming growth factor beta 1 induced epithelial to mesenchymal transition in non-small cell lung cancer cells. Biochim Biophys Acta, Mol Cell Res 1865(5):749–768
35. Abdelmaksoud NM, Abulsoud AI, Abdelghany TM, Elshaer SS, Rizk SM, Senousy MA (2023) Mitochondrial remodeling in colorectal cancer initiation, progression, metastasis, and therapy: a review. Pathol Res Pract 246:154509

Methods in Molecular Biology (2025) 2879: 275–287
DOI 10.1007/7651_2024_538

Published online: 23 April 2024

Assessment of Stiffness-Dependent Autophagosome Formation and Apoptosis in Embryonal Rhabdomyosarcoma Tumor Cells

Serap Sezen, Sevin Adiguzel, Atefeh Zarepour, Arezoo Khosravi, Joseph W. Gordon, Saeid Ghavami, and Ali Zarrabi

Abstract

Remodeling of the extracellular matrix (ECM) eventually causes the stiffening of tumors and changes to the microenvironment. The stiffening alters the biological processes in cancer cells due to altered signaling through cell surface receptors. Autophagy, a key catabolic process in normal and cancer cells, is thought to be involved in mechano-transduction and the level of autophagy is probably stiffness-dependent. Here, we provide a methodology to study the effect of matrix stiffness on autophagy in embryonal rhabdomyosarcoma cells. To mimic stiffness, we seeded cells on GelMA hydrogel matrices with defined stiffness and evaluated autophagy-related endpoints. We also evaluated autophagy-dependent pathways, apoptosis, and cell viability. Specifically, we utilized immunocytochemistry and confocal microscopy to track autophagosome formation through LC3 lipidation. This approach suggests that the use of GelMA hydrogels with defined stiffness represents a novel method to evaluate the role of autophagy in embryonal rhabdomyosarcoma and other cancer cells.

Key words Autophagy, Embryonal rhabdomyosarcoma, GelMA hydrogel, Immunocytochemistry, Stiffness, Tumor microenvironment

1 Introduction

Rhabdomyosarcoma (RMS) is a soft tissue tumor mostly seen in children (70%) and young adults [1]. Although RMS is a rare disease, it accounts for 5% of childhood malignancies and 50% of soft tissue sarcomas [2]. Two types of RMS are common: alveolar RMS (ARMS) and embryonic RMS (ERMS). ERMS is more prevalent compared to ARMS in both adults and children with wide spectrum of morphology of tumor cells in the clinical findings [2].

During tumor development, the biosynthesis of ECM proteins changes resulting in increased tumor stiffness. The increased stiffness of the tumor tissue limits tumor growth but enhances tumor aggressiveness. Through the enzymatic ECM remodeling and

Authors Saeid Ghavami and Ali Zarrabi have senior authorship.

degradation, cells can be free from physical barriers favoring tumor growth and invasion, which is a common feature of most cancers [3]. The increase in ECM stiffness in solid tumor niche is also associated with the deposition of collagen and lysyl oxidase proteins that eventually results in contraction and deposition, associated with the crosslinking of the stroma. The increase in the stiffness of tumors is considered a critical step and leads to change in several molecular mechanisms in most cancers. On the contrary, the softening of the tumor ECM slows down the cancer progression and could be considered targeted therapy for the most tumors [4].

The change in ECM stiffness in tumors also alters autophagy, which is one of the fundamental cellular mechanisms in cancer cells [5, 6]. Autophagy is a physiological event in eukaryotic cells and plays an important role in cellular hemostasis by eliminating damaged or mis-synthesized proteins as well as dysfunctional organelles, mutant proteins prone to aggregation, and intracellular pathogens [7–9]. There are three forms of autophagy: chaperone-mediated, microautophagy, and macroautophagy (hereafter autophagy) [2, 10].

In autophagy, the conversion of microtubule-associated protein 1 light chain 3 (LC3) from the free form (LC3B-I) to the phosphatidyl ethanolamine-conjugated form (LC3B-II) via autophagy-related protein 4 (ATG4) is considered a crucial step for autophagosome formation [11, 12]. Autophagy is regulated through various stressors that define the fate of the cell [13, 14]. In particular, it is regulated by cell surface receptors and through mammalian target of rapamycin (mTOR) signaling, which is directly influenced by the ECM stiffness of the tumor niche [15]. Moreover, autophagy is also involved in ECM mechanotransduction, so it has been suggested that in normal mammalian cells, autophagy increases with increasing matrix stiffness [16]. Besides, matrix stiffness regulates the stromal autophagy and enhances the formation of protumorigenic ECM [17].

The growing tumor cells cause tumor tissues to be up to ten times stiffer than normal tissue [18]. ECM stiffness has an impact on the phenotype of malignant cells which is an important parameter often observed in cell-based experiments and drug discovery research to clarify biological pathways. Although ex vivo experiments have begun to explore the relationship between ECM stiffness, tumor stiffness, and drug activity, a mechanistic-biological link between stiffness and chemotherapy response in RMS has not yet been identified [19]. Although tumor stiffness can be monitored in vivo, it is hard to generate tumor with defined stiffness in animal models. The synthesis of hydrogels and their application in tissue engineering have made it possible to develop 3D cell cultures similar to the tumor microenvironment [20, 21]. The mechanical properties provided by these hydrogels regulate intracellular signaling pathways, and studies similar to natural tissues can be completed with such approaches [22]. Thus, we developed a method to assess the autophagy based on the matrix stiffness in ERMS cells. We generated cell-laden GelMA hydrogels and measured autophagy using immunocytochemistry techniques in conjunction with confocal microscopy (*see* Fig. 1). This approach represents a novel method to evaluate the role of autophagy in embryonal rhabdomyosarcoma and other cancer cells.

2 Materials

2.1 2D Cell Culture

1. Human rhabdoid cell line (A204) (ATCC HTB-82™, Rockville, MD, USA).
2. Mouse muscle cell line (C2C12) (ATCC CRL-1772™ Rockville, MD, USA).
3. Dulbecco's modified Eagle's medium (DMEM) (PAN BIOTECH™ GmbH, Aidenbach, Germany).
4. Penicillin/streptomycin (P/S) (PAN BIOTECH™ GmbH, Aidenbach, Germany).
5. Dulbecco's phosphate buffered saline (DPBS) (PAN BIOTECH™ GmbH, Aidenbach, Germany).
6. Fetal bovine serum (FBS) (Biowest, Nuaillé, France).
7. Trypsin–EDTA (PAN BIOTECH™ GmbH, Aidenbach, Germany).

2.2 3D Cell Culture and Treatments

1. Gelatin methacrylate (GelMA) (CELLINK, Boston, MA, USA).
2. Lithium phenyl-2,4,6-trimethylbenzoylphosphinate (LAP) (Sigma, St. Louis, MO, USA).
3. 96-well and 48-well plates (Greiner BIO ONE, Fisher Scientific, Illkirch, France).
4. Hemocytometer, Counting chambers, Imp. Neubauer (VWR, Radnor, PA, USA).
5. Trypan blue solution 0.4% (Thermo Fisher Scientific, Wilmington, DE, USA).

2.3 3D Cell Culture Staining

1. Calcein, AM, cell-permeant dye (ThermoFisher Scientific, Wilmington, DE, USA).
2. Propium iodide solution (ThermoFisher Scientific, Wilmington, DE, USA).
3. Tetramethylrhodamine methyl ester perchlorate (TMRM) (Sigma Aldrich, St. Louis, MO, USA).
4. Hoechst 33342 (Hellobio, Princeton, NJ, USA).
5. DPBS (PAN BIOTECH™ GmbH, Aidenbach, Germany).

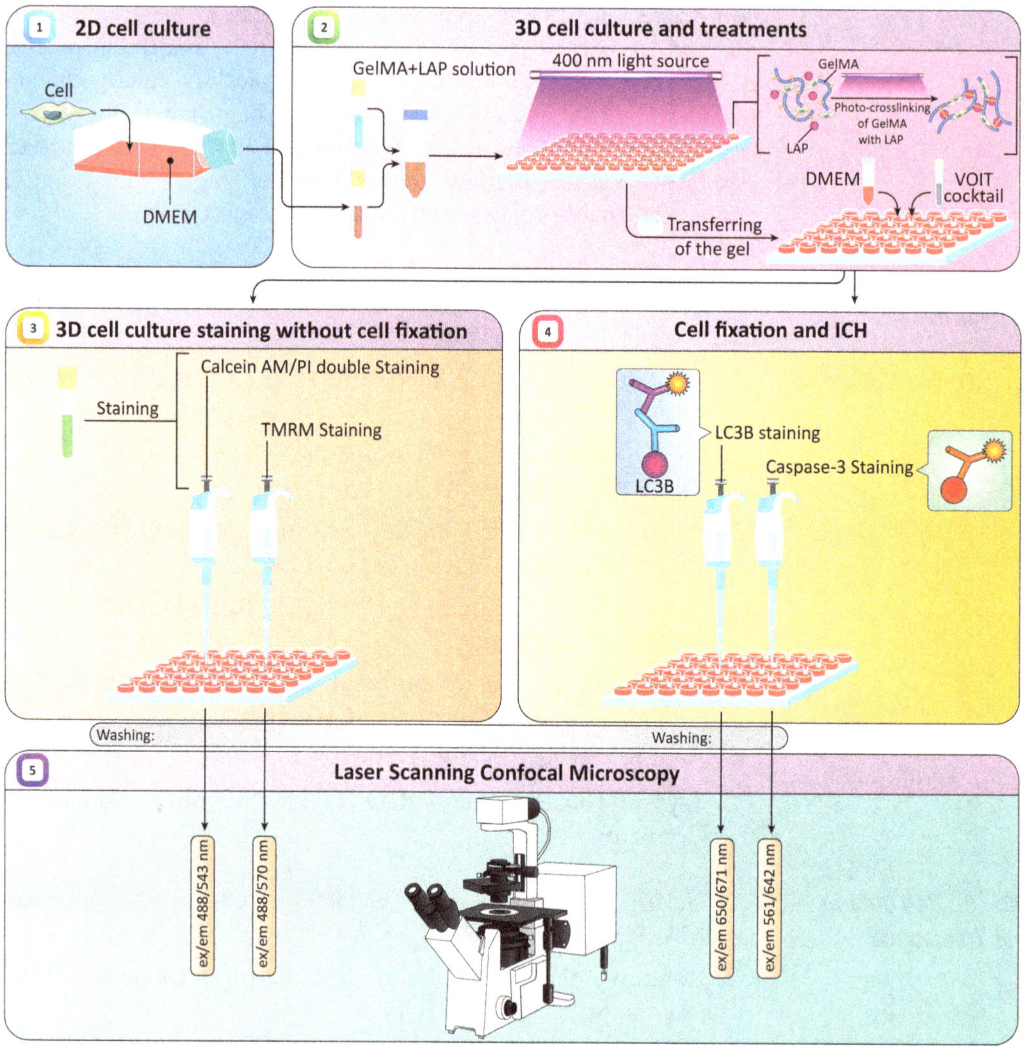

Fig. 1 Workflow

6. DMEM medium (PAN BIOTECH™ GmbH, Aidenbach, Germany).
7. Confocal dish (WVR, Radnor, PA, USA).

2.4 Immunocytochemistry (ICC)

1. Paraformaldehyde powder (Sigma–Aldrich Inc. St. Louis, MO, USA).
2. Triton X-100 (Sigma–Aldrich Inc. St. Louis, MO, USA).
3. Bovine serum albumin powder (BSA) (Sigma–Aldrich Inc. St. Louis, MO, USA).
8. DPBS (PAN BIOTECH™ GmbH, Aidenbach, Germany).
4. Primary Antibody LC3B, (D11) XP® Rabbit mAb (Cell Signaling Technology, Danvers, MA, USA).

5. Seconder Antibody Alexa Fluor® 488 AffiniPure Donkey Anti-Rabbit IgG (Jackson Immunoresearch).
6. Primary antibody cleaved Caspase-3 (Asp175) (D3E9) Rabbit mAb (Alexa Fluor® 647 Conjugate) (Cell Signaling Technology, Danvers, MA, USA).
7. Hoechst 33342 (Hellobio, Princeton, NJ, USA).

3 Methods

3.1 2D Cell Culture

1. Prepare complete growth medium including DMEM supplemented with 10% FBS and 1% P/S and store at 4 °C.
2. Obtain A204 and C2C12 cells. Culture both A204 and C2C12 cells with complete medium at 37 °C in a 5% humidified CO_2 atmosphere.
3. After reaching confluency about 75%, passage cells by removing the medium. Then wash the cells with DPBS. Add 1 mL trypsin–EDTA solution and incubate for 3 min at 37 °C (*see* **Note 1**). Add complete medium and pipette to avoid any cell clumps. Collect the cell suspension and transfer to the falcon tube, and centrifuge at 300 rpm for 5 min. Remove the medium and replace it with fresh medium. Passage the cells 1–3.

3.2 3D Cell Culture

1. Obtain GelMA and LAP. Prepare 5% w/v and 10% w/v GelMA solution in 2 mL DPBS at 50 °C under slow stirring in a water bath. Add 0.5% w/v LAP to GelMA solution and stir for a while. Cover the reaction tube with a piece of aluminum foil.
2. Filter the GelMA+LAP mixture via 0.22 um filter inside the cabinet. Transfer it into a new tube and cover. Keep it at 37 °C water bath until use.
3. Obtain cell pellets by trypsin–EDTA treatment. Count the cell pellets with hemocytometer and trypan blue with standard protocol. Dilute the cell pellets as 1×10^6 cells per mL of GelMA solution. Transfer the diluted cell pellet into 2 mL Eppendorf tubes.
4. Mix cell pellet with GelMA+LAP solution and pipette slowly. Transfer 50 μL of this solution into the wells of 96-well plate. After dividing this mixture, close the lid of the 96-well plate and apply 400 nm light source for 1 min for photo-crosslinking of GelMA with LAP (*see* **Note 2**).
5. After crosslinking, transfer cell-laden gels into the wells of the 48-well plate and label them (*see* **Note 3**). Fill each well of 48-well plate with 250 μL complete DMEM. Grow the cells at

37 °C, in a 5% humidified CO_2 atmosphere overnight before drug treatment.

3.3 3D Cell Culture Staining Without Cell Fixation

3.3.1 Calcein AM/PI Double Staining

1. Prepare Calcein AM dye and prepare a stock solution of Calcein AM in DMSO at 1 mg/mL. The molarity of this stock solution is 1 mM. This stock solution can be stored for short periods of time at −20 °C.
2. Prepare 1 μM working solution of Calcein AM from 1 mM stock solution in DPBS. This working solution should be prepared freshly.
3. Remove the media and wash the cell-laden gels with DPBS several times.
4. Add Calcein AM working solution just enough to cover the gels. Use aluminum foil to cover the plates.
5. Incubate the cells with Calcein AM working solution for 30 min at RT.
6. Remove the Calcein AM solution and wash with DPBS.
7. Obtain 1 mg/mL PI solution. The molarity of this solution is 1.5 mM. Store this solution at 4° C.
8. Prepare 500 nM working solution of PI in DPBS from stock solution.
9. After Calcein AM staining, remove the DPBS and incubate the cell-laden gels with PI solution for 5 min in dark at RT.
10. Discard the PI solution and wash the gels three times with DPBS with 5-min intervals between each wash (*see* **Note 4**).
11. Carefully place the gels on confocal dish and add small amount of DPBS to prevent drying. Observe the cells in laser scanning confocal microscopy. Adjust the parameters for Calcein AM (ex/em 494/517 nm) and PI (ex/em 535/617 nm). Record the images in Z-stack format with 10X objective.

3.3.2 TMRM Staining

1. Obtain TMRM powder. Prepare 5 mg/mL stock solution of TMRM corresponding to 10 mM main stock solution.
2. Dilute 10 mM stock solution into 100 μM stock solution. This stock solution is 1000X of working solution. Both solutions can be stored at −20° C.
3. Prepare working solution by diluting 1:1000 with DPBS. The working concentration for TMRM is 100 nM. This solution should be prepared freshly.
4. Obtain the treated and control cell-laden gels. Remove the media and wash with DPBS.
5. Add 1X TMRM working solution enough to cover the gels. Cover the plate with aluminum foil and incubate at 37° C for 30 min.

6. Remove the TMRM solution and wash the gels three times with DPBS.
7. Carefully place the gels on the confocal dish and add small pieces of DPBS to prevent drying. Adjust the wavelength parameters of TMRM (ex/em 488/570 nm) and objective to 10X. Start imaging in Z-stack format.

3.4 Cell Fixation and ICC

3.4.1 LC3B Staining

1. Obtain paraformaldehyde, BSA, Triton-X-100, and antibody solutions.
2. Prepare 3.7% formaldehyde solution from powder paraformaldehyde with standard protocol in DPBS. This solution can be stored at −20° C.
3. Prepare blocking and antibody dilution solutions from DPBS, BSA, and Triton-X-100. Blocking solution contains 5% BSA + 0.3% Triton-X in DPBS. Antibody dilution buffer includes 1% BSA and 0.3% Triton-X in DPBS. These solutions can be stored at +4° C.
4. Fix cells with 3.7% formaldehyde solution at RT for 40 min.
5. Discard the formaldehyde solution from wells and wash wells with DPBS 3 times. Allow cells to be washed properly by incubating them for 5 min with DPBS in each wash.
6. Add blocking solution enough to cover the gels. Incubate cells with blocking solution for 2 h at RT.
7. Prepare LC3B primary antibody solution using (D11) XP® Rabbit mAb in antibody dilution buffer with a 1:300 dilution. This solution should be prepared freshly (*see* **Note 5**).
8. Incubate the cells with primary antibody solution at 4 °C overnight. Add antibody solution enough to cover the gels.
9. Remove solutions from wells and wash with DPBS three times at 5-min intervals between each wash.
10. Prepare secondary antibody solution containing Alexa Fluor® 488 AffiniPure Donkey Anti-Rabbit IgG in antibody dilution buffer at a 1:300 dilution ratio.
11. Add secondary antibody solution just enough to cover the gels and incubate with seconder antibody solution for 4 h in dark at RT.
12. Discard secondary antibody solution of Alexa Fluor® 488 AffiniPure Donkey Anti-Rabbit IgG and incubate for 5 min during each wash.
13. Obtain Hoechst 33342 (*see* **Note 5**). Prepare 1 mg/mL Hoechst solution in DMSO. This stock can be stored at −20 °C for months. Prepare working solution of Hoechst by diluting stock into 1 μg/mL with DPBS.
14. Add Hoechst working solution to wells enough to cover the gels and incubate for 30 min at RT in dark.

15. Remove Hoechst solution and wash 3 times with DPBS and incubate for 5 min during each wash.
16. Adjust the filters of confocal microscopy for visualization. For LC3B, Alexa Fluor® 488 AffiniPure Donkey Anti-Rabbit IgG (ex/em 495/519 nm) will be imaged with defined wavelengths.

3.4.2 Caspase-3 Staining

1. Perform cell fixation as indicated above.
2. Use the same antibody dilution buffer.
3. Obtain cleaved Caspase-3 (Asp175) (D3E9) Rabbit mAb (Alexa Fluor® 647 Conjugate). This antibody is conjugated with florescence marker and does not require seconder antibody for labeling.
4. Prepare cleaved Caspase-3 primary antibody solution using cleaved (Asp175) (D3E9) Rabbit mAb (Alexa Fluor® 647 Conjugate) in antibody dilution buffer with a 1:300 dilution. This solution should be prepared freshly.
5. Incubate the cells with primary antibody solutions at 4 °C overnight. Add antibody solutions enough to cover the gels. Cover the plate with aluminum foil.
6. Remove solutions from wells and wash with DPBS for 3 times at 5-min intervals between each wash.
7. Adjust the filters of confocal microscopy for visualization of cleaved Caspase-3 (Asp175) (D3E9) Rabbit mAb (Alexa Fluor® 647 Conjugate) (ex/em 650/671 nm).

3.5 Laser Scanning Confocal Microscopy

1. Turn on the device and program (*see* **Note 6**).
2. In the program, adjust the parameters including objective (*see* **Note 7**) and lasers (Table 1).
3. First place the gel containing confocal dish and observe the cells with one laser channel by coarse and fine adjustment. Adjust acquisition parameters including laser intensity, detector gain, and pinhole. Add another channel when needed for multiple imaging.
4. Create a Z-Stack method. Start imaging the slices.
5. Process the data.

3.6 Interpretation of the Results

The viability of the cells was first studied with Calcein AM/PI double staining. This allows us to visualize the effect of matrix stiffness on the viability of the cells. Figure 2 shows the result of Calcein AM/PI double staining of A204 cells grown for 48 h on 5-GelMA and 10-GelMA matrices. Figure 2 also represents the results of TMRM staining. TMRM indicates the mitochondrial

Table 1
Laser excitation and emission wavelengths for florescence indicators

Indicator	Excitation	Emission
Calcein AM	488	543
PI	561	642
TMRM	561	642
Alexa Fluor™ 488	488	543
Alexa Fluor™ 647	561	642
Hoechst 33342	405	642

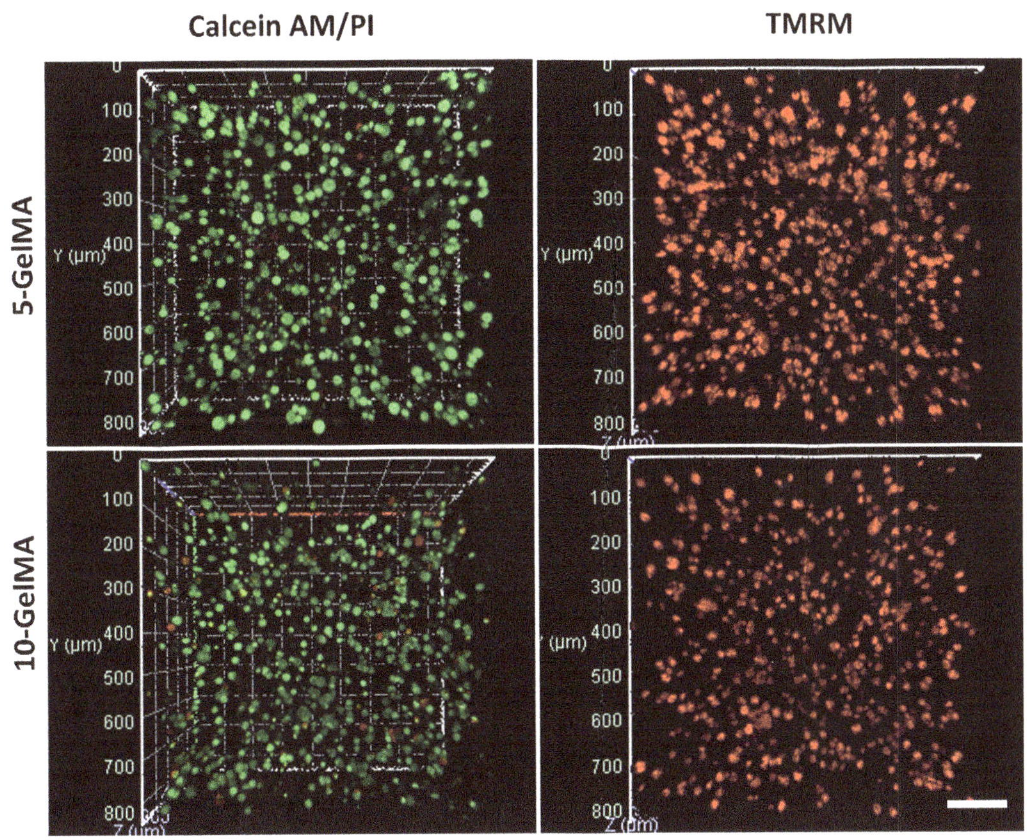

Fig. 2 Confocal images of A204 cells seeded on 5-GelMA and 10-GelMA matrices for 48 h stained with Calcein AM/PI double staining (left) and TMRM (right). The images are taken via 10X objective and represented as Z-Stack format. Scale bar: 100 μm

membrane potential of the cells, which has direct relationship with the apoptosis levels of the cells.

For apoptosis, the level of cleaved Caspase-3 was also imaged with ICC technique. Figure 3 shows the confocal images of A204

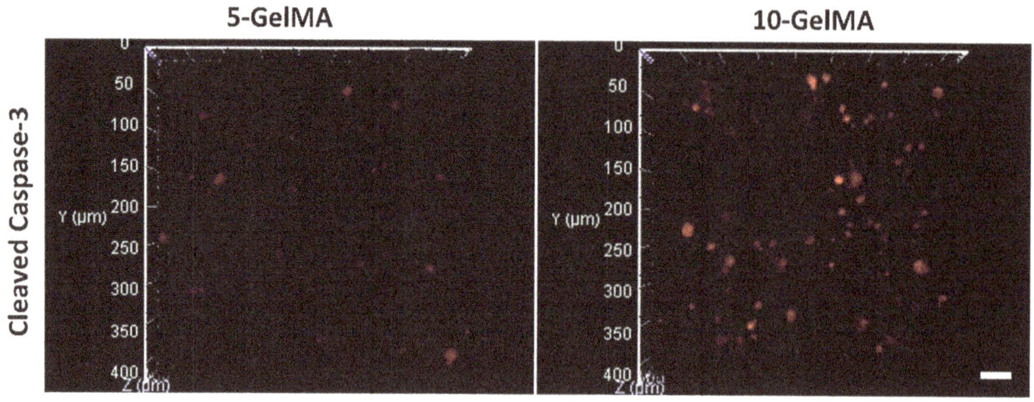

Fig. 3 Confocal images of A204 cells seeded on 5-GelMA and 10-GelMA matrices for 48 h. Cells were fixed and stained for cleaved Caspase-3 with ICC. The images are taken via 20X objective and represented as Z-Stack format. Scale bar: 50 μm

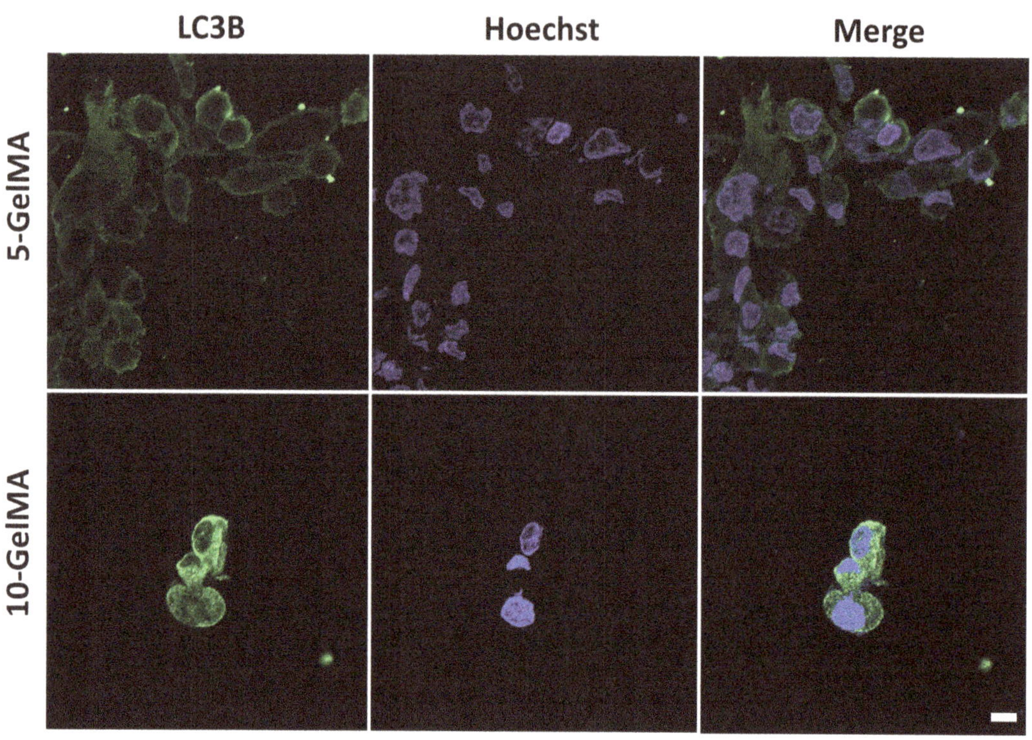

Fig. 4 Confocal images of A204 cells after ICC staining. Cells were seeded on 5-GelMA and 10-GelMA matrices and grown for 48 h. The images were taken with 63X objective with oil. Scale bar: 10 μm

cells seeded on 5-GelMA and 10-GelMA matrices and grown for 48 h.

In the following section, autophagosome was studied. The LC3B puncta and their altered levels were imaged with ICC, followed by nuclear staining. Figure 4 shows ICC of A204 cells seeded on 5-GelMA and 10-GelMA matrices.

4 Notes

1. The amount of the trypsin–EDTA and incubation time are given for the cells seeded to the T25 flask. The amount and time can change according to the flask size.
2. GelMA+LAP mixture should be kept in dark in order to prevent pre-gelation.
3. Once the gelation is complete, the transparent solution turns into blurry gel. These gels are brittle and should be carefully transferred from the 96 wells to the 48 wells filled with the medium. If the gels are dried before transferring, small drop of DMEM medium can be added to keep the gels wet.
4. Before each imaging, the gels should be kept in DPBS. For stained samples, the plates are covered with aluminum foil to prevent photobleaching under light. The imaging should be completed at the same day when staining and ICC protocols are completed.
5. In this study, we used Hoechst 33342 dye. This dye is cell-permeant nuclear counterstain. Instead of this one, DAPI can also be used. DAPI can be suitable for fixed cells, although Hoechst can stain both live and dead cells.
6. As scanning laser confocal microscopy, we used LSM 710, Axio Observer (Carl Zeiss, Oberkochen, Germany), equipped with Zeiss Zen software (version 2010). For each microscopy, the given parameters should be altered to get better images.
7. For 10X imaging, EC Plan-Neofluar 10x/0.30 M27 objective was set. For 63X imaging, we used oil-based objective Plan-Apochromat 63x/1.40 Oil DIC M27.

Acknowledgments

SG (Saeid Ghavami) and AZ (Ali Zarrabi) conceptualized and designed the study. SS (Serap Sezen) and SA (Sevin Adiguzel) performed the experiments. SG and SS analyzed the data. SS and A Kh (Arezoo Khosravi) wrote the manuscript. SS, SG, AZ, SA, Joseph W. Gordon (JWG), Atefeh Zarepour (AtZ), and AKh reviewed and edited the manuscript. AZ funded and SG supervised the experiments.

Funding

This study was funded in part by URGP (University of Manitoba) (SG) and 1002-short term R&D funding program of The Scientific and Technological Research Council of Türkiye (TÜBİTAK, Project No: 221Z107) (AZ).

References

1. Ramadan F, Fahs A, Ghayad SE, Saab R (2020) Signaling pathways in Rhabdomyosarcoma invasion and metastasis. Cancer Metastasis Rev 39(1):287–301. Available from: https://doi.org/10.1007/s10555-020-09860-3
2. Zarrabi A, Perrin D, Kavoosi M, Sommer M, Sezen S, Mehrbod P et al (2023) Rhabdomyosarcoma: current therapy, challenges, and future approaches to treatment strategies. Cancers (Basel) 15(21):5269
3. Gonzalez-Molina J, Kirchhof KM, Rathod B, Moyano-Galceran L, Calvo-Noriega M, Kokaraki G et al (2022) Mechanical confinement and DDR1 signaling synergize to regulate collagen-induced apoptosis in Rhabdomyosarcoma cells. Adv Sci 9(28):2202552. Available from: https://doi.org/10.1002/advs.202202552
4. Liang R, Song G (2023) Matrix stiffness-driven cancer progression and the targeted therapeutic strategy. Mechanobiol Med 1(2):100013. Available from: https://www.sciencedirect.com/science/article/pii/S294990702300013X
5. Liang P, Wang B (2024) An autophagy-independent role of ULK1/ULK2 in mechanotransduction and breast cancer cell migration. Autophagy 1:1–2. Available from: https://doi.org/10.1080/15548627.2023.2300916
6. Gong T, Wu D, Pan H, Sun Z, Yao X, Wang D et al (2023) Biomimetic microenvironmental stiffness boosts stemness of pancreatic ductal adenocarcinoma via augmented autophagy. ACS Biomater Sci Eng 9(9):5347–5360. Available from: https://doi.org/10.1021/acsbiomaterials.3c00487
7. Behrooz AB, Cordani M, Donadelli M, Ghavami S (2023) Metastatic outgrowth via the two-way interplay of autophagy and metabolism. Biochim Biophys Acta (BBA)-Molecular Basis Dis 1870:166824
8. Siapoush S, Rezaei R, Alavifard H, Hatami B, Zali MR, Vosough M et al (2023) Therapeutic implications of targeting autophagy and TGF-β crosstalk for the treatment of liver fibrosis. Life Sci 329:121894
9. Alizadeh J, Kochan MM, Stewart VD, Drewnik DA, Hannila SS, Ghavami S (2021) Inhibition of autophagy flux promotes secretion of chondroitin sulfate proteoglycans in primary rat astrocytes. Mol Neurobiol 58(12):6077–6091
10. Parzych KR, Klionsky DJ (2014) An overview of autophagy: morphology, mechanism, and regulation. Antioxid Redox Signal 20(3):460–473. Available from: https://pubmed.ncbi.nlm.nih.gov/23725295
11. Alizadeh J, Kavoosi M, Singh N, Lorzadeh S, Ravandi A, Kidane B et al (2023) Regulation of autophagy via carbohydrate and lipid metabolism in cancer. Cancers (Basel) 15(8):2195
12. Hajiahmadi S, Lorzadeh S, Iranpour R, Karima S, Rajabibazl M, Shahsavari Z et al (2023) Temozolomide, simvastatin and acetylshikonin combination induces mitochondrial-dependent apoptosis in GBM cells, which is regulated by autophagy. Biology (Basel) 12(2):302
13. Martelli A, Omrani M, Zarghooni M, Citi V, Brogi S, Calderone V et al (2022) New visions on natural products and cancer therapy: autophagy and related regulatory pathways. Cancers (Basel) 14(23):5839
14. Dalvand A, da Silva Rosa SC, Ghavami S, Marzban H (2022) Potential role of TGFB and autophagy in early crebellum development. Biochem Biophys Rep 32:101358
15. D'Agostino S, Tombolan L, Saggioro M, Frasson C, Rampazzo E, Pellegrini S et al (2021) Rhabdomyosarcoma cells produce their own extracellular matrix with minimal involvement of cancer-associated fibroblasts: a preliminary study. Front Oncol 10:600980. Available from: https://pubmed.ncbi.nlm.nih.gov/33585217
16. Li Y, Randriantsilefisoa R, Chen J, Cuellar-Camacho JL, Liang W, Li W (2020) Matrix stiffness regulates chemosensitivity, stemness characteristics, and autophagy in breast cancer cells. ACS Appl Bio Mater 3(7):4474–4485. Available from: https://doi.org/10.1021/acsabm.0c00448

17. Hupfer A, Brichkina A, Koeniger A, Keber C, Denkert C, Pfefferle P et al (2021) Matrix stiffness drives stromal autophagy and promotes formation of a protumorigenic niche. Proc Natl Acad Sci 118(40):e2105367118. Available from: https://doi.org/10.1073/pnas.2105367118
18. Lu P, Weaver VM, Werb Z (2012) The extracellular matrix: a dynamic niche in cancer progression. J Cell Biol 196(4):395–406. Available from: https://pubmed.ncbi.nlm.nih.gov/22351925
19. Schrader J, Gordon-Walker TT, Aucott RL, van Deemter M, Quaas A, Walsh S et al (2011) Matrix stiffness modulates proliferation, chemotherapeutic response, and dormancy in hepatocellular carcinoma cells. Hepatology 53(4):1192–1205. Available from: https://pubmed.ncbi.nlm.nih.gov/21442631
20. Annabi N, Tamayol A, Uquillas JA, Akbari M, Bertassoni LE, Cha C et al (2014) 25th anniversary article: rational design and applications of hydrogels in regenerative medicine. Adv Mater 26(1):85–124. Available from: https://doi.org/10.1002/adma.201303233
21. Stefanek E, Samiei E, Kavoosi M, Esmaeillou M, Roustai Geraylow K, Emami A et al (2021) A bioengineering method for modeling alveolar Rhabdomyosarcoma and assessing chemotherapy responses. MethodsX 8:101473. Available from: https://www.sciencedirect.com/science/article/pii/S2215016121002661
22. Tibbitt MW, Anseth KS (2009) Hydrogels as extracellular matrix mimics for 3D cell culture. Biotechnol Bioeng 103(4):655–663. Available from: https://pubmed.ncbi.nlm.nih.gov/19472329

Methods in Molecular Biology (2025) 2879: 289–300
DOI 10.1007/7651_2024_527

Published online: 28 February 2024

Principles of Hanging Drop Method (Spheroid Formation) in Cell Culture

Mohammad Rasouli, Fatemeh Safari, Mohammad Hossein Kanani, and Hiva Ahvati

Abstract

A type of three-dimensional (3D) cell culture models which is simple and easy is hanging drop method. The hanging drop method emerges as a pivotal technique with diverse applications in cancer research and cell biology. This method facilitates the formation of multicellular spheroids, providing a unique environment for studying cell behavior dynamics. The hanging drop method's theoretical underpinning relies on gravity-enforced self-assembly, allowing for cost-effective, reproducible 3D cell cultures with controlled spheroid sizes. The advantages of this approach include its efficiency in producing cellular heterogeneity, particularly in non-adherent 3D cultures, and its ability to create hypoxic spheroids, making it a suitable model for studying cancer. Moreover, the hanging drop method has proven valuable in investigating various aspects such as tissue structure, signaling pathways, immune activation of cancer cells, and notably, cell proliferation. Researchers have utilized the hanging drop method to explore the dynamics of cell proliferation, studying the effects of mesenchymal stem cells (MSC) secretome on cancer cells. The method's application involves co-culturing different cell lines, assessing spheroid formations, and quantifying their sizes over time. These studies have unveiled intricate cell behavior dynamics, demonstrating how the MSC secretome influences cancer cell growth and viability within a three-dimensional co-culture paradigm.

Key words Hanging drop, 3D cell cultures, Spheroid formation, Cell culture, Cell behavior dynamics

1 Introduction

While current studies have relied heavily on two-dimensional (2D) cell cultures, their inherent limitations like neglecting cell-microenvironment interactions and simplified nutrient gradients hinder accurate portrayal of tumor complexities. Embracing three-dimensional (3D) cell culture models, which better mimic the intricate architecture and communication networks of real tumors [1], is therefore paramount to unraveling autophagy's true influence and ultimately designing precise, personalized cancer therapies (Fig. 1). These 3D cultures provide a more similar environment to *in vivo*, closely mimicking tumor physiology and upregulation of tumor-relevant pathways [2]. The capacity to reproduce the intricate relationships that exist between cancer cells and the stroma around them is one of the main advantages of 3D cell

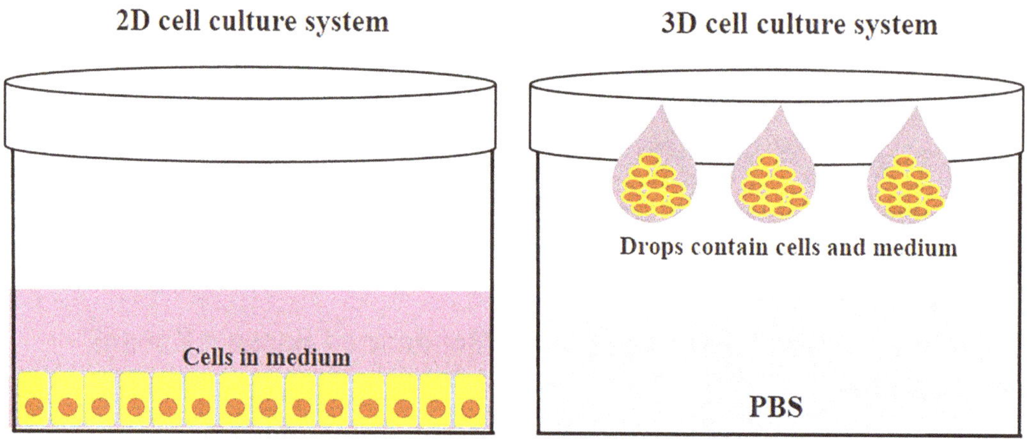

Fig. 1 Schematic models for 2D and 3D (hanging drop method) cell culture system

culture for cancer research. In addition to cancer cells, stromal cells, immune cells, and blood vessels are among the additional cell types seen in tumors. These models provide a practical, scalable, and morally sound substitute for animal models in preclinical research by bridging the gap between conventional 2D cell cultures and animal models. As the science develops, 3D cell cultures have the potential to be extremely important for comprehending cancer biology and expediting the creation of potent anticancer treatments [3].

There are various techniques for 3D cell cultures, which can be broadly classified into two categories: scaffold-based and scaffold-free techniques. Scaffold-based 3D cell culture is a method that relies on a scaffold to provide physical support to cells, allowing them to aggregate, proliferate, and migrate. The scaffolds must possess specific properties to support cell adhesion, proliferation, and activation, such as mechanical properties, degradation kinetics, pore distribution, exposed surface area, and porosity. Various types of scaffolds, including those derived from non-animal sources, have been developed for 3D cell culture applications, with potential implications in cancer research and tissue engineering [4]. Although scaffold-based systems make it possible to manipulate spheroid cultures more precisely in terms of size, shape, and position, they frequently give rise to questions regarding biological compatibility and application in medicine because scaffolds are usually composed of non-human materials [5].

Scaffold-free 3D cell culture is a method of cell culture where cells are grown and organized into three-dimensional structures without the need of an external scaffold. The cells generate 3D tissue-like structures by self-assembling and interacting with nearby cells, as opposed to being immersed within a supporting matrix [3]. The hanging drop method is a 3D cell culture technique used to form spheroids, also known as micro tumors, which are aggregates of cells grown in suspension or embedded in a 3D matrix. The method involves placing a small droplet of cell suspension on the lid of a special culture plate, which is then inverted over a reservoir of media, allowing the cells to hang in the droplet (Fig. 2). Significance of the technique is studying of cell proliferation and motility. The hanging drop method has been shown to be useful for studying cell function in an avascular tumor microenvironment, drug therapies, tumor angiogenesis, and tumor-immune cell interactions [6].

The formation of a 3D milieu that encourages the self-assembly of cells into multicellular spheroids is the theoretical underpinning of cell proliferation in the hanging drop method. Small quantities of a cell suspension are pipetted onto the inside lid of a tissue culture plate, and the drops stay attached to the lid due to surface tension. This approach uses gravity-enforced self-assembly to form spheroids. A single spheroid forms as a result of the cells settling and congregating at the bottom of the drop due to gravity. The hanging drop approach modifies the cell density in each drop to alter spheroid size and cellular composition [7]. The ability of the holding layer to adhere, which is impacted by surface geometry and hydrophilicity, is crucial in preserving the droplet's spherical shape and preventing the hanging drop from leaking [8].

Numerous benefits of this approach include cost effectiveness, spheroid size control, and the ability to co-culture and create spheroids from a variety of cell types. Furthermore, it has been observed that the hanging drop approach produces 100% reproducible 3D cell cultures [9]. The serious concerns surrounding the impacts from 3D structure creation are decreased because this technique is based on cells' natural desire to aggregate to one another rather than relying on the matrices or scaffolds provided. However, the restricted volume of the cell suspension is the indisputable disadvantage of the hanging drop technique. The upside-down surface can only hold up to 50 μL of suspension, including the testing medium, unless there is dripping because there is insufficient surface tension to keep liquids connected. A further drawback is the challenge of modifying the culture media while the spheroids are being cultivated without upsetting them [10]. Three steps are involved in the production of spheroids: (1) integrin-ECM binding forms loose cell aggregation; (2) cadherin expression and accumulation are delayed; and (3) homophilic cadherin-cadherin interactions generate compact spheroids [11].

The duration required for spheroid formation often varies depending on the type of cell, and maintaining 3D spheroids in suspended droplets can present certain difficulties [6]. For instance, in a study involving pMCF7 cells, spheroids were imaged starting from Day 1 and followed up until Day 7 in culture [12]. In another study focusing on ovarian cancer cell lines (A2780 and OVCAR3),

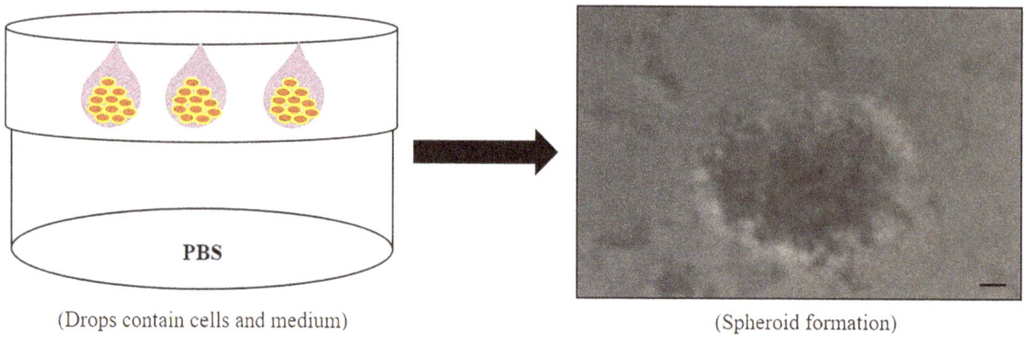

Fig. 2 Schematic model of spheroids formation by using hanging drop technique. A spheroid was shown (magnification: 20, scale bar: 100 μm)

spheroids were observed to form within Day 2 [13]. In separate investigations examining the inhibition of pancreatic cancer cells (Pancl [14–16] and MiaPaca2 [17]), prostate cancer cells (LNCaP) [18], colon cancer cells (HT-29) [19], and breast cancer cells (MDA-MB-231) [20], invasion by hAMSCs secretome, spheroid formation was noted on Day 3.

It is essential to create hanging drops that are both successful and reliable in order to create a steady culture medium for the cells. Dripping, flatting, and even non-response are examples of hanging drop failures that can affect other drops or lower the testing samples needed for additional research [8]. Methyl cellulose is an effective thickening agent that preserves cells that are suspended from shear stresses and facilitates the formation of homogenous 3D cell spheroids. It has been demonstrated that adding methyl cellulose to the hanging drop cultures increases spheroid circularity and compactness, making it an important tool for 3D cell culture research. Methylcellulose facilitates the creation of cell spheroids, which more closely resemble the extracellular geometry found *in vivo* and offer a more physiologically accurate model for researching drug reactions and cell behavior. Moreover, methyl cellulose is an effective thickening agent that preserves cells that are suspended from shear stresses and facilitates the formation of homogenous 3D cell spheroids [21].

The hanging drop method has various applications, including tissue structure, signaling pathways, and immune activation of cancer cells. In a non-adherent 3D culture approach, single tumor cells can create multi-cellular tumor spheroids together with other cell types, which is more efficient in producing cellular heterogeneity. The microenvironment of spheroid cultivation influences the morphologies of these tumor spheroids. Morphologies can create star-shaped spheroids, elongated ovals, entangled bundles, or aggregated circles depending on the culture microenvironment circumstances. Because the spheroid form contains a limited concentration of oxygen at its core, it is the most appropriate model for studying cancer. The main benefit of spheroid culture is that spheroids are hypoxic [9]. Furthermore, cell migration has been studied using the hanging drop approach. By smoothly dripping from the culture system, it has been employed to promote long-term spheroid culture and subsequent examination of tumor spheroids. This technique is helpful for researching cell migration and invasion because it allows the examination of tumor spheroid migration and invasion in an extracellular matrix (ECM) niche [18, 22].

In addition to the previously mentioned applications of the hanging drop method, another significant area of study involves cell proliferation. Researchers have extensively utilized the hanging drop method in investigating cell proliferation dynamics. Specifically, their work focused on co-culturing mesenchymal stem cells (MSC) with cancer cells and exploring the effects of the MSC secretome on the growth and death of cancer cells [14–17, 19–20]. For instance, a study that was conducted recently focused on the application of the hanging drop method for investigating cell proliferation. The experimental design involved the separate culture of human amniotic mesenchymal stem cells (hAMSCs) and MiaPaca2 cells initially. Subsequently, both cell lines underwent co-culturing using 0.4 μm pore-sized transwell membranes at different time points. The primary objective was to examine the impact of hAMSCs secretome on cancer cells. To establish a 3D cell culture model and form spheroids, the hanging drop method was employed. The hAMSCs, reaching approximately 90% confluence, were trypsinized and counted before being subjected to the experiment. Both control groups (cells and medium) and experimental groups (cells with stem cells secretome or conditioned media) were maintained, with media changes occurring every two days. On the third day, the size and number of spheroid formations were assessed. Continuous monitoring of spheroid formation took place using a phase-contrast inverted microscope. Finally, the quantification of spheroid numbers was performed at 3- and 8-days post-seeding, and their diameters were determined using ImageJ software [17]. In these studies, researchers observed a significant decrease in both the number and size of spheroids within the sampled groups. This observed effect was attributed to the influence of the MSCs conditioned media, demonstrating a clear inhibition of cell proliferation and an increase in cell death. These findings show how the hanging drop method can be used to decipher complex cell behavior dynamics and offer important new information about the regulatory function of MSC secretome on cancer cell growth and viability in a three-dimensional co-culture paradigm. In contrast, in another study, it has been demonstrated that a common pattern was observed in hanging drop-based spheroids, which predicts a first phase of diameter reduction (up to day

10) during which the cells compact. This is followed by a second period of growth (from day 10 to 14) during which the cells proliferate. Since cells must first compact during the "spheroidization time" and only then do strong cell-cell adhesions develop, they proliferate, increasing the size of the spheroid. This is a common phenomenon in spheroids formed using the hanging drop method, as was previously described [23].

In a study, researchers by developing mathematical models have discovered that the transient structure of spheroids is independent of their initial size, and the limiting structure can be unrelated to the seeding density. The standard experimental approach often involves comparing spheroid size over time, but the authors argue that assessing spheroid structure as a function of overall size yields more robust results, being less sensitive to variability in spheroid size. The experimental observations are based on two melanoma cell lines, but the proposed modeling framework is suggested to be applicable across a broad range of spheroid culture conditions and cell lines [24]. By comparing the average size of 10–20 (or more) hanging drops between treatment groups, compaction can be evaluated. For image analysis, the program ImageJ is often used. Its thresholds each image and then converts it to Binary Mode. Particle analysis is then used to reveal the entire image area in pixels, which is easily translated to square microns. This method makes it easier to compare the sizes of the sample and control cells [6] and cell viability [5], and growth can be assessed by monitoring the number and diameter of spheroids. The spheroid diameter offers insights into compactness, aggregation, and sphericity. For example, estimating the mean diameter using image analysis functions like Feret's diameter helps evaluate size and shape parameters. Additionally, the sphericity index, indicating how closely the spheroid shape resembles a spherical geometry, can be calculated using the circularity parameter. These measurements are crucial for understanding the impact [1].

To address the drawbacks of the conventional hanging drop technique in 3D cell culture, the PANDA System, Flipped Well-Plate Hanging-Drop Technique, and High-throughput 3D spheroid culture have been developed. The old approach can be labor-intensive and costly because it requires specialized equipment and frequently calls for manually loading cell suspension. Furthermore, these sophisticated procedures have addressed issues such as the hanging drops' evaporation and the requirement for continuous drop creation [12, 25, 26].

PANDA System: The Pressure-Assisted Network for Droplet Accumulation (PANDA) system is a user-friendly chip that consists of a hanging drop cell culture component and a multiple channels fluidic system. It allows for accurate regulation over artificial niches, which in turn allows for the creation of microtissues with different sizes. The PANDA system has produced several consistent 3D glomerulus-like microtissues made up of kidney podocytes and mesenchymal stem cells. Its affordability and speed make it a useful platform for creating microtissues with cellular heterogeneity and tunable 3D environments. This adaptability makes the PANDA system a valuable tool for a wide range of biomedical research applications [12]. Flipped Well-Plate Hanging-Drop Technique addresses the challenge of evaporation of liquid in hanging drops by using conventional 96-well plates and makes 3D tumor growth easier to understand and manage. Preparing the hanging drop is made simple by rotating the plate. The well-plate flip (WPF) technique is a method that produces changeable media droplets that can be used to study cell biology processes, including cancer. Using human colorectal cancer cells (HCT116), researchers were able to produce tremendous scaffold-free 3D tumor spheroids on single-well plates that were larger than 1.5 mm in diameter, demonstrating the method's adaptability. With its versatility, the WPF method provides a useful and effective instrument for fundamental and applied research in the engineering and biology fields [25]. Moreover, high-throughput 3D spheroid culture (384-well hanging drop culture plates) are introduced for high-throughput drug testing and 3D spheroid cultivation. It streamlines the procedure to the same level of simplicity as traditional 2D cultures using currently available high-throughput screening equipment. Compared to the traditional hanging drop method, high-throughput 3D spheroid culture provides a more efficient way to cultivate 3D cells. It is appropriate for high-throughput screening and manufacturing because it allows for the simultaneous creation of many cellular clusters [26].

In fact, the hanging drop method stands out as a robust and versatile approach in the realm of 3D cell cultures, contributing significantly to our understanding of cancer biology and cell behavior dynamics. Its theoretical foundation, grounded in gravity-enforced self-assembly, offers a cost-effective means to produce reproducible 3D cell cultures with controlled spheroid sizes. The method's diverse applications, ranging from tissue structure and signaling pathways to immune activation of cancer cells, highlight its adaptability and relevance in various research avenues. Particularly in non-adherent 3D cultures, the hanging drop method efficiently fosters the creation of multicellular tumor spheroids, providing a unique environment that mirrors the hypoxic conditions of cancer, thereby making it an apt model for cancer studies. Moreover, the hanging drop method has proven instrumental in investigating cell proliferation dynamics, as evidenced by studies exploring the impact of mesenchymal stem cells (MSC) secretome on cancer cells. These studies reveal the intricate interplay between different cell lines in a three-dimensional co-culture paradigm, shedding light on the regulatory role of the MSC secretome in cancer cell growth and viability. The integration of mathematical models to analyze spheroid transient structures and the delineation of practical methodologies for evaluating spheroid characteristics further underscore the method's depth and applicability. By considering factors such as compaction, cell viability, and growth, researchers gain valuable insights into the impact on spheroid size and shape parameters. In essence, the hanging drop method emerges as a crucial tool that transcends the limitations of conventional 2D cell cultures. Its contributions extend beyond the laboratory, offering a unique vantage point to unravel the complexities of cancer biology. As we delve deeper into this methodological landscape, it becomes evident that the hanging drop technique not only enhances our understanding of cellular dynamics but also holds immense potential for shaping the future of precise and personalized cancer therapies.

2 Materials

The necessary materials for performing cell culture by using hanging drop are as follows:

2.1. Cell culture dish.

2.2. PBS (phosphate-buffered saline).

2.3. Trypsin-EDTA (0.05% or 0.25%).

2.4. Cells culture medium.

2.5. FBS (fetal bovine serum: 10 %).

2.6. Antibiotics (penicillin/streptomycin: 50-100mg/mL) (*see* **Note 1**).

2.7. Incubator (at 37 °C, 5% CO2, 95% humidity).

These materials are necessary to establish an atmosphere that is conducive to the development of hanging drop cultures. The pipettor is required to deposit the cell suspension as hanging drops, and the tissue culture dish, PBS, and cover are employed to establish a hydration chamber. Depending on the kind of cells being employed, the procedure's precise details may change.

3 Methods

3.1 Establishing a Single Cell Drop

1. After adherent cell cultures reach 90% confluence (*see* **Note 2**), the monolayers should be washed with PBS. Once well drained, add 1 milliliter (for 60 mm plates) of trypsin-EDTA (*see* **Note 3**), and allow the cells to separate by incubating at 37°C. Add 2 mL of complete media to stop the trypsinization process, then carefully triturate the mixture with a 5-mL

pipette until the cells are in suspension. Fill a 15 mL conical tube with cells.

2. Centrifuge for 5 min at 4000 rpm after a brief vortex.
3. After discarding the supernatant, wash the pellet in 1 mL of full tissue culture media. After doing so, resuspend the cells in two milliliters of the whole tissue culture medium (*see* **Note 4**).
4. Utilizing an automated cell counter or hemacytometer, count the cells and set the concentration to 20×10^3 cells/mL (*see* **Notes 5** and **6**).

3.2 Formation of Hanging Drops

1. Five milliliters of PBS should be added to the bottom of a 60 mm tissue culture plate after the lid has been removed. This will serve as a room for hydration (*see* **Note 7**).
2. After inverting the lid, place 20 μL droplets onto the bottom of the lid. Make sure the droplets are spaced far enough apart to avoid coming into contact. You can put at least twenty drips in each dish.
3. Place the cover back on the bottom chamber filled with PBS and incubate (at 37°C, 5% CO2, 95% humidity). Additionally, the media altered every 2 days and the size and number of spheroids were determined after their formation (around 3 days). Eventually, spheroid formation was monitored with employing a phase contrast inverted microscope (INV100, BEL Engineering, Italy).
4. After taking photo, their diameters were determined using ImageJ software.
5. To apply total cell lysate (TCL) of spheroids into Western blot or quantitative real-time PCR (qRT-PCR), spheroids are collected and centrifuged for 5 min at 4000 rpm. The pellets are stored at –80 °C (*see* **Note 8**).

4 Notes

1. Preventing bacterial contamination in the culture is the goal of using antibiotics, such as penicillin and streptomycin, in cell culture. This is especially crucial when working with primary cells or cell lines, since bacterial contamination can have a negative impact on the health of the cultivated cells as well as the outcomes of the experiments. Antibiotics added to the culture media contribute to the sterility of the environment, lowering the possibility of bacterial contamination and overgrowth and protecting the integrity of the cell culture research.
2. When using the hanging drop method for 3D cell culture, adherent cell cultures should be grown to 90% confluence in

order to guarantee the development of compact and homogenous spheroids. This high confluence encourages cell aggregation into spheroids that resemble normal tissue architecture by having regular sizes and shapes.

3. To maintain cadherin activity, cells can be separated from tissue culture dishes using 0.05% trypsin and 2 mM calcium.
4. To get rid of any remaining trypsin and give the cells fresh growth medium, wash the cell pellet in one milliliter of whole tissue culture fluid and then resuscitate the cells in two milliliters of the entire tissue culture medium. By taking this precaution, trypsin exposure to the cells is prevented from lasting too long, which could compromise their viability and functionality. Additionally, it aids in the elimination of any last remnants of trypsin or FBS, which may obstruct subsequent uses or the creation of 3D spheroids.
5. To guarantee of hanging drop culture, it should be started with a known and constant quality of cells. This makes it possible to standardize the beginning cell population, which is essential for consistency and repeatability when creating 3D spheroids. This step is crucial for experimental control and interpretation since it affects the size and properties of the resultant spheroids as well as maintaining a particular cell concentration.
6. Before hanging drop development, cells can be labeled with fluorescent dyes that intercalate membranes.
7. The moisture from the PBS prevents the evaporation of media from the hanging drops, and the incubation conditions mimic the physiological environment, promoting the aggregation of cells into spheroids. This setup allows the drops to be refreshed by replacing a few microliters of media, maintaining their spherical shape and preventing evaporation due to surface tension.
8. Sheet sizes can be quantified or aggregates can be employed for biochemical or molecular analysis, mechanical testing, or biochemical testing, depending on the experiment.
9. Before hanging drop production, two different cell types can be combined in a 1:1 (or other) ratio and differently stained. These may include embryonic tissues, cancer and stromal cells, or genetically modified cells with particular sticky characteristics.

References

1. Amaral RL, Miranda M, Marcato PD, Swiech K (2017) Comparative analysis of 3D bladder tumor spheroids obtained by forced floating and hanging drop methods for drug screening. Front Physiol 8:605
2. Bingel C, Koeneke E, Ridinger J, Bittmann A, Sill M, Peterziel H, Wrobel JK, Rettig I, Milde T, Fernekorn U et al (2017) Three-dimensional tumor cell growth stimulates autophagic flux and recapitulates chemotherapy resistance. Cell Death Dis 8:e3013

3. Abuwatfa WH, Pitt WG, Husseini GA (2024) Scaffold-based 3D cell culture models in cancer research. J Biomed Sci 31:7
4. Carletti E, Motta A, Migliaresi C (2011) Scaffolds for tissue engineering and 3D cell culture. 3D cell culture. Methods Protoc 695:17–39
5. Vu B, Souza GR, Dengjel J (2021) Scaffold-free 3D cell culture of primary skin fibroblasts induces profound changes of the matrisome. Matrix Biol Plus 11:100066
6. Foty R (2011) A simple hanging drop cell culture protocol for generation of 3D spheroids. J Vis Exp:e2720. https://doi.org/10.3791/2720
7. Achilli T-M, Meyer J, Morgan JR (2012) Advances in the formation, use and understanding of multi-cellular spheroids. Expert Opin Biol Ther 12:1347–1360
8. Cho C-Y, Chiang T-H, Hsieh L-H, Yang W-Y, Hsu H-H, Yeh C-K, Huang C-C, Huang J-H (2020) Development of a novel hanging drop platform for engineering controllable 3D microenvironments. Front Cell Dev Biol 8:327
9. Ryu N-E, Lee S-H, Park H (2019) Spheroid culture system methods and applications for mesenchymal stem cells. Cells 8:1620
10. Velasco V, Shariati SA, Esfandyarpour R (2020) Microtechnology-based methods for organoid models. Microsyst Nanoeng 6:76
11. Lin R-Z, Chang H-Y (2008) Recent advances in three-dimensional multicellular spheroid culture for biomedical research. Biotechnol J Healthc Nutr Technol 3:1172–1184
12. Raghavan S, Mehta P, Horst EN, Ward MR, Rowley KR, Mehta G (2016) Comparative analysis of tumor spheroid generation techniques for differential in vitro drug toxicity. Oncotarget 7:16948
13. Raghavan S, Ward MR, Rowley KR, Wold RM, Takayama S, Buckanovich RJ, Mehta G (2015) Formation of stable small cell number three-dimensional ovarian cancer spheroids using hanging drop arrays for preclinical drug sensitivity assays. Gynecol Oncol 138:181–189
14. Lahiji MAS, Safari F (2022) Potential therapeutic effects of hAMSCs secretome on Panc1 pancreatic cancer cells through downregulation of SgK269, E-cadherin, vimentin, and snail expression. Biologicals 76:24–30. https://doi.org/10.1016/j.biologicals.2022.02.001
15. Safari F, Shafiee Nejad N, Aghaei Nejad A (2022) The inhibition of Panc1 cancer cells invasion by hAMSCs secretome through suppression of tyrosine phosphorylation of SGK223 (at Y411 site), c-Src (at Y416, Y530 sites), AKT activity, and JAK1/Stat3 signaling. Med Oncol 39:28. https://doi.org/10.1007/s12032-022-01649-4
16. Safari F, Dadvar F (2023) In vitro evaluation of autophagy and cell death induction in Panc1 pancreatic cancer by secretome of hAMSCs through downregulation of p-AKT/p-mTOR and upregulation of p-AMPK/ULK1 signal transduction pathways. Tissue Cell 84:102160
17. Rahmani Z, Safari F (2021) Evaluating the in vitro therapeutic effects of human amniotic mesenchymal stromal cells on MiaPaca2 pancreatic cancer cells using 2D and 3D cell culture model. Tissue Cell 68:101479. https://doi.org/10.1016/j.tice.2020.101479
18. Safari F, Shakery T, Sayadamin N (2021) Evaluating the effect of secretome of human amniotic mesenchymal stromal cells on apoptosis induction and epithelial-mesenchymal transition inhibition in LNCaP prostate cancer cells based on 2D and 3D cell culture models. Cell Biochem Funct 39:813–820. https://doi.org/10.1002/cbf.3654
19. Ebadi Zavieh S, Safari F (2022) The antitumor activity of hAMSCs secretome in HT-29 colon cancer cells through downregulation of EGFR/c-Src/IRTKS expression and p38/ERK1/2 phosphorylation. Cell Biochem Biophys 80:395–402. https://doi.org/10.1007/s12013-022-01066-4
20. Shakery T, Safari F (2023) Downregulation of Pinkbar/pAKT and MMP2/MMP9 expression in MDA-MB-231 breast cancer cells as potential targets in cancer therapy by hAMSCs Secretome. Cells Tissues Organs 212(2):155–163. https://doi.org/10.1159/000520370
21. Ware MJ, Colbert K, Keshishian V, Ho J, Corr SJ, Curley SA, Godin B (2016) Generation of homogenous three-dimensional pancreatic cancer cell spheroids using an improved hanging drop technique. Tissue Eng Part C Methods 22:312–321
22. Zhao L, Xiu J, Liu Y, Zhang T, Pan W, Zheng X, Zhang X (2019) A 3D printed hanging drop dripper for tumor spheroids analysis without recovery. Sci Rep 9:19717
23. Cavo M, Delle Cave D, D'Amone E, Gigli G, Lonardo E, Del Mercato LL (2020) A synergic approach to enhance long-term culture and manipulation of MiaPaCa-2 pancreatic cancer spheroids. Sci Rep 10:10192

24. Browning AP, Sharp JA, Murphy RJ, Gunasingh G, Lawson B, Burrage K, Haass NK, Simpson M (2021) Quantitative analysis of tumour spheroid structure. elife 10: e73020
25. Jeong Y, Tin A, Irudayaraj J (2022) Flipped well-plate hanging-drop technique for growing three-dimensional tumors. Front Bioeng Biotechnol 10:898699
26. Tung Y-C, Hsiao AY, Allen SG, Torisawa Y, Ho M, Takayama S (2011) High-throughput 3D spheroid culture and drug testing using a 384 hanging drop array. Analyst 136:473–478

Methods in Molecular Biology (2025) 2879: 301–313
DOI 10.1007/7651_2024_539

Published online: 19 April 2024

Protocol for Testing the Effects of ssDNA Aptamer in HeLa and MCF-7

Haregewoin Bezu Woldekidan, Zandile Nxumalo, Mutsa M. Takundwa, Adugna Abdi Woldesemayat, and Deepak B. Thimiri Govinda Raj

Abstract

Conventional approaches for treating tumors encompass chemotherapy, radiotherapy, and surgery. However, these methods come with their limitations when applied in clinical practice. Aptamers are often referred to as "chemical antibodies" and consist of short DNA or RNA molecules, designed to bind to a wide range of targets, including proteins or nucleic acid structures. They exhibit strong affinities and remarkable specificity for their target molecules, making them capable of functioning as therapeutic agents to directly impede tumor cell proliferation. This approach helps minimize the harm to normal cells, thus reducing toxicity through decreased side effects. Here we report the procedure to develop ssDNA aptamer and investigate its ability to inhibit cancer cell proliferation in HeLa and MCF-7 cancer cell lines.

Key words Aptamer, Cancer cell, Cytotoxic effect

1 Introduction

Intolerable side effects in cancer patients undergoing chemotherapy remain a significant clinical challenge. These side effects can range from mild to life-threatening, affecting vital organs like the heart, lungs, and brain and leading to issues such as drug resistance, carcinogenicity, and infertility. Traditional chemotherapy, a widely used cancer treatment, often lacks selectivity, resulting in immediate toxicity symptoms as well as chronic toxicity [1]. Western treatments are normally used to treat diseased tissues or organs. However, the treatments have failed to achieve the goal due to poor selectivity and resulting "off-target" side effects [2].

To advance cancer therapy, it is imperative to explore new biomarkers, targeted agents, and innovative delivery methods. Achieving tumor-targeted therapy necessitates the development of ligands that can bind directly to cancer cells. Aptamers emerge as a promising class of molecules, often described as antibody's chemical counterparts [3, 4]. For the last two decades, aptamers and the

Authors Haregewoin Bezu Woldekidan and Zandile Nxumalo have equally contributed to this chapter.

SELEX (systemic evolution of ligands by exponential enrichment) technology have garnered significant consideration in the field of life sciences. Aptamers are single-stranded DNA or RNA oligonucleotides, exhibit specific three-dimensional structures, and possess a remarkable ability to bind with high affinity to a diverse array of target molecules, which includes protein, dyes, and even microbial and animal cells or tissues. Aptamers achieve their specificity and high affinity through various molecular interactions. This capacity to interact with such a broad spectrum of targets makes aptamers a promising alternative to traditional antibodies [3, 5, 6]. Monoclonal antibodies (mAbs) are well established as powerful tools in modern medicine for both therapeutic and diagnostic purposes [7]. However, aptamers offer several advantages over traditional antibodies, including high chemical stability, efficiency, cost-effectiveness, and bulk production. They have excellent reproducibility and reliability and can be produced at low cost on a large scale [3, 4]. To mitigate systemic toxicity concerns, aptamers can be directly linked to drugs or utilized in conjunction with nanocarriers. Functionally, they can serve as antagonists, agonists, or targeting ligands [5, 8].

The potential of aptamers as therapeutic agents was demonstrated by Sullenger et al. in 1990, and since then, their utilization has been widely explored in various biomedical domains, including biomarker detection, analysis, imaging, and targeted therapy [9]. Aptamers are classified as free RNA and DNA aptamers, in the realm of cancer therapy, and each is suited to specific molecular targets that represent the disease's hallmark characteristics. As aptamers continue to evolve and expand their applications, they offer a promising platform for developing new diagnostic, therapeutic, and protective treatments across a wide range of diseases [3, 5].

The first aptamer drug, mutagen, received clinical approval several years ago, marking a significant milestone in the field. Currently, numerous aptamer-based agents are undergoing active clinical trials, demonstrating the growing interest in therapeutic potential. Beyond clinical trials, aptamers are also being explored as tumor-targeting ligands, showing promise in significantly improving the outcomes of cancer treatment in animal tumor models [10]. Consequently, numerous aptamer-drug conjugate (AptDC) models have been successfully developed in recent years and are currently undergoing extensive preclinical evaluation [11]. In the pursuit of enhancing therapeutic efficacy while minimizing side effects in cancer treatment, a variety of innovative approaches have been anticipated and studied. For example, a combination of doxorubicin (Dox) and curcumin has been shown to enhance anti-cancer activity, particularly in colorectal cancer cells (CRC) [1]. Another ground-breaking development is the creation of a specially designed aptamer-siRNA chimera, which significantly increased the inhibition of prostate cancer in mice. Additionally, aptamer-conjugated nanoparticles containing docetaxel have demonstrated enhanced cancer-specific cytotoxicity in both in vitro and in vivo settings. Collectively, these findings emphasize the potential of aptamers as outstanding tumor-targeting ligands within targeted therapeutic systems [4, 12].

However, it's worth noting that the development of aptamers using SELEX technology can be a time and labor-intensive process with relatively low cost-efficiency and output. To address these challenges, researchers have explored approaches like HT- SELEX (high-throughput SELEX) or HTS (high-throughput screening), which aim to streamline and optimize the aptamer selection process [10]. HTS-SELEX provides the advantage of sequencing libraries across all selection rounds, making for early identification of improved sequences, timesaving, and avoiding PCR bias. Bioinformatics methods facilitate comprehensive characterization of aptamers, covering aspects such as binding affinity, and aptamer-target interactions, aided by examination of extensive sequence datasets on a global scale [13]. In this study, we report the method to screen and design ssDNA aptamer and evaluated its cytotoxic effects on HeLa and MCF-7 cancer cell lines.

2 Materials

2.1 Reagents

1. 100 uM ssDNA aptamer (WhiteSci, Whitehead Scientific (Pty) Ltd. IDT Integrated Technologies (South Africa)).
2. 1x TE buffer (comprising 10 mM Tris–HCl, 0.1 mM EDTA, pH 8.0) (Merck Life Science (Pty) Ltd. (Merck Group South Africa)).
3. Folding Buffer: 1 mM $MgCl_2$, 1X PBS (137 mM NaCl, 2.7 mM potassium chloride, 8 mM sodium phosphate dibasic, 2 mM potassium phosphate monobasic) (Thermo Fisher Scientific (LTC Tech south Africa Pty LTD)).
4. 100 nM Bleomycin sulfate (BIOCOM Africa (Pty) Ltd., South Africa AP302).
5. 100 nM Metformin (BIOCOM Africa (Pty) Ltd., South Africa 270,432).
6. 100 nM Mitomycin (BIOCOM Africa C(Pty) Ltd., South Africa GR311).
7. HeLa and MCF7 cancer cell line—adherent in DMEM and 10% FBS and 10% DMSO (donated by CSIR SNB laboratory).
8. Dulbecco's modified Eagle medium (DMEM) supplemented with 10% heat-inactivated FBS and 1% antibiotics [100 U/mL penicillin, 100 μg/mL streptomycin, and 250 μg/mL fungizone] (WhiteSci, Whitehead Scientific (Pty) Ltd., South Africa BE12-604F).

9. 0.4% Trypan blue (WhiteSci, Whitehead Scientific (Pty) Ltd., South Africa 17-942E).
10. Presto blue cell viability reagent (WhiteSci, Whitehead Scientific (Pty) Ltd., South Africa).
11. 10X Trypsin (WhiteSci, Whitehead Scientific (Pty) Ltd., South Africa BE02-007E).
12. 0.1% DMSO (Merck Life Science (Pty) Ltd., Merck Group South Africa).
13. 0.1 M BzCl (Merck Life Science (Pty) Ltd., Merck Group South Africa).
14. Conjugate buffer –0.1 M sodium acetate (CH_3COONa), 0.05 M sodium chloride (NaCl), and 5 mM magnesium chloride ($MgCl_2$) (Merck Life Science (Pty) Ltd., Merck Group South Africa).
15. 1X DPBS (Thermo Fisher Scientific, South Africa A1285601).

2.2 In Silico/Computational Tools

1. RCSB PDB (www.rcsb.org).
2. OligoAnalyzer web-based tool (www.eu.idtdna.com/pages/tools/oligoanalyzer).
3. RNA fold web server (www.Vienna RNA web services- rna.tbi.univie.ac.at).
4. RNA composer (www.rnacomposer.cs.put.ponan.pl).
5. PyMOL v. 2.5 (www.pymol.org).

2.3 Equipment

1. Laminar flow hood.
2. Incubator −37 °C and 5% CO_2.
3. Inverted microscope.
4. Centrifuge.
5. Pipettor.
6. Water bath −37 °C.
7. Hemocytometer.
8. Tecan Infinite F500 Microplate Reader.
9. Flasks—T 25 cm^2 and 75 cm^2 (WhiteSci, Whitehead Scientific (Pty) Ltd., South Africa 503501).
10. 96-Well plate (WhiteSci, Whitehead Scientific (Pty) Ltd., South Africa).
11. T 25 cm^2 and T 75 cm^2 culture flask (WhiteSci, Whitehead Scientific (Pty) Ltd., South Africa 708003 & 709003).
12. Centrifuge tubes (15 mL and 50 mL) (WhiteSci, Whitehead Scientific (Pty) Ltd., South Africa).

13. Pipette tips (10 uL–1 mL) (WhiteSci, Whitehead Scientific (Pty) Ltd., South Africa 328001).

3 Methods

3.1 In Silico Aptamer Design and Preparation

3.1.1 Secondary (2D) and Tertiary (3D) Aptamer Structure Prediction

1. Select the RNAfold server from the Vienna RNA secondary structure server.
2. Enter nucleotide sequence into sequence box and use a default.
3. Click the Proceed bar.
4. Next extract the dot-bracket notation from RNAfold and submit it to the RNAcomposer modeling server to get the 3D structure.

3.1.2 OligoAnalyzer Tool

1. Select the OligoAnalyzer tool from any IDT webpage menu.
2. Enter the Sequence into "Sequence box" in the 5′ to 3′ orientation.
3. Select the DNA from the menu under the sequence box.
4. Enter Na+, Mg++, and dNTP concentration, default 50 mM, 0 mM, and 0 nM, respectively.
5. Press the Analyze bar.
6. Visualize the structure using the PyMOL tool.

3.1.3 Aptamer Design

The aptamer used in this study was ssDNA aptamer. Oligonucleotide sequences were randomly selected and characterized using OligoAnalyzer web-based tools. The aptamer GC content, melting temperature (Tm), molecular weight, and other values were computed using this tool. It also determines the potential for a secondary structure. Using the RNAfold web server tool, the secondary structure of this aptamer was also predicted and visualized. Next tertiary structure of the aptamer was identified by RNAcomposer which is a 3D modeling server, and the 3D structure was finally visualized by PyMOL. The model aptamer was synthesized and purchased from WhiteSci IDT Company based on the data gathered from these tools. The aptamer was 5′6, FAM labeled, and had a length of 54 nucleotides (Fig. 1). The aptamer's sequence and dot-bracket notation are as follows:

(5′6, FAM, GCACCACCCTGTGCGGCACCCTGGATTATCTG CCGCCGGAAATGATTGAAGGCC 3′)

```
........(((.((((((............)))))))))........
.......
```

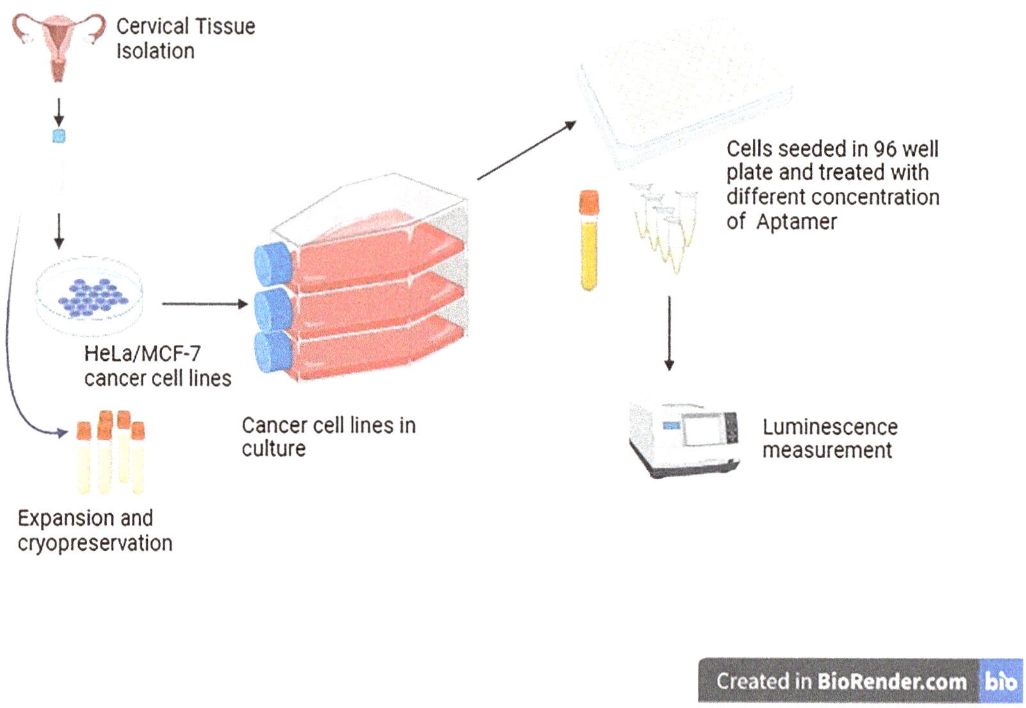

Fig. 1 A summary of methods to evaluate the cytotoxicity effects of aptamer on cell lines

3.2 Preparation of Aptamer and Drugs

Aptamer resuspension procedure—The aptamer was resuspended by the steps briefly described below:

3.2.1 Aptamer Storage Procedure

1. The aptamer was briefly centrifuged in the original storage tube at 2000 rpm for 1 min to ensure that the dried aptamer pelleted at the bottom of the tube.
2. The pelleted aptamer was resuspended in 1x TE buffer (10 mM Tris–HCl, 0.1 mM EDTA, pH 8.0) to achieve 100 uM concentration. For 950 nmol of yielded aptamer, we used 9550uL of 1X TE buffer.
3. The aptamer was incubated at room temperature for 30 min and then vortexed for 20 s.
4. Centrifuge the tube containing the aptamer at 10,000× *g* for 1 min.
5. Make aliquots and store the resuspended ssDNA at −20 °C.

3.2.2 Aptamer Folding Procedure

1. Before experimental usage, the aptamer was diluted to its working concentration in Folding Buffer: 5 mM $MgCl_2$, 1X PBS (137 mM NaCl, 2.7 mM potassium chloride, 8 mM sodium phosphate dibasic, 2 mM potassium phosphate monobasic).

2. Aliquots of the aptamer in folding buffer were prepared and placed onto PCR tubes for post-processing in the thermocycler PCR machine.
3. In the thermocycler, the aptamer solution was heated to 95 °C for 5 min.
4. After 5 min, the aptamer solution was allowed to cool to room temperature (~10 min) and at 37 °C for 15 min.

3.2.3 Preparation of Aptamer

1. 1x TE buffer (Tris–EDTA) is a commonly used buffer solution for resuspending and storing nucleic acids, especially DNA. A stock solution of 100 μM aptamer solution was prepared using 1x TE buffer.
2. A dilution concentration equation was used to prepare various concentrations of aptamer from the stock solution, namely 1 μM, 1.5 μM, 2 μM, 2.5 μM, 3 μM (low concentration), and 20 μM, 40 μM, 60 μM, 80 μM, and 100 μM (high concentration).

$$C1V1 = C2V2 \tag{1}$$

C1 = Initial concentration of stock solution
V1 = Volume of stock solution needed to make a new solution
C2 = Final concentration of new solution
V2 = Final volume of new solution

3. The aptamer was diluted in 1x DPBS comprising 5 mM $MgCl_2$. To maintain the folded structure and most accessible and three-dimensional structure, the aptamer was heated in thermocycler heating protocol (denaturation at 95 °C for 5 min), followed by slow cooling to 22 °C, over a 10-min period and then maintained at 37 °C for 15 min [14].

3.2.4 Preparation of Aptamer-Drug Mixture

1. Before the aptamer-drug conjugation experimental work, target cells (HeLa and MCF_7) were seeded in 96-well plates and incubated overnight (see Subheading 3.4).
2. After 24 h of cell incubation, aptamer was prepared according to Subheading 3.2.2 Aptamer-Folding Procedure as indicated above.
3. This was followed by preparing aptamer-drug mixture in molar ratio of 1.0, 1.2, 1.4, 1.6, and 1.8 in conjugate buffer (0.1 M CH_3COONa, 0.05 M NaCl, and 5 mM $MgCl_2$).
4. Once the aptamer-drug conjugates were prepared, the mixture was incubated for 1 h at 37 °C and 5% CO_2 incubator.

   Note: To demonstrate the synergistic and drug-loaded effect of the aptamer, this known anti-cancer drug was mixed with the aptamer. The aptamer was prepared at a molar ratio of aptamer:drug 1.0, 1.2, 1.4, 1.6, and 1.8. The drugs were separately mixed with the aptamer. The aptamer was combined with a fixed drug concentration (100 nM) in conjugation buffer (0.1 M CH_3COONa, 0.05 M NaCl, and 5 mM $MgCl_2$). The mixture was incubated for 1 h at 37 °C to permit the intercalation of the drug [1].

3.3 Cytotoxicity Assay

3.3.1 Cell Lines and Cultures

1. Decontaminate and prepare laminar air flow hood.
2. Incubate the cell culture media in warm water bath at 37 °C.
3. Thawing HeLa and MCF-7 cells at 37 °C until a small ice crystal remains.
4. Transfer the thawed cells to a sterile centrifuge tube containing a pre-warmed culture medium. Centrifuge the cells at a speed of 2000× *g* for 5 min. Carefully aspirate supernatant.
5. Resuspend the cell pellet in fresh and prewarmed culture medium. Then seed the cells into a tissue culture flask (T 25 cm^2/75 cm^2).
6. Place the tissue culture vessels in a cell culture incubator set at 37 °C with 5% humidity.
7. Allow the cells to adhere and proliferate for 48–72 h.
8. When the cells reach approximately 70–80% confluency, they are ready for subculturing.
9. Prepare the sub-culture and aspirate the old medium from the culture vessel.
10. Wash the cell two times with 2 mL phosphate-buffered saline (PBS).
11. Add 2 mL trypsin–EDTA solution to detach from the surface and incubate them at 37 °C for 5 min.
12. Stop the trypsin activity by adding an equivalent volume of culture medium.
13. Then determine the cell concentration using a hemocytometer or an automated cell counter by staining the cell with 4% 1 uL trypan blue.
14. Seed the cells into 96-well plates and test the percentage cell viability assay.
15. Add aptamer and aptamer drug mixture and evaluate its effect in a time- and concentration-dependent manner as mentioned below.
16. To evaluate the % cell viability effect of the aptamer and aptamer-drug conjugation, we used HeLa (human cervical cancer) and MCF-7 (human breast cancer). The cell lines were grown and maintained in culture with Dulbecco's modified Eagle medium (DMEM) supplemented with 10% heat-inactivated FBS and 1% antibiotics [100 U/mL penicillin,

100 μg/mL streptomycin, and 250 μg/mL fungizone]. The cells were maintained at 37 °C in a 5% CO_2 humidified atmosphere in T25[2] cm culture flasks.

3.3.2 Effect of Free Aptamer on the Viability of the Cell Lines

1. Treated HeLa and MCF-7 cancer cell line response to free aptamer were studied by Presto blue assay.
2. About 1.5×10^5 cells were suspended in a culture medium and seeded into 96-well plates.
3. Following 24 h of incubation, the old medium was replaced with a fresh medium. The cells were treated with 20 uL of free aptamer solution at varied concentrations as mentioned above, for 1, 4, 24, 48, and 72 h which were manually printed in triplicate [15].
4. Thereafter, each well was treated with viability stain of about 10uL Presto Blue solution and incubated for a further 10 min at 37 °C, 5% CO_2.

3.3.3 Cell Viability of Drug and Aptamer on Cell Growth

1. Approximately 1.5×10^5 cells were seeded into 96-well plates in the same manner as previously described, and the cells were then incubated for an additional 24 h.
2. The cells were then exposed to 20 uL of aptamer-drug mixture at various concentrations. The treatment was carried out after 24, 48, and 72 h of incubation. Then, the cell viability was determined using presto blue assay according to the typical protocol drawn by the manufacturer. A negative control of 0.1% DMSO and a positive control of 0.1 M BzCl were made (*see* **Notes 1–3**).
3. The cell viability were measured using the Tecan infinite F500, Luminescence with (λEx = 535 nm, λEm = 612 nm). All results were averaged from biological triplicates and values are reported as mean ± SEM.

3.4 Anticipated Result and Discussion

In the context of cell line studies, aptamers can be utilized for various purposes such as targeting cell surface receptors, interfering with intracellular processes, or delivering therapeutic payloads. Aptamers have gained significant attention as potential therapeutic agents due to their ability to target various biomolecules and even whole cells.

When investigating the effects of aptamer treatment in cell lines, it is important to understand how the concentration and duration of exposure to aptamer influence their efficacy and biological responses. This is because aptamer binding and subsequent cellular responses often exhibit concentration- and time-dependent properties. Therefore, conducting thorough concentration- and time-dependency studies is essential for characterizing the effects of aptamer-based treatments accurately (*see* **Notes 4** and **5**).

3.5 Conclusion

A variety of previously identified aptamers have shown good anticancer (cytotoxic) properties [12, 20]. Aptamers are characterized by their strong affinities and remarkable specificity toward target molecules. This unique property makes aptamer potential therapeutic agents to directly inhib the proliferation of tumor cells. Furthermore, they hold significant promise in the realm of targeted drug delivery systems that can be harnessed for tumor diagnosis and the specific targeting of tumor cells, thus reducing toxicity to healthy cells. Particularly, some aptamers have demonstrated superior antitumor activity compared to antibodies [21–23]. Overall, developing methods to utilize aptamers for testing cancer cells holds great promise for advancing the field of oncology and improving outcomes for patients with cancer and other diseases.

4 Notes

1. Concentration-dependent effects: the efficacy of aptamer binding and subsequent biological effects often increase with increasing aptamer concentration until reaching a plateau, where further increases in concentration may not significantly enhance responses. By researching the concentration-dependent effects, researchers can determine the minimum effective concentration required to achieving a desired biological outcome and assess the potential toxicity or off-target effects associated with higher aptamer concentration. To observe concentration-dependent effects, a range of aptamer concentrations should be tested across multiple experimental replicates to ensure statistical significance and reproducibility [16, 17].
2. Time-dependent effects: the duration of aptamer exposure can influence the kinetic of aptamer binding, internalization, and downstream cellular responses. Short-time exposure to aptamer may elicit immediate responses such as receptor modulation, signaling pathway activation or inhibition, or changes in gene expression. However, long-term exposure may lead to sustained effects such as cell proliferation inhibition, apoptosis induction, or alteration in cell phenotype. Time-course experiments involving multiple time points are necessary to capture the dynamic changes induced by aptamer treatment over time and to determine the optimal duration of treatment for achieving desire therapeutic outcomes [17, 18].
3. Generation of linear standard curves: to quantify aptamer binding or assess aptamer-mediated effects, it is essential to establish

a standard curve relating aptamer concentration to the observed response. A linear standard curve allows for the interpolation of unknown concentration based on their corresponding response values within the linear range of detection. Thus, to generate a linear standard curve, various concentration of aptamer should be tested against a constant target or cell line under standardized experimental conditions. The responses (e.g., binding signal, cellular viability, and gene expression) at each aptamer concentration are measured, and a plot of response versus concentration is constructed. The linear range of the standard curve is determined by the concentration at which the response exhibits a linear relationship with aptamer concentration, typically at lower concentration where saturation effects are minimal. This standard curve serves as a reference for quantifying aptamer concentration in subsequent experiments and for comparing the efficacy of different aptamer formulation or experimental conditions [17, 19].

4. Generally understanding the concentration and time-dependent effects of aptamer treatment in cell lines is crucial for optimizing their therapeutic potential. Generating linear standard curves facilitates accurate quantification and comparison of aptamer-mediated responses, thereby advancing the development and application of aptamer-based therapeutics in various biomedical contexts.
5. Certain studies reported that aptamers have displayed superior antitumor activity as compared to antibodies, highlighting their potential as innovative alternatives in oncological interventions. Using aptamers for cancer cell testing offers a compelling path forward for improving patient outcomes and oncology research. This underscores the importance of continued efforts in developing methodologies to leverage the therapeutic potential of aptamers, with the ultimate goal of revolutionizing cancer treatment strategies and addressing unmet medical needs.

Acknowledgments

This work was supported by Addis Ababa Science and Technology University (AASTU), Ethiopia, the Council for Scientific and Industrial Research (CSIR), South Africa, and the Organization for Women in Science for the Developing World (OWSD).

Author Contribution Haregewoin B. Woldekidan: Conceptualization, methodology, performed the analysis, and the main writer of the article; Deepak B. Thimiri Govinda Raj: funding acquisition, writing, reviewing and editing, and supervision; Adugna

A. Woldesemayat: writing, reviewing and editing, and supervision; Zandile Nxumalo: performed data collection and analysis; Mutsa Takundwa: funding acquisition, reviewing, and editing.

References

1. Lohlamoh W, Soontornworajit B, Rotkrua P (2021) Anti-proliferative effect of doxorubicin-loaded AS1411 aptamer on colorectal cancer cell. Asian Pac J Cancer Prev 22(7):2209–2219. https://doi.org/10.31557/APJCP.2021.22.7.2209
2. Zhu G, Niu G, Chen X (2015) Aptamer-drug conjugates. Bioconjug Chem 26(11): 2186–2197. https://doi.org/10.1021/acs.bioconjchem.5b00291
3. Aljohani MM, Cialla-May D, Popp J, Chinnappan R, Al-Kattan K, Zourob M (2022) Aptamers: potential diagnostic and therapeutic agents for blood diseases. Molecules 27(2):1–21. https://doi.org/10.3390/molecules27020383
4. Weaver SD, Whelan RJ (2021) Characterization of DNA aptamer-protein binding using fluorescence anisotropy assays in low-volume, high-efficiency plates. Anal Methods 13(10): 1302–1307. https://doi.org/10.1039/d0ay02256j
5. Fan R et al (2023) Application of aptamer-drug delivery system in the therapy of breast cancer. Biomed Pharmacother 161:114444. https://doi.org/10.1016/j.biopha.2023.114444
6. Thevendran R, Citartan M (2022) Assays to estimate the binding affinity of aptamers. Talanta 238:122971. https://doi.org/10.1016/j.talanta.2021.122971
7. Mahmuda A et al (2017) Monoclonal antibodies: a review of therapeutic applications and future prospects. Trop J Pharm Res 16(3): 713–722. https://doi.org/10.4314/tjpr.v16i3.29
8. Sun H, Zu Y (2015) A highlight of recent advances in aptamer technology and its application. Molecules 20(7):11959–11980. https://doi.org/10.3390/molecules200711959
9. Sullenger BA, Gallardo HF, Ungers GE, Gilboa E (1990) Overexpression of TAR sequences renders cells resistant to human immunodeficiency virus replication. Cell 63(3):601–608. https://doi.org/10.1016/0092-8674(90)90455-N
10. Buglak AA, Samokhvalov AV, Zherdev AV, Dzantiev BB (2020) Methods and applications of in silico aptamer design and modeling. Int J Mol Sci 21(22):1–25. https://doi.org/10.3390/ijms21228420
11. Wu X, Chen J, Wu M, Zhao JX (2015) Aptamers: active targeting ligands for cancer diagnosis and therapy. Theranostics 5(4):322–344. https://doi.org/10.7150/thno.10257
12. Luo S, Wang S, Luo N, Chen F, Hu C, Zhang K (2018) The application of aptamer 5TR1 in triple negative breast cancer target therapy. J Cell Biochem 119(1):896–908. https://doi.org/10.1002/jcb.26254
13. Woldekidan HB, Woldesemayat AA, Adam G, Tafesse M, Thimiri Govinda Raj DB (2023) Aptamer-based tumor-targeted diagnosis and drug delivery. Adv Exp Med Biol 1409 (September):173–192. https://doi.org/10.1007/5584_2022_732
14. Macdonald J, Henri J, Goodman L, Xiang D, Duan W, Shigdar S (2017) Development of a Bifunctional Aptamer targeting the transferrin receptor and epithelial cell adhesion molecule (EpCAM) for the treatment of brain cancer metastases. ACS Chem Neurosci 8(4): 777–784. https://doi.org/10.1021/acschemneuro.6b00369
15. Niu W, Teng IT, Chen X, Tan W, Veige AS (2018) Aptamer-mediated selective delivery of a cytotoxic cationic NHC-Au(i) complex to cancer cells. Dalton Trans 47(1):120–126. https://doi.org/10.1039/c7dt02616a
16. Emahi I, Gruenke PR, Baum DA (2015) Effect of Aptamer binding on the electron-transfer properties of redox cofactors. J Mol Evol 81(5–6):186–193. https://doi.org/10.1007/s00239-015-9707-7
17. Odeh F et al (2019) Aptamers chemistry: chemical modifications and. Molecules 25(3): 1–51
18. Ogurcovs A, Kadiwala K, Sledevskis E, Krasovska M, Plaksenkova I, Butanovs E (2022) Effect of DNA Aptamer concentration on the conductivity of a water-gated Al:ZnO thin-film transistor-based biosensor. Sensors 22(9):1–12. https://doi.org/10.3390/s22093408
19. Yu H, Zhu J, Shen G, Deng Y, Geng X, Wang L (2023) Improving aptamer performance: key factors and strategies. Microchim Acta

190(7):255. https://doi.org/10.1007/s00604-023-05836-6

20. Henri JL, Nakhjavani M, McCoombe S, Shigdar S (2023) Cytotoxic effects of aptamer-doxorubicin conjugates in an ovarian cancer cell line. Biochimie 204:108–117. https://doi.org/10.1016/j.biochi.2022.09.005
21. Han J, Gao L, Wang J, Wang J (2020) Application and development of aptamer in cancer: from clinical diagnosis to cancer therapy. J Cancer 11(23):6902–6915. https://doi.org/10.7150/JCA.49532
22. Doherty C, Wilbanks B, Khatua S, James L, Iii M (2024) Aptamers in neuro-oncology: an emerging therapeutic modality. Neuro-Oncology 26(2023):38–54
23. Kim DH, Seo JM, Shin KJ, Yang SG (2021) Design and clinical developments of aptamer-drug conjugates for targeted cancer therapy. Biomater Res 25(1):42. https://doi.org/10.1186/s40824-021-00244-4

Index

Kursad Turksen (ed.), *Autophagy in Development and Disease: Methods and Protocols*, Methods in Molecular Biology, vol. 2879, https://doi.org/10.1007/978-1-0716-4268-9,

www.ingramcontent.com/pod-product-compliance
Lightning Source LLC
Chambersburg PA
CBHW052006180325
23653CB00034B/76
9781071642672